W9-BVY-537

DESIGNING TECHNICAL REPORTS

DESIGNING TECHNICAL REPORTS

Writing for Audiences in Organizations

J. C. Mathes

Dwight W. Stevenson

BOBBS-MERRILL EDUCATIONAL PUBLISHING
Indianapolis

To Rosemary and Mary Ann, and to John, Ann, and Melissa for the many hours we weren't home—and for putting up with us the few hours we were.

Printed in the United States of America

The Bobbs-Merrill Company, Inc.
4300 West 62nd Street
Indianapolis, Indiana 46268

First Edition

Third Printing—1979

Designed by Viki Webb

Library of Congress Cataloging in Publication Data

Mathes, John C
 Designing technical reports.

 Includes index.
 1. Technical writing. I. Stevenson, Dwight W.,
1933– joint author. II. Title.
T11.M36 808'.066'6021 75-44249
ISBN 0-672-61367-0

Contents

List of Figures

Acknowledgments

To our colleagues in the Humanities Department of the College of Engineering at the University of Michigan we want to express our thanks for their help and encouragement. Particularly we want to thank W. Earl Britton, Peter R. Klaver, Dorothy Lambert Mack, Leslie A. Olsen, Thomas M. Sawyer, and Richard E. Young.

To our colleagues in other departments of the College of Engineering we want to express our appreciation for their advice and assistance on technical matters. Especially we want to thank Harry Benford, Professor of Naval Architecture and Marine Engineering; Kan Chen, Professor of Electrical and Computer Engineering; Amelio D'Archangelo, Professor of Naval Architecture and Marine Engineering; Donald Gray, Professor of Civil Engineering; Robert Harris, Professor of Civil Engineering; Herman Merte, Professor of Mechanical Engineering; Wilbur Nelson, Professor of Aerospace Engineering; Brymer Williams, Professor of Chemical and Metallurgical Engineering; John Young, Director, Engineering Placement Service. Also, in other units of the University, Maurita Holland, Head, Engineering and Transportation Library; and Karl Lagler, Professor of Natural Resources.

To our associates without whose contributions there would be no book we also express our thanks. Especially Vi Benner who designed and typed our prototype manuscript; Cy Barnes and Harry Willsher who prepared the illustrations; Sharon Finton, Vicki West and Dorothy Strand who saw us through several revisions; to our editor, Larry Ligget, and to Ray Warden, who gave much editorial assistance to the book.

We also thank the College of Engineering for giving us the released time and sabbatical leaves to do the research, our students, and the participants in the Engineering Summer Conference for Engineers, Scientists and Technical Writers and in the Conference on Teaching Technical and Professional Writing for allowing us to try out our material on them.

To the many engineers, scientists, managers, editors, and technical writers and illustrators to whom we are indebted, we thank them for their contributions and suggestions. Among them particularly:

Timothy Adama
Paul Ainslie
John Ambrose
David Armstrong

Kenneth Arnold
Daniel Atkins
John Baldwin
Janis Baron
Howard Barlow
David Baumgartner
Karen Bilich
Richard Bollinger
David Boyd
Robert Bradley
Gordon Braidwood
Richard Bratcher
Lee Carpenter
Donald Cooke
Gary Cousins
Raymond Daniels
Robert DiGiovanni
Keith Higgins
John Eder
John Elinger
Donald Eschman
Raymond Fales
James Fell
Donald Fleming
Leo Fitzpatrick
Charles Hale
Robert Hasse
James Higgins
John Holland
William Hunley
Michael Huston
Thomas Johnson
Roland Jones
Ronald Kane
Dale Kempf
Jack Keyte
Kenneth Kits
David Krause
Stephen Lusky
John Mackovjak
Gary Madison
Ernest Mazzatenta
Michael Meisel
John Milewski
Jay Miner
Jeffrey Mundth
Hugh Munroe
James Ogonowski
Gerhard Ohlhaver
Craig Oppenlander
John Oyer
William Parker
Jack Partridge
Kenneth Petty
Richard Polmear
Ann Resetar
George Robinson
Keith Roe
Stephanie Rosenbaum
Lester Rosenblatt
Richard Rykowski
John Sacher
David Saunders
Richard Schaadt
Mark Shahly
Kirk Sherhart
Fred Shippy
William Sikula
Raymond Smit
Patricia Smith
Michael Steer
Melvin Takata
David Tarsi
David Terrell
Craig Tessmer
Donald Thompson
Margaret Vukelich
George Waller
Patrice Walsh
Gary Was
Huey Wegner
Carol Wissel
David Witmer
Steve Wright

To the following organizations who permitted us to use their reports and helped us with our research, our deepest appreciation:

ADP Network Services, Cyphernetics Division
American President Lines
Ayers, Lewis, Norris, and May, Consulting Engineers
Chrysler Corporation
Commission on Public Hospital Administration
Community Systems Foundation, Ltd.
Consumer's Power Co.

Dames and Moore Consultants
Detroit Edison Co.
Dow Chemical Co.
Dresser Manufacturing Corp.
Environmental Research Institute of Michigan
FMC Corp.
General Motors Research Laboratory
Institute of Electrical and Electronic Engineers, Inc.
Institute of Marketing: Construction Industry and Marketing Group (England)
Maritime Administration
McNamee, Porter, and Seeley, Consulting Engineers
Monsanto Research Corp.
M. Rosenblatt and Son, Inc.
National Highway Traffic Safety Administration
Naval Ship Engineering Center
Richards Rosen Press
USDA Forest Service
W. A. Kraft Corp.

Preface

This book began with a paradox: practicing engineers need to write efficiently and effectively; yet despite that obvious need, the writing experiences of most engineering students in college do not prepare them to write as they must in industry and government. They may have learned to write as students in the classroom, but they have not learned how to write as professionals employed in complex organizations. After they graduate, they must learn how to write as professionals. Put as simply as possible, that is what this book is all about.

Every engineering student has heard—and every practicing engineer knows the essential truth of—dramatic pronouncements about his or her need for communication skills. An engineer in management asserts, "an engineer who can't communicate is in trouble; a manager who can't communicate is finished." Another engineer warns, "if you can't tell them what you are doing, they'll get someone else to do it." This is a familiar theme—and a true one.

Yet engineers are not trained well by their college writing experiences. Despite the usual freshman English themes on "Justice" or "Civil Disobedience," the term papers in history or American literature, the numerous lab reports, and despite the occasional technical report or design project report, engineering graduates are usually ill-equipped to cope with the communication needs of professional engineers. In fact, the writing most engineering students do in college bears so little resemblance to the writing required of professional engineers that it is even appropriate to say students are trained to communicate poorly.

In college, students write for an audience of one person—a professor; in industry, they must learn to write for a large, diverse audience in an organization. In college they write for a reader who knows the field and probably knows more about their technical material than the writer does; in industry, they must learn to write to people who perhaps do not know the field or who almost certainly know less about the material than the writer. In college, they write for pedagogical purposes—to demonstrate to a professor their mastery of concepts, processes, and information; in industry, their mastery is assumed, and they must learn to write for instrumental purposes—to help people in an organization make judgments and act upon the results they present. In short, even though engineering students may have written quite a bit by the time of graduation, little in their college writing experiences prepares them for the communication situations they face as professionals.

The result is that the graduating student, the student in design courses, and most obviously the young practicing engineer are likely to feel frustrated. The placement service manual at one university even warns beginning engineers to expect that feeling. It says, "Since success in engineering school does not depend much on communication abilities, you may not have developed your skills and therefore may resent the absolute dependence you will have on them for accomplishment in an organization." "Absolute dependence" on communication skills—an engineer's own words.

So what to do about it? What can the engineering student and young professional do?

We believe the engineer must learn to approach the design of technical communication systematically, just as he or she has learned to approach technical design problems. We assume that if an engineer can use rigorous, systematic procedures to resolve technical problems, he or she should be equally able to apply systematic procedures to resolve communication problems. Neither process, we believe, should depend upon innate ability, intuition, or blind luck. The purpose of this book, therefore, is to present a systematic procedure which will enable the engineer to approach and solve the problem of report design confidently and effectively.

A word about our method is necessary. This book begins with basic design principles and concentrates on them. It therefore differs substantially from most technical writing texts, which usually begin with questions of report format, technical style, sentence structure, or mechanics—and often never go beyond such secondary concerns. This book focuses on the questions a writer must answer first: who is to read the report, what do they want to know, what does the writer want to accomplish, and how should the report be structured to meet these needs? We do this because we believe one cannot design a tool, a bridge, an inventory system, or a report unless one first knows a good deal about such fundamental questions as how the finished product will be used, by whom, and for what purposes. We begin, in short, at the beginning.

J. C. Mathes and Dwight W. Stevenson
Ann Arbor, Michigan
1976

ONE
Determining the Function of the Report in the System

THE TECHNICAL COMMUNICATION PROCESS

AUDIENCE ANALYSIS: THE PROBLEM AND A SOLUTION

THE PROBLEMATIC CONTEXT:
THE PURPOSE OF THE REPORT

CHAPTER 1

The technical communication process

No matter what technical investigations the engineer conducts, the organization in which he or she functions continues to operate without modification until the engineer writes reports whose objective is to modify the processes of the system.

When an engineer first has to write a technical report, he or she must design a piece of writing quite unlike anything he has written before. The technical report is distinctive because it is prepared within a professional context. It is not just a piece of writing from one individual to another, like a letter or class assignment, or from one individual to some vague segment of our society, like an essay or a magazine article. The technical report is written by an engineer for a specific purpose within a specific organization—such as a corporation or governmental institution. And it is written solely because segments of that organization need reports to accomplish their missions. The technical report is *an act of communication by a professional in an organizational system to transfer information necessary for the system to continue to function.*

Because technical reports have specific organizational purposes, they necessarily have distinctive design features. That is, they have design features other types of writing lack because reports are used differently. When you read an essay or a magazine article, you read straight through, top to bottom. You expect it to unfold before you, and you expect to read it all. A report, however, is seldom designed to be read straight through. It is structured to be used by many readers who are cramped for time and by people with widely differing needs and interests. Thus to write reports, you must learn new methods. You cannot expect to write technical reports in the same manner you wrote papers in college.

How, then, should you write reports? If the technical report has distinctive characteristics, where does that leave most engineers? Did you, for instance, as most college students, have six hours of freshman composition? Did you have high school English? What kinds of writing were you taught? It is unfortunate, but very likely, that none of your writing instruction taught you how to write effective reports.

If your experiences were similar to those of most students, you were asked to write essays. For the most part, though, the purposes of these essays were not really to communicate; the purpose of this type of writing was essentially humanistic. That is, you wrote to improve the ability of your reasoning and the quality of your perceptions and experiences. Seldom did your essays have some real communication purpose. Your professor assigned grades to you on the basis of your essays, but you did not really expect him to modify his behavior or to act on the information you presented. Thus from earlier writing experiences you bring few skills to help you write technical reports.

As engineers you have had other, somewhat more appropriate, types of writing experiences. You have had homework in physics and math and have had laboratory reports. In technical courses you have produced design project reports. These writing experiences may be more useful to you as you start writing technical reports. However, even these experiences contain pitfalls that continue to trap engineers when they start writing technical reports on the job. Many of your writing activities as an engineering student lacked an appropriate communication purpose. The "system" survives if a student produces a poorly written laboratory report; the only thing that happens is the student flunks. More importantly, when students write in their roles as students to individuals in their roles as professors, this is not true communication in the sense we intend here. The "better" your lab reports, the less effect you had because your purpose was to produce a form corresponding to some existing template in the teacher's syllabus. Even design project reports can develop poor report writing habits, because—for pedagogical purposes—students must emphasize the details of how they derived their designs rather than focus on the useful results of their designs.

In contrast to these college writing experiences, technical reports are defined and measured by their effects on an organizational system—that is, by changes they cause in the organization's operations. To write effective reports you must first learn to design reports in terms of their distinctive purposes. You do not discard everything you have learned from other types of writing experiences, but you must learn what the technical communication process requires you to do so that you know what skills to keep and what new skills you should develop.

In this book we explain a design approach addressing the technical report as a distinct form of communication tailored to the engineer's role in an organizational system. We will explain how you can design a report so that most of your readers will need to read only segments to be able to use the report. (And they will make more use of it than the few readers who will read the entire report.) Such a report will indeed be quite different from the college English essay.

THE ENGINEER'S ROLE WITHIN AN ORGANIZATIONAL SYSTEM

To develop the skills necessary to write good reports, you need to know how these reports figure in your role as a professional engineer. Your organizational role—your daily activities—consists of processing information that comes to you from the organization, of responding by performing technical and intellectual activities, and of transmitting information into the organization to affect its operation. The most important means of transmitting information into the organization is the report. The effects your report has upon the organization in turn can modify or change your activities, and can even be used to evaluate your performance as an engineer.

We can illustrate the engineer's role as in Figure 1-1, *The engineer's role related to system input and output.* Most of the engineer's role within the system involves

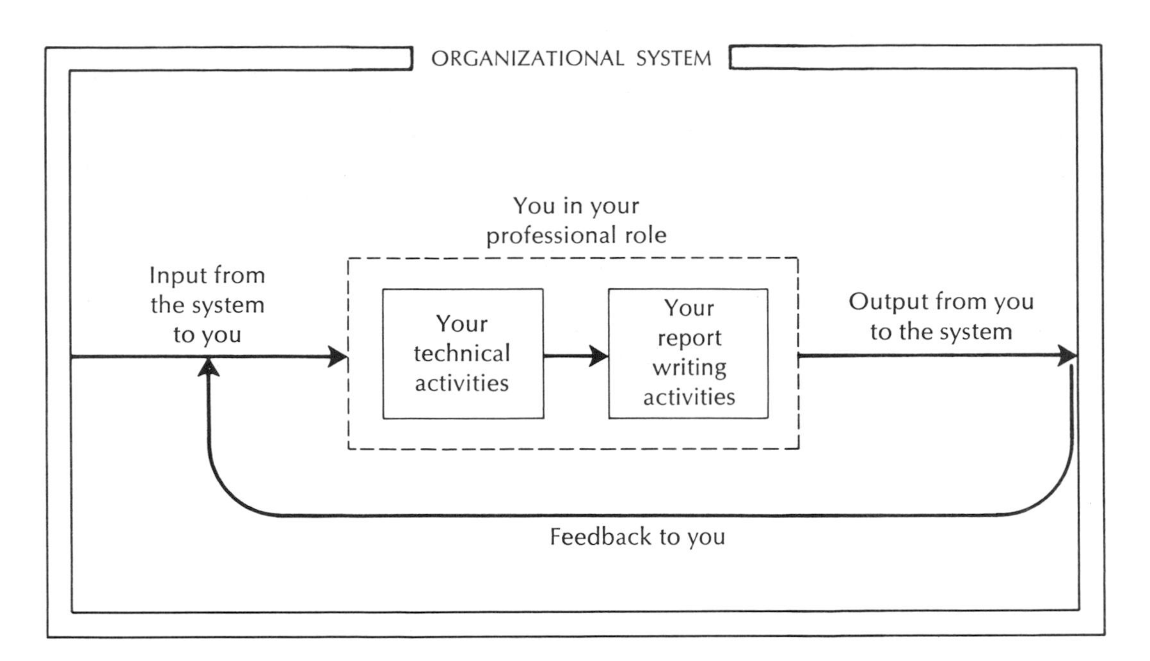

Figure 1–1. The engineer's role related to system input and output.

processing information. His output to the system consists of various types of reports. Although most of his education has prepared him to perform numerous mental and technical acts, these acts acquire meaning in an organization only when they are transformed into information for the organization. The system itself responds only to this output. This output is what causes machinery to be tooled, ground to be excavated, boxcars to be loaded, sales to increase, and people to live better.

At this point we need to make an important distinction. The role of the engineer distinguishes his reports from those written by professional technical writers. In many large organizations some formal reports and articles for persons outside the organization are written by technical writers whose sole function is to edit reports and write articles. The technical writer, however, does not process information technically. He transforms or rearranges information without changing the technical content. In contrast, the role of the engineer is to develop, modify, and change the information he processes—to generate additional information for the organization. We concern ourselves with the reports you write in your role as an engineer.

The influence of the system is the primary factor to consider when you design a report. What is this system? How can it influence your writing? The organization consists of patterns of relationships marking the paths through which information is processed. These paths are communication routes, primarily the routes by which all kinds of reports travel. They are also the routes of responsibility and influence within the organization. Although superficially these paths are described by the organization chart, in fact they are much more complex than they might first appear because every engineer interacts in numerous ways not specified on an organization chart. The engineer's role marks the intersection of numerous paths, and his role is defined by the manner in which he processes the information transmitted on these paths. Information, perhaps in the form of an assignment, comes to him. He performs technical and intellectual tasks. Then he transmits the results to accomplish some purpose related to his role and the assignment.

As an example, assume the organization has a problem with excessive noise on a production line. The problem is transmitted as a report assigning a technical task to an industrial engineer. He receives this information, analyzes the noise situation, and derives recommendations. He transmits the recommendations to the persons in production who requested the study. He also reports to the decision maker who can authorize changes in the production line. If incorporated, these changes will affect the entire system. Consequently, his report is transmitted to other areas of the organization: to purchasing, labor relations, and the legal department, for example. Furthermore, the report could have numerous additional secondary effects by indirectly influencing the decisions and even the attitudes and values of other persons in their professional roles. The custodian's, shipping clerk's, and the production worker's routines may be affected. The meaning of the report is in the various responses it stimulates in the system; the original analysis of the noise problem has no intrinsic value itself.

When you write reports, therefore, you must think in terms of the concrete needs of specific persons in the organization and of the various effects the report will have on the organization. In terms of these needs and responses, you determine the purposes for your report, and design your report accordingly. Part of your writing task is to identify these purposes. You must design your report to affect the system in ways you intend.

We may then modify our definition of the report. The technical report is *the processing of information by an engineer in his or her professional role, the processing designed in response to a stimulus from the organizational system and embodying designs to modify the behavior of the system in purposeful ways.*

Some technical reports involve very few relationships in the organization and transfer very limited types of information (brief memoranda, short letters, simple instructions, etc.). Reports such as these are simplified because they have a homogeneous audience and a single purpose. Such technical reports can pose few problems for most writers, although they should embody most of the basic principles we develop. Many technical reports, however, involve numerous relationships in the organization and transfer complex types of information, although such report writing situations receive relatively little attention in many books on technical writing. In this book we develop our principles for report design on the basis of the complex report writing situation. As we develop principles for report design, we explain how to apply them to complex reports and when to apply them to simple reports.

THE REPORT DESIGN PROCESS

As you can see from the importance of the processing of information to the organization, the technical report is not just a reflex action performed at the conclusion of a technical investigation. It originates somewhere in the organization and terminates somewhere in the organization. As an engineer, you must view your communication endeavors as related to the organization as well as to your technical activities. You should adopt this view in order to effect the transition from student writing to professional writing. The inexperienced report writer usually writes the report only to describe his technical activities, as if he were still a student. The experienced engineer designs the report to serve the needs of the organization.

The report design process can be illustrated in terms of the engineer's role in the organization, as in Figure 1-2, *The report design process.* When you have completed your technical investigation in response to an assignment, you are immersed in the details of your investigation. You now must remove yourself from the investigation in order to write your report. You need to design your report in terms of its function in the system, not in terms of the investigation. You design the report in terms of your technical assignment, and of your conclusions and recommendations to the organization. You start the report design process by analyzing your audiences and purposes, that is, by determining the output to the system.

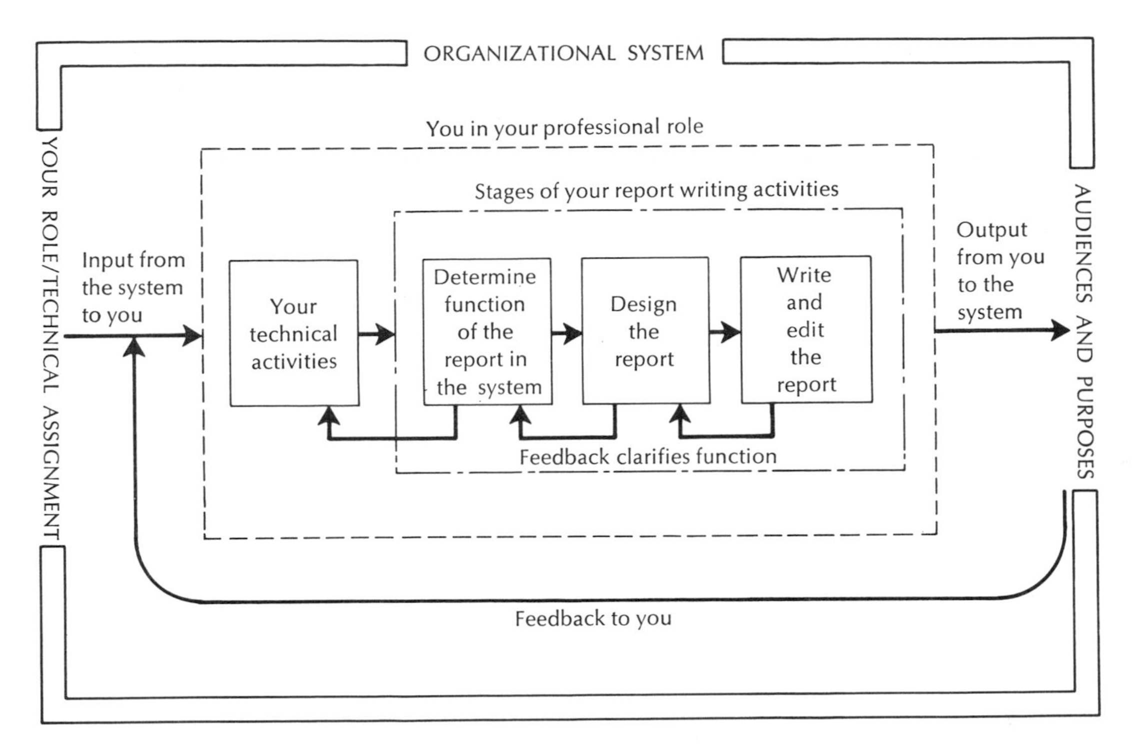

Figure 1–2. The report design process.

We have arranged this book to reflect the three stages of your report writing activities, as illustrated in Figure 1-2. Chapters 1 through 10 correspond to the process as follows:

Determining the function of the report in the system
1. The technical communication process
2. Audience analysis: the problem and a solution
3. The problematic context: the purpose of the report

Designing the report

4. Designing the basic report structure
5. Designing the opening component
6. Designing the discussion component

Writing and editing the report

7. Arranging report segments and units
8. Editing sentences
9. Additional design features: layout and visual aids
10. Report design: guide and checklist

In the chapters that follow, we explain how the audiences affect the design of a report and how to determine the nature and the needs of the specific audiences for any individual report. Then we explain how to define the purposes for a report and how purposes control design.

Next we explain how you establish a complex, two-component structure for a report in order for it to meet multiple organizational needs. The first component addresses audiences who have no need to know the details of the technical investigation but who need to know both the problem and the solution in relatively general terms. The second component addresses audiences who primarily need a more detailed explanation of the problem, technical investigation, and solution.

Finally, we explain how to implement your report design in terms of segments and units (clusters of related paragraphs), appropriate sentence structures, and additional design features such as figures. We conclude by presenting a guide to help you design the report and a checklist to help you revise and edit it before you execute your professional responsibilities by sending the report into the system.

With these explanations as a basis, you should be able to write instrumentally useful reports—even those of you who always had trouble in English classes. Although we will have provided you with a number of procedures by which to design and write your reports, perhaps most importantly we will have provided you with a practical approach to report writing.

Our approach to report design is axiomatic: *The needs of the audiences in the organizational system determine the design of a report.* When you understand this axiom, you can write reports confidently.

CHAPTER 2

Audience analysis: The problem and a solution

Before designing a report, the engineer must identify and analyze the backgrounds and needs of his primary and secondary audiences so that he can design the report to meet the different needs of these audiences. To do this he should use a systematic procedure rather than a piecemeal method for audience analysis.

Every communication situation involves three fundamental components: a writer, a message, and an audience. However, many report writers treat the communication situation as if there were only two components: a writer and his message. Writers often ignore their readers because writers are preoccupied with their own problems and with the subject matter of the communication. The consequence is a poorly designed, ineffective report.

As an example, a student related to the class her first communication experience on a design project during summer employment with an automobile company. After she had been working on her assignment for a few weeks, her supervisor asked her to jot him a memo explaining what she was doing. Not wanting to take much time away from her work and not thinking the report very important, she gave him a handwritten memo and continued her technical activities. Soon after, the department manager inquired on the progress of the project. The supervisor immediately responded that he had just had a progress report, and thereupon forwarded the engineer's brief memo. Needless to say, the engineer felt embarrassed when her undeveloped and inadequately explained memo became an official report to the organization. The engineer thought her memo was written just to her supervisor, who was quite familiar with her assignment. Due to her lack of experience with organizational behavior, she made several false assumptions about her report audience, and therefore about her report's purpose.

The inexperienced report writer often fails to design his report effectively because he makes several false assumptions about the report writing situation. If the writer would stop to analyze the audience component he would realize that:

1. It is false to assume that the person addressed is the audience.
2. It is false to assume that the audience is a group of specialists in the field.
3. It is false to assume that the report has a finite period of use.
4. It is false to assume that the author and the audience always will be available for reference.
5. It is false to assume that the audience is familiar with the assignment.
6. It is false to assume that the audience has been involved in daily discussions of the material.
7. It is false to assume that the audience awaits the report.
8. It is false to assume that the audience has time to read the report.

Assumptions one and two indicate a writer's lack of awareness of the nature of his report audience. Assumptions three, four, and five indicate his lack of appreciation of the dynamic nature of the system. Assumptions six, seven, and eight indicate a writer's lack of consideration of the demands of day-by-day job activity.

A report has value only to the extent that it is useful to the organization. It is often used primarily by someone other than the person who requested it. Furthermore, the report may be responding to a variety of needs within the organization. These needs suggest that the persons who will use the report are not specialists or perhaps not even technically knowledgeable about the report's subject. The specialist is the engineer. Unless he is engaged in basic research, he usually must communicate with persons representing many different areas of operation in the organization.

In addition, the report is often useful over an extended period of time. Each written communication is filed in several offices. Last year's report can be incomprehensible if the writer did not anticipate and explain his purpose adequately. In these situations, even within the office where a report originated, the author as well as his supervisor will probably not be available to explain the report. Although organizational charts remain unchanged for years, personnel, assignments, and professional roles change constantly. Because of this dynamic process, even the immediate audience of a report sometimes is not familiar with the writer's technical assignment. Thus, the report writer usually must design his report for a dynamic situation.

Finally, the report writer must also be alert to the communication traps in relatively static situations. Not all readers will have heard the coffee break chats that fill in the details necessary to make even a routine recommendation convincing. A report can arrive at a time when the reader's mind is churning with other concerns. Even if it is expected, the report usually meets a reader who needs to act immediately. The reader usually does not have time to read through the whole report; he wants the useful information clearly and succinctly. To the reader, time probably is the most important commodity. Beginning report writers seldom realize they must design their reports to be used efficiently rather than read closely.

The sources of the false assumptions we have been discussing are not difficult to identify. The original source is the artificial communication a student is required to perform in college. In writing only for professors, a student learns to write for audiences of one, audiences who know more than the writer knows, and audiences who have no instrumental interests in what the report contains. The subsequent source, on the job, is the writer's natural attempt to simplify his task. The report writer, relying upon daily contact and familiarity, simply finds it easier to write a report for his own supervisor than to write for a supervisor in a different department. The writer also finds it easier to concentrate upon his own concerns than to consider the needs of his readers. He finds it difficult to address complex audiences and face the design problems they pose.

AUDIENCE COMPONENTS AND PROBLEMS THEY POSE

To write a report you must first understand how your audience poses a problem. Then you must analyze your audience in order to be able to design a report structure that provides an optimum solution. To explain the components of the report audience you must do more than just identify names, titles, and roles. You must determine who your audiences are as related to the purpose and content of your report. "Who" involves the specific operational functions of the persons who will read the report, as well as their educational and business backgrounds. These persons can be widely distributed, as is evident if you consider the operational relationships within a typical organization.

Classifying audiences only according to directions of communication flow along the paths delineated by the conventional organizational chart, we can identify three types of report audiences: *horizontal, vertical,* and *external.* For example, in the organization chart in Figure 2-1, *Part of organization chart for naval ship engineering center,*[1] horizontal audiences exist on each level. The Ship Concept Design Division and the Command and Surveillance Division form horizontal audiences for each other. Vertical audiences exist between levels. The Ship Concept Design Division and the Surface Ship Design Branch form vertical audiences for each other. External audiences exist when any unit interacts with a separate organization, such as when the Surface Ship Design Branch communicates with the Newport News Shipbuilding Company.

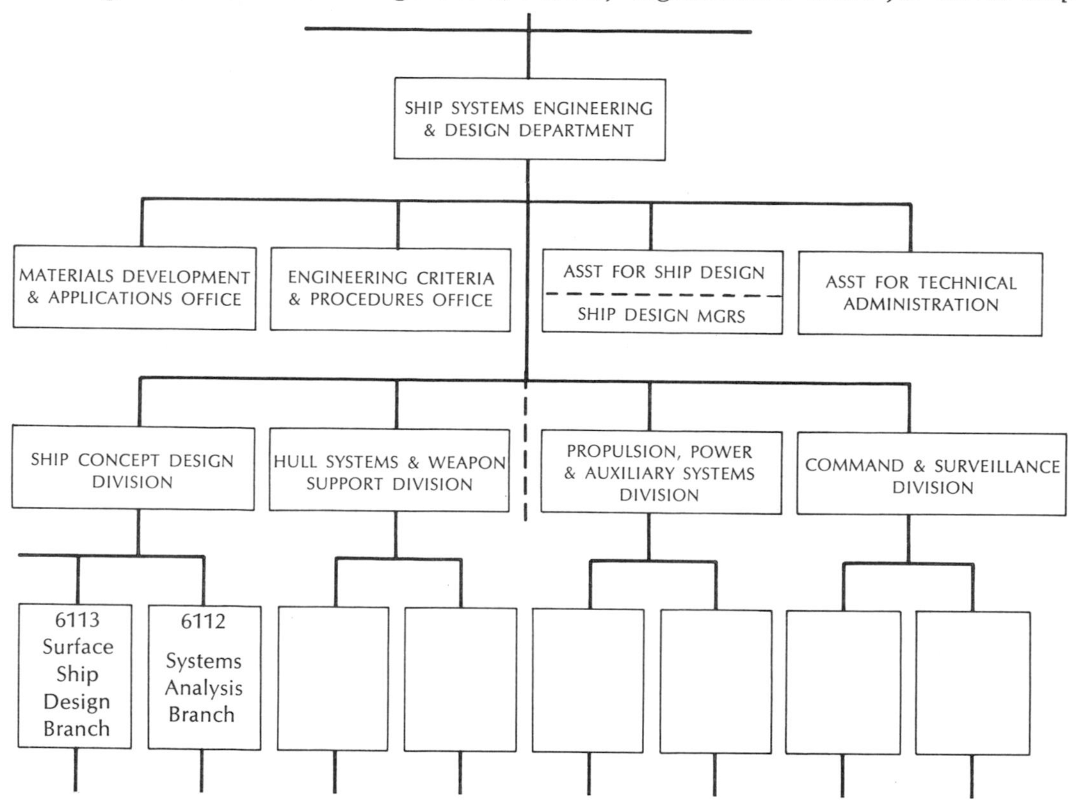

Figure 2–1. Part of organization chart for naval ship engineering center.

[1] A reference in H. B. Benford and J. C. Mathes, *Your Future in Naval Architecture,* Richards Rosen, New York, 1968.

What the report writer first must realize is the separation between him and any of these three types of audiences. Few reports are written for horizontal audiences within the same unit, such as from one person in the Surface Ship Design Branch to another person or project group within the Surface Ship Design Branch itself. Instead, a report at least addresses horizontal audiences within a larger framework, such as from the Surface Ship Design Branch to the Systems Analysis Branch. Important reports usually have complex audiences, that is, vertical and horizontal, and sometimes external audiences as well.

An analysis of the problems generated by horizontal audiences—often assumed to pose few problems—illustrates the difficulties most writers face in all report writing situations. A systems engineer in the Systems Analysis Branch has little technical education in common with the naval architect in the Surface Ship Design Branch. In most colleges he takes only a few of the same mathematics and engineering science courses. The systems engineer would not know the wave resistance theory familiar to the naval architect, although he could use the results of his analysis. In turn, the naval architect would not know stochastics and probability theory, although he could understand systems models. But the differences between these audiences and writers go well beyond differences in training. In addition to having different educational backgrounds, the audiences will have different concerns, such as budget, production, or contract obligations. The audiences will also be separated from the writer by organizational politics and competition, as well as by personality differences among the people concerned.

When the writer addresses a horizontal audience in another organizational unit, he usually addresses a person in an organizational role. When addressed to the role rather than the person, the report is aimed at a department or a group. This means the report will have audiences in addition to the person addressed. It may be read primarily by staff personnel and subordinates. The addressee ultimately may act on the basis of the information reported, but at times he serves only to transfer the report to persons in his department who will use it. Furthermore, the report may have audiences in addition to those in the department addressed. It may be forwarded to other persons elsewhere, such as lawyers and comptrollers. The report travels routinely throughout organizational paths, and will have unknown or unanticipated audiences as well.

Consequently, even when on the same horizontal organizational level, the writer and his audience have little in common beyond the fact of working for the same organization, of having the same "rank," and perhaps of having the same educational level of attainment. Educational backgrounds can be entirely different; more important, needs, values, and uses are different. The report writer may recommend the choice of one switch over another on the basis of a cost-efficiency analysis; his audiences may be concerned for business relationships, distribution patterns, client preferences, and budgets. Therefore, the writer should not assume that his audience has technical competence in the field, familiarity with the technical assignment, knowledge of him or of personnel in his group, similar value perspectives, or even complementary motives. The differences between writer and audience are distinctive, and may even be irreconcilable.

The differences are magnified when the writer addresses vertical audiences. Reports directed at vertical audiences, that is, between levels of an organization chart, invariably have horizontal audience components also. These complex report writing situations pose significant communication problems for the writer. Differences between writer and audience are fundamental. The primary audiences for the reports, especially informal reports, must act or make decisions on the basis of the reports. The reports thus have only instrumental value, that is, value insofar as they can be used effectively. The writer must design his report primarily according to how it will be used.

In addition to horizontal audiences and to vertical audiences, many reports are also directed to external audiences. External audiences, whether they consist of a few or many persons, have the distinctive, dissimilar features of the complex vertical audience. With external audiences these features invariably are exaggerated, especially those involving need and value. An additional complication is that the external audience can judge an entire organization on the basis of the writer's report. And sometimes most important of all, concerns for tact and business relationships override technical concerns.

In actual practice the writer often finds audiences in different divisions of his own company to be "external" audiences. One engineer encountered this problem in his first position after graduation. He was sent to investigate the inconsistent test data being sent to his group from a different division of the company in another city. He found that the test procedures being used in that division were faulty. However, at his supervisor's direction he had to write a report that would not "step on any toes." He had to write the report in such a manner as to have the other division correct its test procedures while not implying that the division was in any way at fault. An engineer who assumes that the purpose of his report is just to explain a technical investigation is poorly prepared for professional practice.

Most of the important communication situations for an engineer during his first five years out of college occur when he reports to his supervisor, department head, and beyond. In these situations, his audiences are action-oriented line management who are uninterested in the technical details and may even be unfamiliar with the assignment. In addition, his audiences become acquainted with him professionally through his reports; therefore, it is more directly the report than the investigation that is important to the writer's career.

Audience components and the significant design problems they pose are well illustrated by the various audiences for a formal report written by an engineer on the development of a process to make a high purity chemical, as listed in Figure 2-2, *Complex audience components for a formal report by a chemical engineer on a process to make a high purity chemical.* The purpose of the report was to explain the process; others would make a feasibility study of the process and evaluate it in comparison to other processes.

The various audiences for this report, as you can determine just by reading their titles, would have had quite different roles, backgrounds, interests, values, needs, and uses for the report. The writer's brief analysis of the audiences yielded the following:

> He could not determine the nature of many of his audiences, who they were, or what the specifics of their roles were.
>
> His audiences had little familiarity with his assignment.
>
> His report would be used for information, for evaluation of the process, and for evaluation of the company's position in the field.
>
> Some of his audiences would have from a minute to a half hour to glance at the report, some would take the report home to study it, and some would use it over extended periods of time for process analysis and for economic and manufacturing feasibility studies.
>
> The useful lifetime of the report could be as long as twenty years.
>
> The report would be used to evaluate the achievements of the writer's department.

The report would be used to evaluate the writing and technical proficiencies of the writer himself.

This report writer classified his audiences in terms of the conventional organization chart. Then to make them more than just names, titles, and roles he asked himself what they would know about his report and how they would use it. Even then he had only partially solved his audience problem and had just begun to clarify the design problems he faced. To do so he needed to analyze his audiences systematically.

A METHOD FOR SYSTEMATIC AUDIENCE ANALYSIS

To introduce the audience problem that report writers must face, we have used the conventional concept of the organization chart to classify audiences as *horizontal, vertical,* and *external.* However, when the writer comes to the task of performing an instrumentally useful audience analysis for a particular report, this concept of the organization and this classification system for report audiences are not very helpful.

First, the writer does not view from outside the total communication system modeled by the company organization chart. He is within the system himself, so his view is always relative. Second, the conventional outsider's view does not yield sufficiently detailed information about the report audiences. A single bloc on the organization chart looks just like any other bloc, but in fact each bloc represents one or several human beings with distinctive roles, backgrounds, and personal characteristics. Third, and most importantly, the outsider's view does not help

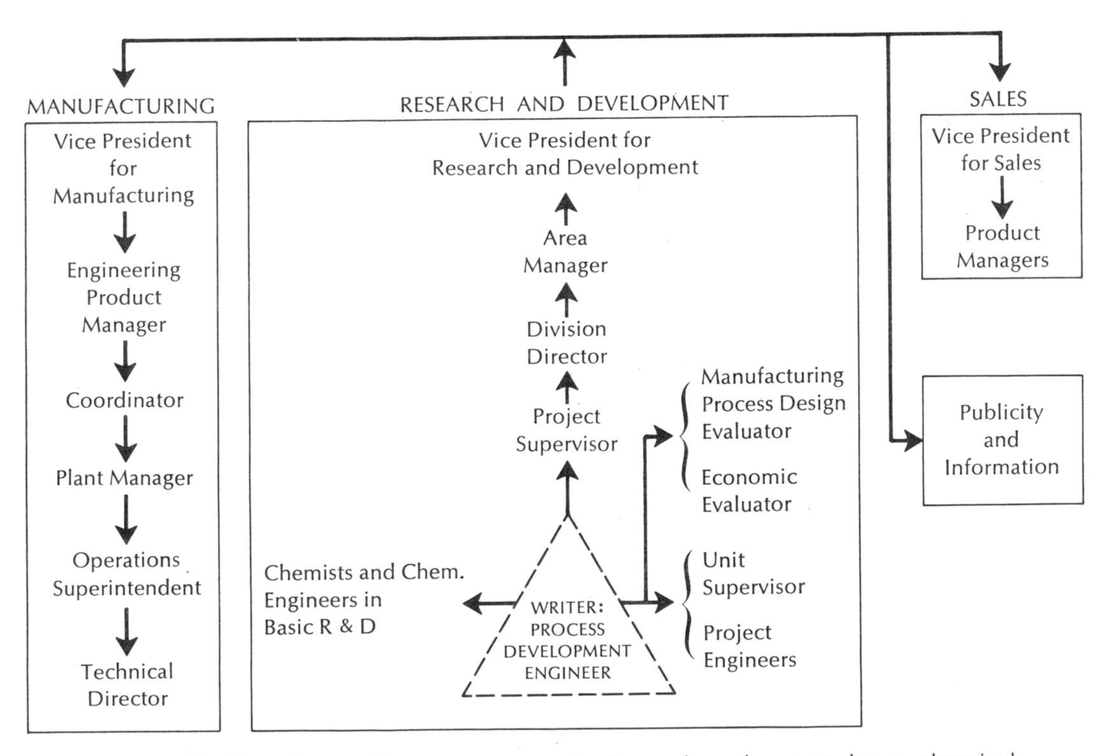

Figure 2–2. Complex audience components for a formal report by a chemical engineer on a process to make a high purity chemical.

much to clarify the specific routes of communication, as determined by audience needs, which an individual report will follow. The organization chart may describe the organization, but it does not describe how the organization functions. Thus many of the routes a report follows—and consequently the needs it addresses—will not be signaled by the company organization chart.

In short, the conventional concept of report audiences derived from organization charts is necessarily abstract and unspecific. For that reason a more effective method for audience analysis is needed. In the remaining portion of this chapter we will present a three-step procedure. The procedure calls for preparing an egocentric organization chart to identify individual report readers, characterizing these readers, and classifying them to establish priorities. Based upon an egocentric view of the organization and concerned primarily with what report readers need, this system should yield the information the writer must have if he is to design an individual report effectively.

Prepare an Egocentric Organization Chart

An egocentric organization chart differs from the conventional chart in two senses. First, it identifies specific individuals rather than complex organizational units. A bloc on the conventional chart may often represent a number of people, but insofar as possible the egocentric chart identifies particular individuals who are potential readers of reports a writer produces. Second, the egocentric chart categorizes people in terms of their proximity to the report writer rather than in terms of their hierarchical relationship to the report writer. Readers are not identified as organizationally superior, inferior, or equal to the writer, but rather as near or distant from the writer. We find it effective to identify four different degrees of distance as is illustrated in Figure 2-3, *Egocentric organization chart.* In this figure, with the triangle representing the writer, each circle is an individual reader identified by his organizational title and by his primary operational concerns. The four degrees of distance are identified by the four concentric rings. The potential readers in the first ring are those people with whom the writer associates daily. They are typically those people in his same office or project group.

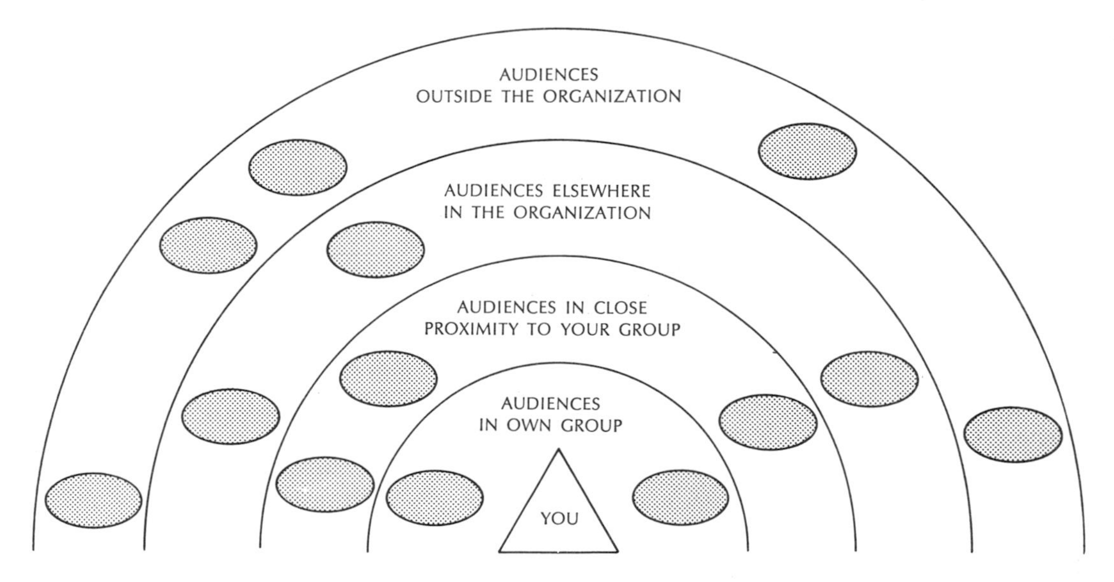

Figure 2–3. Egocentric organization chart.

The readers in the second ring are those people in other offices with whom the writer must normally interact in order to perform his job. Typically, these are persons in adjacent and management groups. The readers in the third ring are persons relatively more distant but still within the same organization. They are distant management, public relations, sales, legal department, production, purchasing, and so on. They are operationally dissimilar persons. The readers in the fourth ring are persons beyond the organization. They may work for the same company but in a division in another city. Or they may work for an entirely different organization.

Having prepared the egocentric organization chart, the report writer is able to see himself and his potential audiences from a useful perspective. Rather than seeing himself as an insignificantly small part of a complex structure—as he is apt to do with the conventional organizational chart—the writer sees himself as a center from which communication radiates throughout an organization. He sees his readers as individuals rather than as faceless blocs. And he sees that what he writes is addressed to people with varying and significant degrees of difference.

A good illustration of the perspective provided by the egocentric organization chart is the chart prepared by a chemical engineer working for a large corporation, Figure 2-4, *Actual egocentric organization chart of an engineer in a large corporation.* It is important to notice how the operational concerns of the persons even in close proximity vary considerably from those of the development engineer. What these people need from reports written by this engineer, then, has little to do with the processes by which he defined his technical problems.

The chemical engineer himself is concerned with the research and development of production processes and has little interest in, or knowledge of, budgetary matters. Some of the audiences in his group are chemists concerned with production—not with research and development. Because of this they have, as he said, "lost familiarity with the technical background, and instead depend mostly on experience." Other audiences in his group are technicians concerned only with operations. With only two years of college, they have had no more than introductory chemistry courses and have had no engineering courses.

Still another audience in his group is his group leader. Rather than being concerned with development, this reader is concerned with facilities and production operations. Consequently, he too is "losing familiarity with the technical material." Particularly significant for the report writer is that his group leader in his professional capacity does not use his B.S.Ch.E. degree. His role is that of manager, so his needs have become administrative rather than technical.

The concerns of the chemical engineer/report writer's audiences in close proximity to his group change again. Instead of being concerned with development or production operations, these audiences are primarily concerned with the budget. They have little technical contact, and are described as "business oriented." Both Manager I and Manager II are older, and neither has a degree in engineering. One has a Ph.D. degree in chemistry, the other an M.S. degree in technology. Both have had technical experience in the lab, but neither can readily follow technical explanations. As the chemical engineer said, both would find it "difficult to return to the lab."

The report writer's department head and the other persons through whom the group communicates with audiences elsewhere in the organization, and beyond it, have additional concerns as well as different backgrounds. The department head is concerned with budget, personnel, and labor relations. The person in contact with outside funding units—in this case, a government agency—has business adminis-

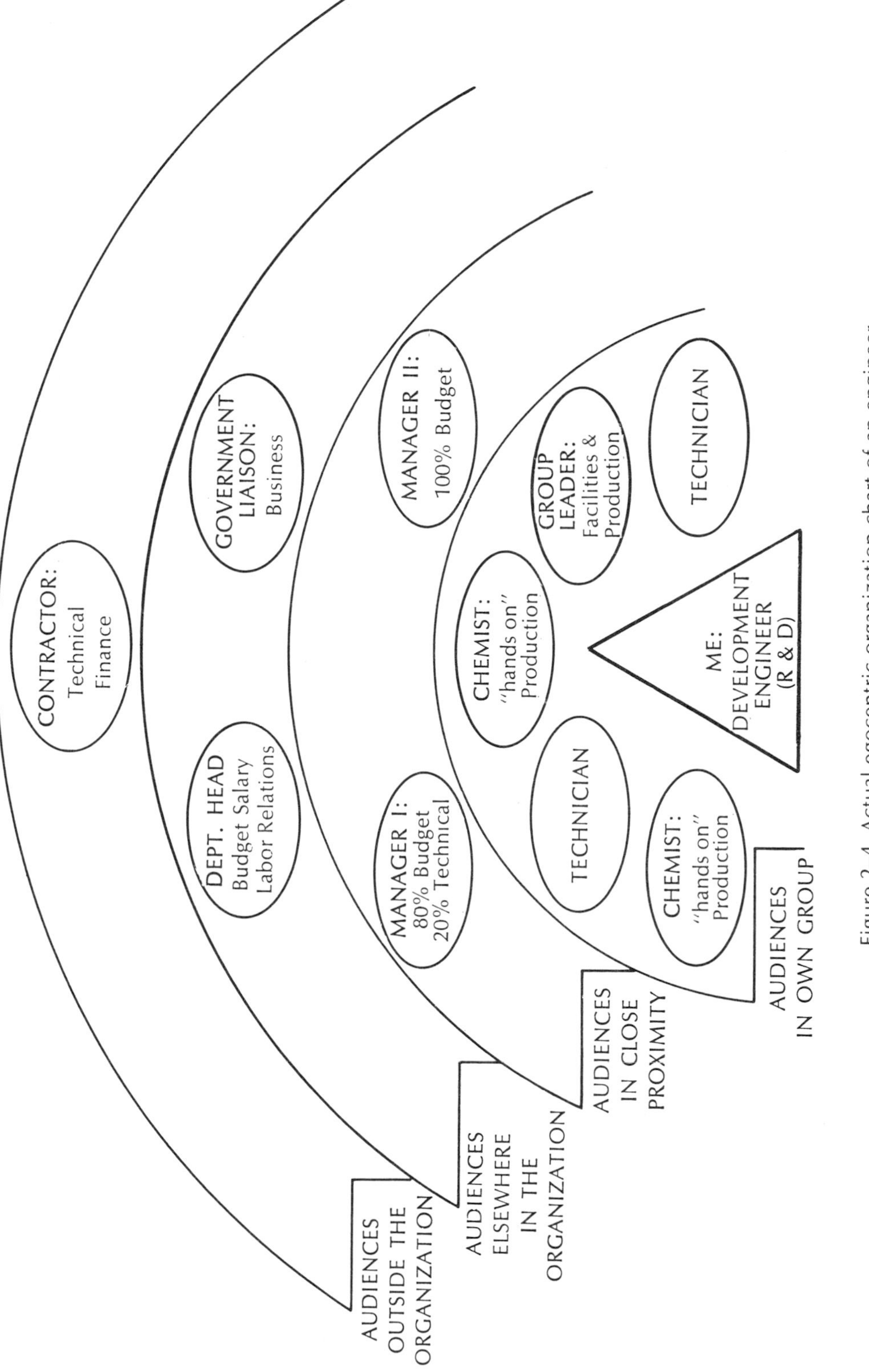

Figure 2–4. Actual egocentric organization chart of an engineer in a large corporation.

tration degrees and is entirely business oriented. The person in contact with subcontractors has both technical and financial concerns.

Notice that when this writer examined his audiences even in his own group as well as those in close proximity to him, he saw that the natures, the backgrounds, and especially the operational concerns of his audiences vary and differ considerably. As he widened the scope of his egocentric organization chart, he knew less and less about his audiences. However, he could assume they will vary even more than those of the audiences in close proximity.

Thus, in the process of examining the audience situation with an egocentric organization chart, a report writer can uncover not only the fact that audiences have functionally different interests, but also the nature of those functional differences. He can proceed to classify the audiences for each particular report in terms of audience needs.

Preparation of the egocentric organization chart is the first step of your procedure of systematic audience analysis. Notice that this step can be performed once to describe your typical report audience situation but must be particularized for each report to define the audiences for that report. Having prepared the egocentric chart once, the writer revises his chart for subsequent reports by adding or subtracting individual audiences.

Characterize the Individual Report Readers

In the process of preparing the egocentric organization chart, you immediately begin to think of your individual report readers in particular terms. In preparing the egocentric chart discussed above, the report writer mentioned such items as a reader's age, academic degrees, and background in the organization as well as his operational concerns. All of these particulars will come to mind when you think of your audiences as individuals. However, a systematic rather than piecemeal audience analysis will yield more useful information. The second step of audience analysis is, therefore, a systematic characterization of each person identified in the egocentric organization chart. A systematic characterization is made in terms of *operational, objective,* and *personal* characteristics.

The *operational characteristics* of your audiences are particularly important. As you identify the operational characteristics for a person affected by your report, try to identify significant differences between his or her role and yours. What are his professional values? How does he spend his time? That is, will his daily concerns and attitudes enable him to react to your report easily, or will they make it difficult for him to grasp what you are talking about? What does he know about your role, and in particular, what does he know or remember about your technical assignment and the organizational problem that occasioned your report to come to him? You should also consider carefully what he will need from your report. As you think over your entire technical investigation, ask yourself if that person will involve staff personnel in action on your report, or if he will in turn activate other persons elsewhere in the organization when he receives the report. If he should, you must take their reactions into account when you write your report.

In addition, you should ask yourself, "How will my report affect his role?" A student engineer recently told us of an experience he had during summer employment when he was asked to evaluate the efficiency of the plant's waste treatment process. Armed with his fresh knowledge from advanced chemical engineering courses, to his surprise he found that, by making a simple change in the process, the company could save more than $200,000 a year. He fired off his report with great anticipation of glowing accolades—none came. How had his report affected the roles of some of his audiences? Although the writer had not considered the report's consequences when he wrote it, the supervisor, the manager, and related personnel now were faced with the problem of accounting for their waste of $200,000 a year. It should have been no surprise that they were less than elated over his discovery.

By *objective characteristics* we mean specific, relevant background data about the person. As you try to identify his or her educational background, you may note differences you might have otherwise neglected. Should his education seem to approximate yours, do not assume he knows what you know. Remember that the half-life of engineering education today is about five years. Thus, anyone five to ten years older than you, if you are recently out of college, probably will be only superficially familiar with the material and jargon of your advanced technical courses. If you can further identify his past professional experiences and roles, you might be able to anticipate his first-hand knowledge of your role and technical activities as well as to clarify any residual organizational commitments and value systems he might have. When you judge his knowledge of your technical area, ask yourself, "Could he participate in a professional conference in my field of specialization?"

For *personal characteristics,* when you identify a person by name, ask yourself how often the name changes in this organizational role. When you note his or her approximate age, remind yourself how differences in age can inhibit communication. Also note personal concerns that could influence his reactions to your report.

A convenient way to conduct the audience analysis we have been describing and to store the information it yields is to use an analysis form similar to the one in Figure 2-5, *Form for characterizing individual report readers.* It may be a little time-consuming to do this the first time around, but you can establish a file of audience characterizations. Then you can add to or subtract from this file as an individual communication situation requires.

One final point: This form is a means to an end rather than an end in itself. What is important for the report writer is that he think systematically about the questions this form raises. The novice usually has to force himself to analyze his audiences systematically. The experienced writer does this automatically.

Classify Audiences in Terms of How They Will Use Your Report

For each report you write, trace out the communication routes on your egocentric organization chart and add other routes not on the chart. Do not limit these routes to those specifically identified by the assignment and the addressees of the report. Rather, think through the total impacts of your report on the organization. That is, think in terms of the first, second, and even some third-order consequences of your report, and trace out the significant communication routes involved. All of these consequences define your actual communication.

When you think in terms of consequences, primarily you think in terms of the uses to which your report will be put. No longer are you concerned with your technical investigation itself. In fact, when you consider how readers will use your report, you realize that very few of your potential readers will have any real interest in the details of your technical investigation. Instead, they want to know the answers to such questions as "Why was this investigation made? What is the significance of the problem it addresses? What am I supposed to do with the results of this investigation? What will it cost? What are the implications—for sales, for production, for the unions? What happens next? Who does it? Who is responsible?"

It is precisely this audience concern for nontechnical questions that causes so much trouble for young practicing engineers. Professionally, much of what the

NAME: TITLE:

A. OPERATIONAL CHARACTERISTICS:

1. His role within the organization and consequent value system:

2. His daily concerns and attitudes:

3. His knowledge of your technical responsibilities and assignment:

4. What he will need from your report:

5. What staff and other persons will be activitated by your report through him:

6. How your report could affect his role:

B. OBJECTIVE CHARACTERISTICS:

1. His education—levels, fields, and years:

2. His past professional experiences and roles:

3. His knowledge of your technical area:

C. PERSONAL CHARACTERISTICS:

Personal characteristics that could influence his reactions—age, attitudes, pet concerns, etc.

Figure 2–5. Form for characterizing individual report readers.

engineer spends his time doing is, at most, of only marginal concern to many of his audiences. His audiences ask questions about things which perhaps never entered his thoughts during his own technical activities when he received the assignment, defined the problem, and performed his investigation. These questions, however, must enter into his considerations when he writes his report.

Having defined the communication routes for a report you now know what audiences you will have and what questions they will want answered. The final step in our method of audience analysis is to assign priorities to your audiences. Classify them in terms of how they will use your report. In order of their importance to you (not in terms of their proximity to you), classify your audiences by these three categories:

> *Primary audiences*—who make decisions or act on the basis of the information a report contains.
>
> *Secondary audiences*—who are affected by the decisions and actions.
>
> *Immediate audiences*—who route the report or transmit the information it contains.

The *primary audience* for a report consists of those persons who will make decisions or act on the basis of the information provided by the report. The report overall should be designed to meet the needs of these users. The primary audience can consist of one person who will act in an official capacity, or it can consist of several persons representing several offices using the report. The important point here is that the primary audience for a report can consist of persons from any ring on the egocentric organization chart. They may be distant or in close proximity to the writer. They may be his organizational superiors, inferiors, or equals. They are simply those readers for whom the report is primarily intended. They are the top priority users.

In theory at least, primary audiences act in terms of their organizational roles rather than as individuals with distinctive idiosyncracies, predilections, and values. Your audience analysis should indicate when these personal concerns are likely to override organizational concerns. A typical primary audience is the decision maker, but his actual decisions are often determined by the evaluations and recommendations of staff personnel. Thus the report whose primary audience is a decision maker with line responsibility actually has an audience of staff personnel. Another type of primary audience is the production superintendent, but again his actions are often contingent upon the reactions of others.

In addition, because the report enters into a system, in time both the line and staff personnel will change; roles rather than individuals provide continuity. For this reason, it is helpful to remember the words of one engineer when he said, "A complete change of personnel could occur over the lifetime of my report." The report remains in the file. The report writer must not assume that his primary audience will be familiar with the technical assignment. He must design the report so that it contains adequate information concerning the reasons for the assignment, details of the procedures used, the results of the investigation, and conclusions and recommendations. This information is needed so that any future component of his primary audience will be able to use the report confidently.

The *secondary audiences* for a report consist of those persons other than primary decision makers or users who are affected by the information the report transmits into the system. These are the people whose activities are affected when a primary audience makes a decision, such as when production supervision has to adjust to management decisions. They must respond appropriately when a primary audience acts, such as when personnel and labor relations have to accommodate production line changes. The report writer must not neglect the needs of his secondary audiences. In tracing out his communication routes, he will identify several secondary audiences. Analysis of their needs will reveal what additional information the report should contain. This information is often omitted by writers who do not classify their audiences sufficiently.

The *immediate audience* for a report are those persons who route the report or transmit the information it contains. It is essential for the report writer to identify his immediate audiences and not to confuse them with his primary audiences. The immediate audience might be the report writer's supervisor or another middle management person. Yet usually his role will be to transmit information rather than to use the information directly. An information system has numerous persons who transmit reports but who may not act upon the information or who may not be affected by the information in ways of concern to the report writers. Often, a report is addressed to the writer's supervisor, but except for an incidental memo report, the supervisor serves only to transmit and expedite the information flow throughout the organizational system.

A word of caution: at times the immediate audience is also part of the primary audience; at other times the immediate audience is part of the secondary audience. For each report you write you must distinguish those among your readers who will function as conduits to the primary audience.

As an example of these distinctions between categories of report audiences, consider how audiences identified on the egocentric organization chart, Figure 2-4, can be categorized. Assume that the chemical engineer writes a report on a particular process improvement he has designed. The immediate audience might be his Group Leader. Another would be Manager I, transmitting the report to Manager II. The primary audiences might be Manager II and the Department Head; they would ask a barrage of nontechnical questions similar to those we mentioned a moment ago. They will decide whether or not the organization will implement the improvement recommended by the writer. The Department Head also could be part of the secondary audience by asking questions relating to labor relations and union contracts. Other secondary audiences, each asking different questions of the report, could be:

The person in contact with the funding agency, who will be concerned with budget and contract implications.

The person in contact with subcontractors, determining how they are affected.

The Group Leader, whose activities will be changed.

The "hands on" chemist, whose production responsibilities will be affected.

The technicians, whose job descriptions will change.

In addition to the secondary audiences on the egocentric organization chart, the report will have other secondary audiences throughout the organization—technical service and development, for example, or perhaps waste treatment.

At some length we have been discussing a fairly detailed method for systematic audience analysis. The method may have seemed more complicated than it actually is. Reduced to its basic ingredients, the method requires you, first, to identify all the individuals who will read the report, second, to characterize them, and third, to classify them. The *Matrix for audience analysis,* Figure 2-6, is a convenient device

Characteristics / Types of audiences	Operational	Objective	Personal
Primary	①	④	⑦
Secondary	②	⑤	⑧
Immediate	③	⑥	⑨

Figure 2–6. Matrix for audience analysis.

for characterizing and classifying your readers once you have identified them. At a glance, the matrix reveals what information you have and what information you still need to generate. Above all, the matrix forces you to think systematically. If you are able to fill in a good deal of specific information in each cell (particularly in the first six cells), you have gone a long way towards seeing how the needs of your audiences will determine the design of your report.

We have not introduced a systematic method for audience analysis with the expectation that it will make your communication task easy. We have introduced you to the problems you must account for when you design your reports—problems you otherwise might ignore. You should, at least, appreciate the complexity of a report audience. Thus, when you come to write a report, you are less likely to make false assumptions about your audience. To develop this attitude is perhaps as important as to acquire the specific information the analysis yields. On the basis of this attitude, you now are ready to determine the specific purpose of your report.

CHAPTER 3

The problematic context: The purpose of the report

The technical task *and the* rhetorical task *are two very different things. To solve a technical problem requires one sense of purpose; but to describe the solution for someone else, or defend it against alternatives, or sell it to management, or foresee its consequences—these require a new sense of purpose.*

The first task in technical communication is to discover precisely who your audiences are. The first question to yourself should be, "With whom am I trying to communicate?" Until you have asked yourself that question, and thoroughly answered it, you are not ready to make any of the subsequent choices you will have to make about the design of your piece of communication. You will not know where to begin, what to include, what to leave out, how the report should be organized, or even what stylistic level to use. You will not know any of the dozens of things you must decide upon before you can put your report onto paper and into someone else's hands.

The central purpose of the last chapter was to describe a method that you can use to discover who your audiences are. If you make careful use of the method, you should be able to get a good sense of the demands imposed upon you by your particular audiences, and you should be able to move on to defining your rhetorical purpose. The central purpose of this chapter, therefore, is to describe this next step: formulating the purpose statement of the report. Specifically, we will try to answer three questions in this chapter: Why is it imperative that you state your purpose? How should you state your purpose? Where should you state your purpose?

WHY IS IT IMPERATIVE THAT YOU STATE YOUR PURPOSE?

There are four reasons. First, if you do not state your purpose, you cannot assume that the audiences we have discussed in the previous chapter will know what you intend to accomplish with your report. Second, if you do not formulate your purpose, you may not be forced to turn aside from your technical activities to report writing activities. Third, if you do not formulate your purpose, you may not yourself understand the organizational aspects of the technical problem you have investigated. Finally, if you do not formulate your purpose, you are not behaving in a way fundamentally consistent with your discipline. In the following pages we will discuss each of these four points in turn.

A novice writer often fails to consider the nature of his audiences. He will sometimes argue, "It isn't necessary for me to state my purpose because it will certainly be clear to the man who assigned the task. After all, he asked for it; why wouldn't he understand it?" Yet as we saw in the previous chapter, the man who assigned the task is almost certainly not the only person who needs the results. Your primary audience may be elsewhere in the organization. Moreover, even if the man who assigned the task were the only one who needed the results, he is a busy man who cannot be expected to remember the precise details of a task he assigned six days or six weeks ago. The report may have been the most important thing in your recent experience, but it is almost certainly not that large in his eyes. Thus you will simply have to tell him the purpose you are setting out to accomplish and why, or risk his understanding neither the technical problem you were investigating nor the reasons you are now sending a report to him.

As you can further see on the basis of what we explained in the previous chapter, your report will have a number of readers—not just one—over its useful life of weeks, months, or even years. Some of the readers will never have heard of you or your assignment. Others may be only vaguely familiar with your technical activities. The purpose of your report will never be apparent to all of the readers unless they are explicitly told.

Each of the readers brings to your report his own interests, his own values, and his own immediate concerns. Readers never see quite the same things in a report simply because they do not play the same roles in an organizational structure. The company comptroller may be far more interested in the costs section of your report than in any other. The head of Research and Development will worry first about the technical merit of your proposal and then about its cost. Meanwhile, the foreman out in the shop will worry about the floor space required for all that hardware you request. In short, different readers bring to your report different perceptions. To assume that they all have the same needs, share the same values, and possess the same information is simply to ignore the differences that exist among professionals in organizations and to invite them to distort your purpose.

Your report is *your report*, not simply a mass of raw data from which each reader assembles his own version of your work. You want each of your readers to be perfectly clear on what you are attempting to do with this particular report—what *your* specific rhetorical purpose is. Leaving it up to the reader to supply his own sense of your purpose is risky at best, and in most cases downright foolish.

The second reason you must formulate a purpose statement is that doing so forces you to turn aside from your technical activity to your rhetorical activity. The *technical task* and the *rhetorical task* are two very different things. To solve the technical problem required one sense of purpose. To describe the solution for production personnel, to defend the solution against alternatives, to examine the costs involved, to sell it to management, or to foresee the consequences of the solution requires a new sense of purpose. The technical objective is antecedent to, and separate from, the rhetorical objective. Putting aside his "technical hat" and replacing it with his "reporter's hat," the investigator assumes a new role. Previously he sought to design a solution to a technical problem; now he seeks to design a report that will tell a complex audience what they need to know about the technical solution. Without a specific redefinition of objective when he begins to design the report, and without a conscious recognition that his role has changed, the writer is likely to make his report a wasteful, ineffectively arranged recitation of past action. If there were no other reasons for stating the purpose of a report, that would be sufficient justification.

The failure of the writer to change hats—to turn toward his or her audience and define his or her rhetorical purpose—accounts for the incomprehensibility of many reports that seem to have a subject, a structure, and even acceptable writing. Although you cannot quite pinpoint what is wrong, you still cannot understand the report. Here is the beginning of one such report, an absolutely classic example of a writer's talking essentially to himself:

negative example

> The symmetrically spiraled curve program was designed and written to compute the basic characteristics of a symmetrically spiraled circular curve. In addition to those characteristics, the program will also compute the deflection angles required to set stakes at quarter stations (every 25 feet) along the curve.
>
> **Data Cards**
>
> Two data cards are required by the symmetrically spiraled curve program for every curve that is to be computed . . . .

The report continues in this manner for six more pages, which we will spare you. The writer communes with himself to the very end, immersed in the particulars of his technical task. He plunges in without first giving his readers an overview of the purpose of the report, the "why." Perhaps that is natural enough. He is just forgetting that his readers do not know as much as he does about the context of the report and the problem it addresses. Furthermore, he is forgetting that some of his readers do not share his technical interests. But most importantly, he is forgetting his rhetorical purpose—addressing audience needs. It is imperative that he state his purpose so the audiences can see why they must continue reading. In the example above, unless the reader already knows all about the symmetrically spiraled curve program, the memo makes no sense whatever. And if the reader does already know, what is the sense of his going on? It would have been an easy matter for this writer to explain the technical problem he was attacking and the specific rhetorical purpose of his report. Yet this writer did not apparently realize that the effort was necessary. The following rewritten version makes apparent how easy it would have been to explain the technical problem and rhetorical purpose of his report. Chances are you will find this version a good deal clearer:

> Symmetrically spiraled curves accommodate the natural driving path of the motorist. When properly designed, these curves produce a more comfortable and safer ride. However, engineers have hesitated to use these curves because of the difficulty in calculating them. Consequently, the symmetrically spiraled curve program was designed and written to quickly compute the basic characteristics of the curve. This memo explains how to arrange the necessary data on computer cards so that highway engineers can use the symmetrically spiraled curve program to design a curve.

In this version the writer introduces the rhetorical purpose of the report. It is to explain to someone how to do something; that is, how to apply the results of the investigation. In the earlier version he simply assumed that his purpose was to introduce only the technical investigation itself. In other words, he failed to change hats.

While we are on the subject of this example, notice also that the purpose statement in the rewritten version probably helped the writer design his report as much as it helped his readers understand it. Formulation of a purpose statement helps the writer set up a conscious control over what goes into his report and how it is arranged. As we will explain in Chapter six, your technical investigation itself does not help you to do this. By stating the purpose, you are providing yourself with a focus, a means of delineating the relevant material. If you fail to set up this conscious control you risk being like the traveler who does not specify his destination precisely and who is consequently never really sure of his route or how far he has progressed at any given moment. If you articulate for yourself—and obviously for your reader—precisely what you intend to do in a piece of reportage, you stand a much better chance of doing it.

The third reason for formulating an explicit purpose statement is that it forces the writer to examine the organizational context that gives rise, first to the assignment, and then to the report. To this point in our analysis of purpose statements we have been discussing the writer as if he always understood his own purpose. But unfortunately much technical work is done by investigators who do not understand the problem. They may understand the specific technical questions, the "what," but they may be blind to the "why," the need or organizational issue lying behind those questions. Now it may strike you that when an engineer has worked on a project for two or three weeks—or months—he could hardly be uncertain about what he has done. Thus you may feel that there is no need at this point for him to formulate a purpose statement *for his own benefit.* Yet that is precisely what we are saying he should do. Buried in the particulars of a technical task, the "what" of engineering, you can easily lose sight of your motives, the "why."

As an example of this blindness to the problematic context, we remember a student who was trying to design a two-man earth auger in a mechanical engineering design course. Upon receiving his assignment, the student immediately immersed himself in the details of the type of clutch to be used, the power of the engine, the size of the auger flutes, and so on. He soon found himself totally lost. Only when we pushed him to examine why anyone would want to design a two-man earth auger (aside from the fact that he was required to do it to pass a course), did he realize how deficiencies in previous designs gave rise to the need and in fact provided the criteria for a new design. The problem was not what kind of clutch, engine, or flutes to use; the problem was to design a two-man earth auger that is lighter, safer, and more maneuverable than the cumbersome ones presently on the market. As soon as the student understood the problem in that sense, he had no difficulty in designing the earth auger and writing the proposal for his design.

The problem this particular student was tackling in his design project was not basically a technical problem. Instead, it was a marketing problem with a technical solution. He just had trouble in seeing it that way at first. And his was a difficulty common in the classroom because technical professors often do not explain the problems that give rise to their design assignments. This difficulty is also common in industry, where supervisors do not explain and engineers do not ask what the nature of the problem is.

As an example of this industrial situation, we recall the assignment of a young engineer working for an automobile manufacturer. He was asked to examine the aluminum panels on a number of older foreign automobiles for any evidence of wear and damage. He dutifully listed all the cars he examined, itemized the defects in the aluminum panels, and submitted his report. Never once in the whole process did he consider the reason for the technical investigation. The reason turned out to be quite specific: it was to determine which of the steel panels on current automobiles might reasonably be replaced with aluminum panels in order to save weight. Thus his report, which could have made clear recommendations for specific design changes, ended up being a somewhat purposeless catalogue of observations. He had left it to someone else to be the problem solver.

Both of these examples show how the engineer sometimes must force himself to define the organizational needs and issues behind his assignment. The needs and issues are certainly there, but frequently are not provided. Although the engineer can proceed meticulously, he cannot proceed purposefully until he knows the problem.

The three reasons previously outlined should be justification enough for explicitly stating your purpose. Both your reader's understanding and your own are advanced when you state your purpose. But there is yet another reason for technical report writers to state their purpose. The nature of the technical problem solver's discipline makes clarity a fundamental principle. That seems obvious, yet judging from many report writers' behavior, the principle is not as obvious or as consistently observed as it should be.

Imprecision is undesirable in technical professions. Scientists and engineers should be accurate, precise, methodical, and direct. Any inaccuracy or uncertainty will produce questionable results. A report without a clear purpose or which can be interpreted as having several apparent purposes is a bad report. A report which primarily intends to defend a hypothesis obviously differs from one which primarily intends to describe something objectively. Yet sometimes it is almost impossible to tell whether the author of a report saw his function as persuasive or expository. Is he advocating something, or not? If so, precisely what is he advocating? The reader must know immediately what the author intends. Thus we may say that because your aim as a scientist or engineer is accuracy, precision, and certitude, to fail to state the most basic aim of a report is fundamentally unprofessional.

Before we finish with our discussion of the importance of stating your purpose, let us point out that an explicit statement of purpose goes against the old "freshman English" grain for some writers. They feel, as some students have told us, that explicitly telling your reader what you are up to is somehow "too obvious." They say, "If I state the purpose for him, the reader won't be interested enough to go on. I'll lose all the suspense." And of course that is an understandable attitude. We would all like to be the writer who can grab the reader's attention and fascinate him by sheer cleverness. But the fact is, ambiguity, suspense, and surprise are not appropriate characteristics of report writing. Engineering should not be a mystery. So, even if it goes against the grain to state your purpose, the very nature of the discipline in which you work demands it in every piece of technical discourse you produce.

If you are faithful to your discipline, then, you state your purpose. When you do, you help your reader and yourself as well. Any writer performs his task best if he consciously articulates his rhetorical purpose for himself before he ever picks up the pencil. Any reader understands most clearly when he is told directly what he is supposed to learn from a report. The report, like the technical task itself, must be a deliberate piece of design work which fulfills a specific function. It is not simply a record of what has happened.

HOW SHOULD YOU STATE YOUR PURPOSE?

We come now to the second of the three questions we posed at the beginning of this chapter: *how* should you state the purpose of a report? [1] It is here that many writers have difficulty. Yet it is here that the greatest care must be taken, for if a writer cannot state his purpose well, it matters very little that he theoretically realizes the importance of purpose statements. It is the formulation of purpose statements that really counts. In the following pages of this section of the chapter, we will explain a method which, in effect, constitutes an algorithm for problem statement.

The most important thing to realize about stating the purpose of a report is that your purpose never exists in a vacuum. For every "what," there is a "why." The organizational context of technical reportage contains a problem and an audience who have an interest in it. As far as the organization is concerned, it may be a problem still not fully understood, much less solved. It may be a problem for which tentative solutions are just being considered. It may be a problem which has already been met and successfully resolved. In short, it may be an ill-defined, future problem; a partially-defined, present problem; or a well-defined, past problem. But most of the time there is a problem. There is an organizational problem which led to your investigation and which is now leading to your report.

Thus, your rhetorical purpose cannot be stated unless you first explain the organizational and technical problems at issue. To put it differently, if there were no problem there would be no need for your report. To write the report without explaining the problem that gave rise to it—as many writers curiously do—is to leave out the reason for the whole thing.

The organization has a problem. Perhaps you discover it yourself; perhaps it is assigned to you. In any event, your role as an engineer is to be a problem solver. Your function is to perform one, several, or all of these tasks: to articulate and define the problem; to explore the problematic data; to formulate hypotheses; to test hypotheses; to find workable solutions. Graphically we might represent the initial situation as in Figure 3-1, *Function of the engineer as technical problem solver.*

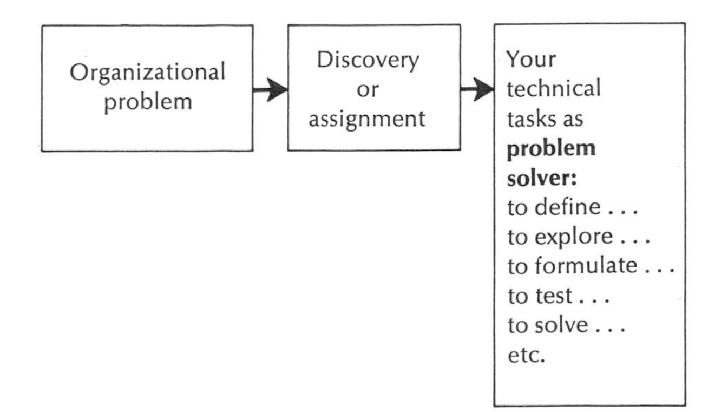

Figure 3–1. Function of the engineer as technical problem solver.

Yet clearly you are not the only one with an interest in the problem, not the only one for whom it has significance and consequence. Therefore, your function is not

---

[1] Recent scholarship on the subject of problem statement is helpful here. See Richard E. Young, Alton L. Becker, Kenneth L. Pike, *Rhetoric: Discovery and Change* (New York: Harcourt, Brace and World, 1970), especially Chap. 5.

solely to be a problem solver. Your role is to interact with the organization; thus you must be a writer as well. Here you put on the report writer's hat. On paper you must tell people in the organization what you have done for them. Thus we must introduce your audience into our graphic representation. And we must also redefine your function, taking account of the fact that you have both a *technical task* and a *rhetorical task*. Graphically, then, we have this situation in Figure 3-2, *Functions of the engineer as problem solver and report writer.*

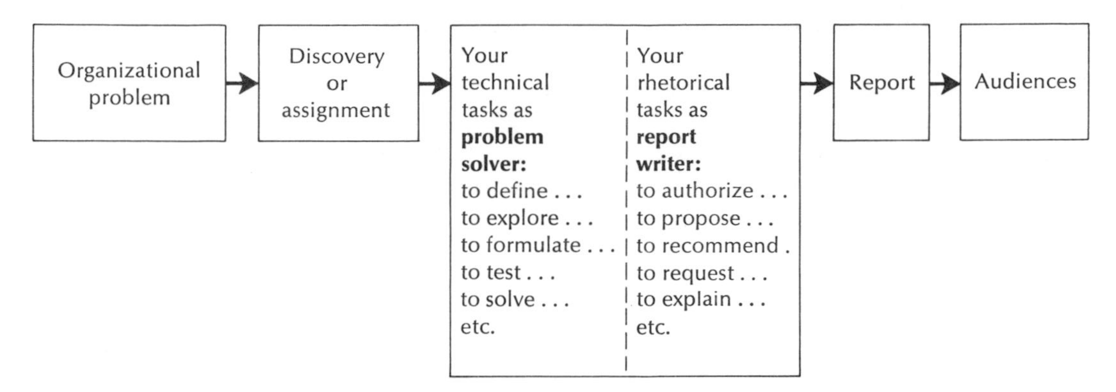

Figure 3–2. Functions of the engineer as problem solver and report writer.

As this figure indicates, you perform what might be termed "transform functions" between the problem and those other people interested in and affected by it. And thus neither of your two functions, technical or rhetorical, is sufficient by itself to assure the kind of understanding and organizational response you seek. Only when you perform both of your functions successfully do you enable your audiences to perceive both the problem and solution and thereby respond appropriately. Graphically, the relationship in its complete form becomes something like that in Figure 3-3, *Understanding and organizational response as the consequences of the engineer's two functions.* Here your report functions as a tool, a means by which your audiences can respond as your purpose would have them respond. If you are successful, the understanding of the problem that your

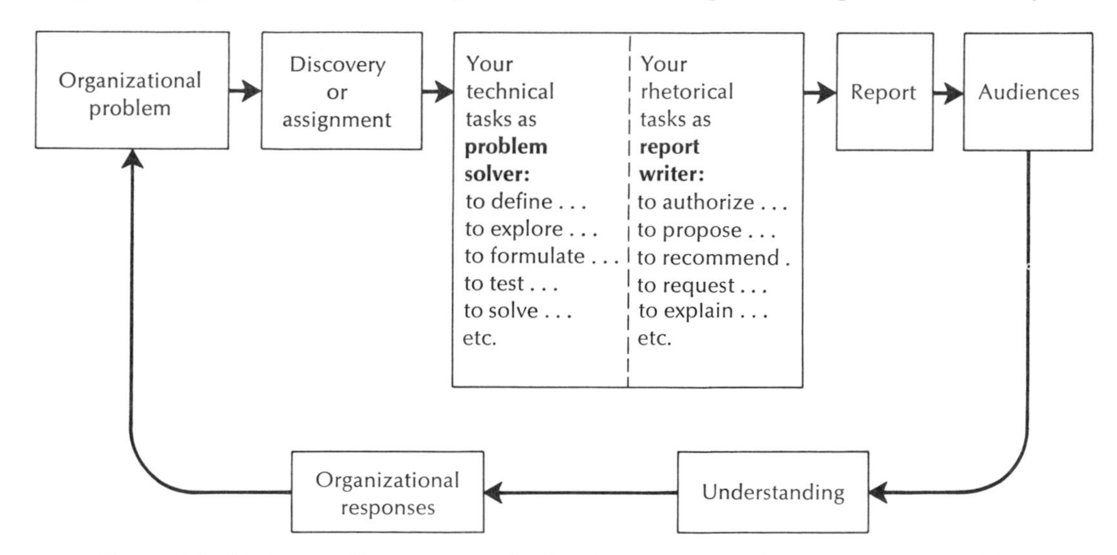

Figure 3–3. Understanding and organizational response as the consequences of the engineer's two functions.

audiences have should be roughly equal to your own understanding. On the basis of this understanding and your recommendations, the audiences will act; that is, your audiences will create the organizational responses required to implement a solution to the initial problem. The report you write is only the means to that end. Its purpose is to effect those responses.

That, then, is the situation. As we said, your purpose never exists in a vacuum. Your rhetorical purpose, whatever it is, must ultimately be understood and stated in terms of the problem that initially gave rise to the technical tasks, and then the report. From this explanation of the situation we can derive an algorithm by which you can state the purpose for a report. This algorithm has three necessary elements which are derived from the initial components of the situation we described above. The first derives from the organizational problem; the second from the assignment and subsequent technical tasks; the third from the writer's rhetorical tasks. This is illustrated in Figure 3-4, *The three elements of the purpose statement derived from the organizational context of the engineer's work.*

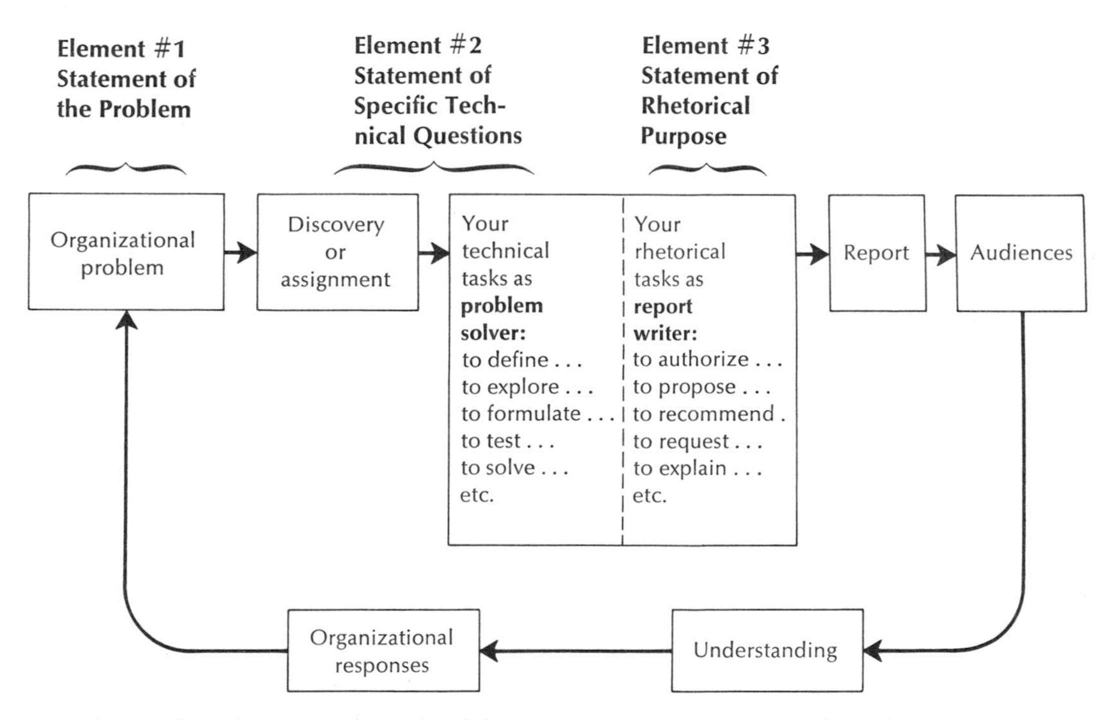

Figure 3–4. The three elements of the purpose statement derived from the organizational context of the engineer's work.

A complete purpose statement begins with a statement of the problem as perceived by the organization, identifies the specific technical questions or tasks addressed by the technical investigation, and ends with an explicit statement of the writer's rhetorical intention in relation to that problem. A good purpose statement, then, includes at least the following elements:

1. A statement of the problem, i.e., the conflict at issue in the organization.
2. The posing of specific technical questions or tasks arising out of that problem, and addressed by the technical investigation.
3. The statement of the rhetorical purpose, i.e., a statement of what the report is designed to do in relation to the organizational problem and the consequent technical questions and tasks.

Perhaps these elements would be more evident if we looked at some examples. Here is an example from a research engineer concerned with process development:

> The investigation discussed in this report is part of our study of efficient methods for drying high concentration liquid food extracts. The investigation was undertaken because although vacuum drying of liquid extracts is potentially less costly than freeze drying, current techniques for vacuum drying do not permit retention of volatile materials in the extract as well as does the freeze drying technique. Since there is an apparent relationship between volatile materials and flavor, unless we can develop new techniques for vacuum drying, we appear to be committed to freeze drying.
>
> The objective of our investigation was to develop a vacuum drying technique that retains volatile materials as well as the freeze drying technique does. Freeze drying establishes a solid structure which controls diffusion of volatiles during the drying process. We believed, therefore, that if a vacuum drying technique could be developed which quickly dried surface layers of the extract, vacuum drying could achieve high retention of volatiles at relatively low drying temperatures. To that end we sought to answer the following questions:
>
> 1. Would high initial temperatures in the vacuum drying process reduce loss of volatiles by drying surface layers quickly?
> 2. Would extremely rapid drying of thin films reduce the loss of volatiles in the vacuum drying process?
>
> The purpose of this report is to present the results of our investigation. These results indicate that vacuum drying techniques do not permit retention of volatiles which compares favorably with retention of volatiles during freeze drying. The failure of experimental vacuum drying techniques indicates that future investigation in this direction is unwarranted. Furthermore, based upon the results of our investigation, we believe future study should be aimed at increasing dryer capacity in the freeze drying process.

The basic elements in this purpose statement should be easy to identify. The first states *the problem:*

> Vacuum drying of liquid extracts is potentially less costly than freeze drying; however, current techniques for vacuum drying do not permit retention of volatile materials in the extract as well as does the freeze drying technique. . . . Unless we can develop new techniques for vacuum drying, we appear to be committed to freeze drying.

Notice here that from the organizational perspective the problem involves a conflict between cost and product quality. Vacuum drying is cheaper than freeze drying; unfortunately, however, the product produced by vacuum drying does not taste as good as it might. Obviously the organization has a conflict of interests here. It wants low cost *and* good taste. In his first paragraph, the writer has pinpointed this organizational problem, this conflict, by the way he has phrased the problem. "A is true, but B, which conflicts with A, is also true."

The second basic element of a good purpose statement states the *specific questions or tasks arising out of the problem:*

> The objective of our investigation was to develop a vacuum drying technique that retains volatile materials as well as the freeze drying technique does. That is, how is it possible quickly to dry out the surface layers of the extract and thus retain volatiles at relatively low drying temperatures? Would high initial temperatures in the vacuum drying process reduce the loss of volatiles by drying surface layers quickly? Would extremely rapid drying of thin films reduce the loss of volatiles in the vacuum drying process?

Notice that phrased as a question the overall objective of the study is open-ended: it simply asks how can vacuum drying be done? The subsequent questions are much more sharply focused. They are the two specific hypothesis questions addressed by the investigation. They are answerable by "yes" or "no." As here, the specific objectives of the investigation are often best phrased as questions, but these questions may be either open-ended or hypothesis questions.

The third basic element of a good purpose statement states the *rhetorical purpose:*

> The purpose of this report is to present the results of our investigation. These results indicate that vacuum drying techniques do not permit retention of volatiles which compares favorably with retention of volatiles during freeze drying. Future investigation in this direction is unwarranted. Furthermore, we believe future study should be aimed at increasing dryer capacity in the freeze drying process.

This is what the report writer seeks to accomplish by writing the report. The first sentence states his purpose in its most general terms. The subsequent three sentences make his rhetorical objectives much more particular. He asserts three generalizations which have emerged in the investigation and of which he will seek to convince his readers.

Now obviously the purpose statement above comes from a report in which the author has reached a series of clear conclusions. Many reports are written when the writer has not progressed that far in the problem solving process. For example, some writers will wish only to define the likely consequences of a problem, i.e. clarifying questions without attempting to resolve them. Thus the statement of a conclusion or even the statement of hypothesis questions is not always appropriate. Nevertheless, in most reports the three elements of problem statement which we have identified will work.

Here is another example. This one is from the introduction to a report by a nuclear engineer conducting tests on an experimental power source for cardiac pacemakers:

> Until recently it was not possible to repair or replace the human heart. At the present time the only way to replace the human heart is with another human heart. The results of these transplants have shown only modest success. However, some hearts not functioning normally do not need to be replaced, and can instead be repaired. Those persons suffering from heart block or any interruption of the normal stimulus to the heart can receive aid from a cardiac pacemaker.
>
> A cardiac pacemaker is an artificial means of electrically and rhythmically stimulating the heart to ensure a constant pumping action and to maintain an uninterrupted flow of blood through the body. Powered by batteries with a lifetime of two or three years, conventional pacemakers are effective. However, because the pacemaker is surgically implanted in the body, surgery is required whenever it is necessary to replace the batteries. This is expensive, inconvenient, and—to a degree—hazardous for the recipient. Therefore a power supply with a longer lifetime than batteries is desirable for pacemakers.
>
> If the batteries in the conventional pacemaker were replaced by a radioisotope-powered electrical generator, it would be possible to decrease the frequency of power supply replacement and, hence, decrease the number of surgeries required. However, a radioisotope-powered unit presents some problems not encountered in battery-powered units:
>
> 1. A very low radiation level (five mrem/hour) must be maintained at the outer surfaces of the pacemaker due to its proximity to living body tissue.
> 2. The absolute integrity of the power-source containment materials must be

> maintained throughout the life of the source because of the radioactive materials contained therein.
>
> 3. The wattage of the source must be small enough so that dissipating any excess heat produced from the decay of the radioisotope does not become a problem.
>
> In addition to these problems, of course, a radioisotope-powered electrical generator —like the conventional battery-powered pacemaker—must be small enough to be used in the body, fabricated from containment materials compatible with the body fluids, and reliable enough to produce rhythmic electrical impulses during the life of the pacemaker.
>
> After careful evaluation by concerned agencies, 238_{Pu} was the isotope chosen for the power source in an experimental cardiac pacemaker. It was chosen because:
>
> 1. 238_{Pu} has a long half-life (87.5 years).
> 2. 238_{Pu} has a low specific neutron-emission rate.
> 3. 238_{Pu} has known heat source fabrication characteristics.
>
> This laboratory developed, fabricated, and for three years tested a 238_{Pu}-powered electrical generator to see if it met design criteria stated above.
>
> This report describes the progress we have made to date in loading and welding the 238_{Pu}-powered heat sources to be utilized in thermoelectric-powered cardiac pacemakers. This information has been requested by the American Nuclear Society because it represents the latest knowledge in the field.

Here again the three essential elements of the purpose statement are easy to identify. The first states *the problem.*

> Powered by batteries with a lifetime of two or three years, conventional pacemakers are effective. However, because the pacemaker is surgically implanted in the body, surgery is required whenever it is necessary to replace the batteries. This is expensive, inconvenient, and—to a degree—hazardous for the recipient.

These three sentences establish the basic need which is explicitly stated in the last sentence of the paragraph (a power supply with a longer lifetime than batteries is desirable for pacemakers). They pinpoint the problem. Notice that the problem in this instance is a large scale research problem, not a small, internal organizational problem. Nevertheless, the pattern for stating the problem is precisely the same as in the first example: A is true, but B, which conflicts with A, is also true.

The second element of the purpose statement states the *specific questions or tasks arising out of the problem*:

> A radioisotope-powered unit presents some problems not encountered in battery-powered units:
>
> 1. A very low radiation level (five mrem/hour) must be maintained at the outer surface of the pacemaker due to its proximity to living body tissue.
> 2. The absolute integrity of the power-source containment materials must be maintained throughout the life of the source because of the radioactive materials contained therein.
> 3. The wattage of the source must be small enough so that dissipating any excess heat produced from the decay of the radioisotope does not become a problem.
>
> In addition to these problems, of course, a radioisotope-powered electrical generator —like the conventional battery-powered pacemaker—must be small enough to be used in the body, fabricated from containment materials compatible with the body fluids, and reliable enough to produce rhythmic electrical impulses during the life of the pacemaker.

Here the writer establishes the specific design objectives. Notice that he has categorized them into two groups: those peculiar to the isotope-powered unit and those characteristic of any pacemaker which is implanted in the body. Notice also that in this case the writer states design objectives; he does not pose them as questions, as the writer in the first example did. Again, however, the essential function of specifying the nature of the technical investigation is served.

The third element of a purpose statement states the *rhetorical purpose*:

> This report describes the progress we have made . . . [because] . . . this information has been requested. . . .

In the last paragraph the writer establishes the rhetorical purpose. It is a progress report whose purpose is primarily informative. The information has general utility because it expands the frontiers in an area of on-going basic scientific research.

Here is an example from a report written by an industrial engineering student for a design project. Although the assignment was for a course, his report, made on the basis of extensive technical investigation at his local hospital,[2] was written to the hospital and effected significant changes in its operation:

> Federal regulations require that hospitals treating patients under the Medicare or Medicaid program obtain from the treating physician statements certifying the patient's need for hospitalization. For patients whose initial certification has run out, the hospital must obtain statements recertifying their need for hospitalization. If these conditions are not met, the hospital will not be reimbursed by the federal government.
>
> The procedure for certification and recertification presently used in Centerville Hospital does not always provide the proper certification and recertification information. Certification cards are often missing from patients' charts or are only partially completed. The result is that cards are sometimes illegally completed by unauthorized personnel in the Medical Records-office or Credit offices.
>
> The obvious conclusion is that the present system is defective. The question is, then, where does the present system break down? Further, what modifications might be made to assure proper certification/recertification of all patients under Medicare or Medicaid?
>
> The purpose of this report is to review the present procedure, consider alternative procedures, and recommend a specific revision of the system by which the hospital obtains certification or recertification of patients under treatment in either the Medicare or Medicaid programs.

The elements of this purpose statement are again the essential three: the problem, the specific technical questions or tasks arising from that problem, and the rhetorical purpose. First, *the problem*:

> Federal regulations require completed certification or recertification forms for all patients treated under Medicare or Medicaid programs; however, the present system used by Centerville Hospital does not always provide properly completed forms.

In the report itself, notice how this conflict is presented in two paragraphs, each stating one side of the conflict.

Second, the *specific technical questions or tasks arising out of the problem*:

> Where does the present system break down? What modifications might be made?

---

[2] The names of people, places, and companies throughout this text are real. In a few places, however, pseudonymns have been supplied for reasons of confidentiality or to avoid embarrassing someone. Unless an obvious psuedonymn such as *Centerville* or *John Doe* or *Doe Shipbuilding Company* is used the reader can assume that the names of people, places, and companies are real.

Notice how this step forms an independent paragraph unit.
Third, the *rhetorical purpose*:

> The purpose of this report is to review the present procedure, consider alternatives, and recommend a specific revision of the system by which the hospital can obtain proper certification or recertification of patients under treatment in either the Medicare or Medicaid programs.

Again, this step forms an independent paragraph unit.

The examples we have seen so far all come from reasonably lengthy or complex technical reports. Thus the statements are fairly detailed statements of the problems under discussion and the specific purposes of the reports. Not all purpose statements must be so detailed, though. In fact, some statements of purpose require no more than a few sentences. Here, for example, is a very short specimen which nonetheless incorporates the essential principles we have seen in the longer purpose statements. It is from a letter report addressed to the Maritime Administration:

> On August 8, 1969, American President Lines authorized specification changes on five Seamasters through the issuance of Change Order No. 16 (ref.e). The Owner in making this economic decision relied upon a preliminary cost estimate given by Doe Shipbuilders, the Contractor. The Contractor now submits a final estimate exceeding the preliminary estimate by 600%. The Owner considers this final estimate unreasonable, and in accordance with standard contractual procedure asks Marad to establish a fair and reasonable cost. The Owner requests Marad to review and adjudicate this final estimate.

The elements of this purpose statement are the essential three. First, *the problem*:

> The Owner in making this economic decision relied upon a preliminary cost estimate given by the Contractor. The Contractor now submits a final estimate exceeding the preliminary estimate by 600%.

Second, the *specific questions or tasks arising out of the problem*:

> The Owner considers this final estimate unreasonable, and in accordance with standard contractual procedure asks Marad to establish a fair and reasonable cost.

Third, the *rhetorical purpose*:

> The Owner requests Marad to review and adjudicate this final estimate.

Because the purpose statement is so important, it might be helpful if we present several additional examples. These may serve as useful models when you come to write a report. They may also suggest how a problem can be clarified for the writer himself by careful statement. As we have observed previously, engineers can often clarify the nature of their technical investigation itself when they follow our suggestions for statement of the problem and derivation of the report purpose. As you recall from the example of the two-man earth augur, too often students have the problem handed to them by their professors. Thus they miss the valuable training of having to identify a problem and to specify the technical questions that need to be answered. Experienced engineers know that often the most difficult part of their job is to identify the problem precisely and to determine what the relevant technical questions are. In each of the four examples which follow, the pattern of problem—specific technical questions or tasks—rhetorical purpose is effectively thought out and defined, as our marginal annotations indicate.

From a report by a mechanical engineer:

Introduction

In order to meet federal exhaust emissions standards for unburned hydrocarbons (HC) and carbon monoxide (CO), it has become necessary to use exhaust afterburners in the form of thermal and/or catalytic reactors. Due to the large amount of air required by these devices, it is generally necessary to inject fresh air into the exhaust manifold at the port. In the past this was accomplished through the use of an externally mounted air manifold, feeding air into the exhaust ports from a belt driven pump.

Due to the high initial and installation costs of the air manifold, we redesigned our new 2600 series V-8 cylinder heads with cored-in air passages to eliminate the external manifolds. The design includes small stainless steel tubes to direct the reactor air from the air passages to the rear of the exhaust valve.

Elimination of these port air tubes would result in a substantial savings in materials, machining, and production time. Consequently, Mr. Relander assigned me to determine if these tubes could be eliminated from the design and still have it meet federal exhaust emissions standards.

The specific questions I addressed were:

1. Can the reactor air tubes be eliminated completely without severe exhaust emissions penalty?
2. What if any is the relationship between air tube length and exhaust emissions (HC, CO, and NO_x)?

I investigated these questions according to federal exhaust emissions test procedures. This report shows that it may be possible to eliminate air tubes from the design if the system is used with a catalytic converter able to handle the slightly higher CO concentration which would result. I also establish a relationship between air tube length and exhaust CO levels.

initial problem

initial solution

new problem

new solution, i.e., present design

problem with the present design (implied)

assignment to address the problem

specific technical questions arising from the problem, i.e., the technical task clarified

rhetorical purpose based upon completion of the assignment

additional purpose implied, based upon answers to technical questions

From a report by a civil engineer:

Project Background

The Surveying Department of Tri-State Associates is presently working on a project for the Edison Power Company which involves the survey of a 35 mile corridor in St. Clair County. This survey is being used for legal descriptions, land acquisition, and coordinate mapping.

The finished product of such a survey requires accuracy and speed. As portions of the line are finished by the field crews, the information is to be reduced, corrected, and mapped as quickly as possible. Each portion of the survey is connected to adjoining portions so that the entire line will be described by one coordinate system.

Present Situation

Currently the Surveying Field Office does not have a system which is capable of meeting our rate of input. The volume of data amassed in the process of surveying a 35 mile corridor is enormous. All the measurements taken in the field must be reduced and corrected with maximum accuracy and efficiency. Yet I have only one person working full time at a desk calculator, trying to handle the data of two field crews. This operation is much too slow and allows great opportunity for human error to be introduced.

A system is needed which will be accurate and efficient so that we can keep up with the data input and so that we can use the operator in other capacities. Consequently, we have decided to install a computing system to handle the data. To choose a system for installation requires that our exact needs be specified and available systems be compared.

problem introduced—notice how the conflict and specific technical tasks are implied

direct statement of the problem with the conflict specified

basic need and hypothesis yield the specific technical tasks

Study Objective

rhetorical purpose stated

This report describes our computing needs, examines and compares those systems currently available for immediate use, and recommends a system to be installed. The system requested must be installed as soon as possible.

From a report by an industrial engineer:

Background

problem

assignment

Our report on "Housekeeping Department Staffing Overview," dated November 22, 1973, indicates a standard staffing requirement of 191 full-time employees at Sunnybrook Medical Centre in Toronto, Ontario. The actual staffing is 200 employees, indicating an overstaffing of 9 people. Your recommendations for accomplishing a staffing reduction include improved cyclical scheduling techniques and policies, and reduced cleaning frequencies.

additional problem and task

The Sunnybrook Administration, realizing this overstaffing and acknowledging your recommendations, has requested CSF Ltd. to perform an analysis of the housekeeping cyclical schedules. In conjunction with this analysis, the Administration is also concerned with replacement of workers who are absent due to illness, vacation, and holidays. At present, there is no regulative provision for these occurrences, resulting in a decreased quality of housekeeping.

Purpose

specific technical tasks

The purposes of this study are: (1) to evaluate the present cyclical schedules, (2) to develop new schedules, where possible, that would reduce staffing without jeopardizing the quality standards, and (3) to develop a staffing replacement method to handle absenteeism.

rhetorical purpose

I have received the present housekeeping schedules from Sunnybrook Medical Centre and have performed a preliminary analysis of them. The purpose of this report is to summarize my findings to date for your information and to specify additional data I need to complete the analysis.

From a report by a metallurgical engineer:

problem

There were a number of instances of radiator hose clamp failure on our cars produced during the period between February 1–21, 1974. The clamps were snapping and allowing loss of engine coolant only days after the cars came off the production line. Yet there did not seem to be any immediately obvious reason for the clamp failures. The supplier, Doe Corporation, is a longtime supplier of all our hose clamps, and we have never had any problems before. In fact, since February 21 there have been no additional instances of clamp failures.

specific technical tasks

In order to investigate the cause of the clamp failures, Mr. Carl Page, Department Manager, Production Quality Control, sent me to visit Doe Corporation. I was instructed to: (1) pinpoint the cause of the clamp failures, (2) make certain that the clamps coming from the supplier now and in the future are reliable.

rhetorical purpose

The purpose of this report is to show that the clamp failures were due to a temporarily defective plating process, and to demonstrate that the clamps now coming from Doe meet our specifications.

These purpose statements have a completely natural pattern which signals the writer's recognition that he has two functions in an organization: on the one hand, he is a technical investigator; on the other hand, he is the writer who seeks to affect the behavior of his audiences in purposeful ways. Those are two different things. Unless he realizes the difference, the report writer is likely to produce a mediocre report no matter how successfully he answers the technical questions.

WHERE SHOULD YOU STATE YOUR PURPOSE?

Now that we have discussed why you must state the purpose for every report you write and have explained how to do it, we come to the last of the three questions we posed at the beginning of the chapter: Where should you state the purpose of a report?

In a basic sense we have already answered that question. A purpose statement should be first both for the writer and for the reader. The writer needs to develop the purpose statement so he knows exactly what he seeks to accomplish by writing his report. The reader needs the purpose statement so that he knows ahead of time whether or not he needs to read the report, and if he does, what to look for. This seems an obvious point. However, judging from the evidence of example, the natural tendency of many writers is to bypass the purpose statement and to jump right into the details. They begin like this:

> Before we get into the comparison of one fish to another, we have to define some terms. As far as the Icthyologist is concerned, a fish has three regions: the *head,* the *body,* and the *tail.* I will consider the fish as two: the *entrance* and the *run.* The entrance is the volume from the nose to the maximum sectional area, and the run is the volume from the maximum sectional area to the tail. I will also define a fish as any fishlike creature (thereby including mammals like whales and porpoise) whose major means of propulsion is his tail. I define the coefficient of finness as the area of the maximum section divided by the area of the circumscribing circle at this point . . . .

negative example

Without so much as a word about the problem the report addresses, the specific questions the technical investigation explored, or the rhetorical purpose of the report, this writer has jumped right into a series of curious definitions that would probably amuse more than inform his readers. In fact, except for a few naval architects and marine biologists, most audiences would have no idea what this report is talking about. They would just have to be patient and hope things clear up.

In this second example one doubts that even the specialists would be able to understand what the report is about. Although this is the first paragraph from a proposal to the Department of the Navy for the funding of a ship design, the writer immediately jumps into a catalog of subcontractors. He gives no hint of his purpose:

> **Section 1**
>
> **Approach**
>
> The R. J. Fisher Company's proposal is based on the selection of subcontractors who possess specialized expertise in the development of air cushion and planing vehicles. The subcontractors selected for each type of vehicle are:

Designator No.	Subcontractor
P 60–25 (Planing)	Pacific Research Co.
C 60–40 (Multi-Cell Air Cushion)	Ramjet Corp.
C 60–40 (Peripheral Jet Air Cushion)	Land and Sea Design Co.
P 50–60 (Planing)	Pacific Research Co.
P 40–60 (Planing)	MacMartin Associates, Inc.

negative example

We raise this point about the natural tendency to bypass purpose statement because it would seem from the evidence of illustration that you may have to force yourself to state your purpose explicitly. Theoretically understanding that and actually doing it—in the press of the day to day routine of your job—may be two very different things. Just keep this axiom in mind: Every piece of technical discourse you produce should begin with a clear and explicit statement of purpose. That holds for everything from a half-page letter to a one-hundred page report.

There is one more point to be made about beginning with an explicit statement of purpose. The full presentation of the purpose statement will usually appear at the beginning of the discussion section of your report and is often labeled "Introduction" or "Purpose." In most technical reports, however, various preliminaries precede the discussion. This is true whether the report is in memorandum form or full formal report form. Therefore, the introduction to the discussion is not in an absolute sense the beginning. The memorandum has a heading including a subject line; the formal report has a title, foreword, and summary. These preliminaries should also give the reader a sense of the report's purpose, although obviously these preliminary glimpses of the purpose will not be nearly so detailed as the full statement of purpose in the introduction to the discussion.

To clarify how the purpose statement is embodied in various segments of the report in progressively greater detail, consider the following example. Each segment takes the reader by gradual degrees into the report, yet each clearly signals a sense of purpose:

> *Title:* A Lifting Cradle and Platform Support Structure for the Towing Tank: A Design Proposal.

This serves to announce the topic and to give a very general sense of the report's purpose. Some readers already familiar with the project would find this indication of the report's purpose sufficient to allow them to turn to specific sections of the discussion—others would not.

> **Foreword**
>
> Propulsion tests conducted with ship models in the University Towing Tank often require more than one day to complete. However, the present hoist system is inadequate to lift ballasted models from the tank for temporary storage. Mr. Fred Miner, Director of Testing Operations, has requested that I design a means of temporarily storing large ship models out of the water in the towing tank. He has also requested that I study the possibility of a new supporting structure for the platform around the ballasting basin.
>
> This report proposes a design for the construction of a new lifting cradle for model storage and a redesign of the platform supports.

This segment provides an explanation of the technical assignment. In this case it briefly identifies the problem and stipulates the technical objectives of the design. It repeats the notion that this report is an organizational design proposal. It also implies the actions to ensue, acceptance of the proposal, and consequent authorization to construct the cradle and supports.

> **Summary**
>
> To lift ballasted models from the towing tanks for temporary storage, the present hoist system can be supplemented by a lifting cradle. The cradle requires a new platform support structure to handle the increase in loads.
>
> The lifting cradle is designed to handle models up to 16 feet long and 500 lbs. in weight, and is designed to fit between the side platforms of the basin. The lifting cradle is made from welded aluminum I-beams . . . [and so on for two paragraphs].

This segment briefly repeats the design objective and presents the solution to the problem. It then summarizes the results of the technical investigation, in this case, the design. Thus, it explains in general how the assignment was completed. Notice how this segment coupled with the foreword provides an overview of the whole problem and solution.

Introduction

Propulsion tests conducted with ship models in the towing tank often require more than one working day to complete. However, the models cannot be left in the water overnight due to necessary leakage at the stern bearing. Therefore, the models are hoisted clear of the water for overnight storage.

When ballasted, many of the larger models are heavier than the capacity of the present hoist system. Deballasting to store the models consumes valuable working time. As an alternative to the deballasting procedure, the Director proposed a lifting cradle to supplement the present hoist system by lifting models clear of the water for temporary storage.

Associated with the cradle design is a redesign of the under-structure of the platform surrounding the ballasting basin. The present structure is inadequate for accepting the loads of the proposed cradle and is only marginally sound for its present use. The question is, can the present structure be reinforced or must it be replaced?

In the sections that follow, I present a design for first, the construction of a lifting cradle, and second, a new supporting structure for the platform around the ballasting basin.

This segment explains the problem in technical terms and introduces a hypothesis, which forms the basis of the assignment. The segment clarifies the technical aspects of the problem in considerable detail. It does so in order to establish the specific technical objectives of the design presented in the discussion. It does not add to your understanding of the organizational background or the rhetorical purpose; instead, it focuses on the assignment and specific technical tasks and considerably increases your understanding of them.

As you can see from this example, the preliminaries of a report serve primarily to clarify the report's purpose. The purpose statement, therefore, is not simply a block of prose labeled, "Introduction." It is an idea which permeates the whole report.

When the purpose statement permeates the whole report, notice especially how you enable your audiences to read the report efficiently. The deliberate redundancy[3] you see in the example above has a function. That function should be apparent on the basis of what you saw in the chapter on audience. Different readers will read different segments of the report. Few will read all of it. Or at least, few will read straight through from beginning to end. Thus, by being deliberately redundant you are taking account of the real nature of your audiences. You are allowing readers to begin—as they will—anywhere. You are allowing them to start and stop where they will, without sacrificing a general understanding of the purpose of the whole. Redundancy it is, but redundancy with a purpose.

In the opening of this chapter we said an engineer is always a person who wears two hats: a problem solver's hat and a report writer's hat. As we have shown, to meet the needs of the organization, both technical tasks and rhetorical tasks are

[3] The term *redundancy* is used in a sense approximating the engineering meaning of the word. *Redundancy* is deliberate duplication to assure greater reliability.

necessary. But we have also suggested that the two tasks are really parts of a whole. The engineer who can define the problem clearly, who therefore really understands his technical task, unquestionably performs his technical investigations with optimum success. Similarly, the writer who can define his purpose clearly, and therefore really understands his rhetorical task, is able to design his report so that the reader can obtain the information efficiently. Or, to put it differently, the technical task and the rhetorical task are the complementary parts of one whole. There may be two hats, but there is only one head.

TWO
Designing the Report

DESIGNING THE BASIC REPORT STRUCTURE
DESIGNING THE OPENING COMPONENT
DESIGNING THE DISCUSSION COMPONENT

CHAPTER 4

Designing the basic report structure

Almost every report has a complex audience. To meet the needs of these audiences efficiently, the writer must design a report structure with two components. The first component addresses audiences interested in general information; the second component addresses audiences interested in particulars.

When you have analyzed your audience and formulated a purpose statement for your report, you have clarified the rhetorical situation. This analysis of the rhetorical situation, not the technical task itself, must yield the criteria by which you design your report. Your analysis makes it evident that you must present more material than the record of your technical investigation contains. Your analysis of audience indicates that many of your readers are searching for essentially nontechnical information. Your analysis of purpose indicates that your report must address not only the specific technical questions of your investigation but also the organizational problem which gave rise to those questions. If you are alert to the evidence, your analysis tells you what must be in the report and how it must be organized. In this chapter we explain how to design the basic structure of a report in terms of audience and purpose rather than in terms of subject matter.

A superficial glance at the rhetorical situation might dismay a writer and lead him to establish an ineffective basic structure. Some audiences have only five minutes to find the budgeting implications of a technical design; others will take several days to analyze necessary production modifications. The reams of data, results, and even conclusions from the technical investigation itself—what the engineer spent all his time doing wearing his engineering hat—do not enable him to select and organize his material. Faced with this sort of complication, he may be tempted to throw away his rhetorical hat and plunge into a recitation of his technical material. He will put everything out cafeteria-fashion and let the clients pick up what they want. His assumption will be to explain everything and there will be something for everyone. That approach may be the easiest for the writer, but on the basis of what we have seen in our discussion of the rhetorical situation, it is totally ineffective for the reader. Another approach is essential.

There seem to be three possibilities. The first approach, suggested by many technical writing texts, asks the writer to write to "the common denominator." He must present his material in such a way that it is simultaneously clear to all his readers. In other words, this approach posits a general audience of educated laymen. Yet our analysis of report audiences has shown that this is clearly not the case. Some audiences have very specific interests and all audiences are particular audiences. The production supervisor and the legal counsel have little in common beyond working for the same organization. For that reason we reject this first approach.

The second approach, also suggested by some technical writing texts, asks the writer to write more than one version of a report so that he can address his different audiences. Instead of one report, he must produce two, three, or more reports. This may work in some situations, but most report writing situations call for a report to travel complex routes within the organization. In fact, the report functions by helping immediate, primary, and secondary audiences to interact with each other. Reports which travel only a portion of the communication route do not help audiences to interact. For that reason, we regard this second alternative as inappropriate most of the time.

The third approach is the one we explain in this chapter. We believe that rather than trying to create a homogeneous audience where in fact there is none or to write different reports for different audiences, a report writer can design a report so that different readers can read different components. In addition, we believe he can arrange those components so that audiences uninterested in the particulars of the technical investigation can immediately get what they need from the report. Furthermore, we believe the writer should not hesitate to repeat material so that different readers can read selectively and efficiently.

We have three suggestions to help you design the basic structure of your reports. As we do subsequently, we put our advice in the form of suggestions. They are not inflexible rules, but they form the basis of a systematic methodology for report design. Our three suggestions are:

1. Think in terms of two basic components for your report structure.
2. Design your report so that it moves from general to particular between the two components.
3. Make the components self-sufficient and selectively redundant.

These three suggestions are arranged in the order in which they are applied.

First suggestion: Think in terms of two basic components for your report structure.

Why are there *two* components? The logic for this derives from our previous discussion. You do not have a homogeneous audience. Usually, you have primary audiences who need to make decisions or to act in order to resolve organizational problems. You also have secondary and immediate audiences who need to familiarize themselves with various aspects of the report's contents or to understand why actions and decisions are necessary.

Given the fact that you do not have a homogeneous audience, it makes sense to design your report with two clearly distinct components. One component explains the problem and solution for those audiences who do not need the details of your technical investigation. The second component provides the details other audiences seek. Figure 4-1, *Two basic structural components of a report and their functions,* illustrates a report's two basic structural components and explains their functions.

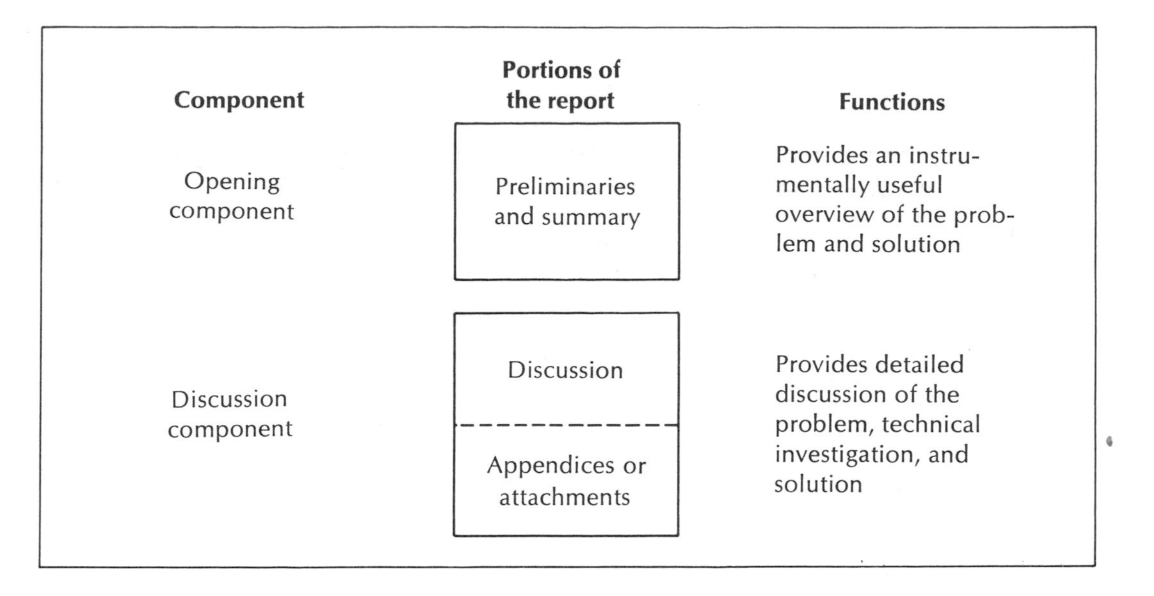

Figure 4–1. Two basic structural components of a report and their functions.

In the majority of reports, the first component addresses decision makers. (Their staff personnel will read the second component.) Although the decision maker will not read the bulk of the report (the second component), he needs certain types of information to implement the conclusions and recommendations. He needs to know the issue of organizational concern, to be reminded of the nature of your assignment, and to know what you conclude. And he needs to be told how he can use those conclusions. The decision maker has no interest in the details of the technical investigation itself. Thus, the first component of the report provides the instrumentally useful overview he needs. The rest of the report, the second component—the discussion plus appendices or attachments—provides what other audiences need. These other audiences, for the most part secondary audiences, need to know details about the problem and about your technical investigation. Some will need to be convinced of your conclusions. Others will need to implement your recommendations. Your supervisor may want to evaluate your work or to make sure that his department is safe. In reports aimed primarily at decision makers, secondary audiences have considerable, although selective, interest in your technical investigation. The second component of these reports provides this detailed information. As you can see, a report organized into two basic components can meet primary and secondary audience needs more effectively than a report with a one-component structure.

The reason for designing your reports to have a two-component structure may not be apparent to you at first. Most students are too accustomed to the reading habits of professors to realize that complex audiences in industry read differently than professors do. However, if you talk to report users in industry about how they actually read reports you will quickly understand the logic behind the two-component structure. For example, a department head in a major automobile company told us he reads a report in the following manner: first the title, then the purpose statement, then the summary. After that, he may spend a few minutes leafing through the report, as he said, "looking at the pictures" (figures, charts, graphs), and reading captions. Then if he is interested enough, he said, "I put the report in my briefcase to take home and read. If not, I put it aside for my secretary to file for future reference." For this management reader, if the report contains important information, that information must be easily accessible right away or the report is likely to get filed before it gets read.

The vice president of a large consulting firm emphasized the same point. He said, "I keep trying to impress on my engineers that I don't have time, and nobody has time, to read through a whole report to find out what it is all about. I want the important stuff right there in the first few paragraphs." A division head in a research organization—with the reputation as someone who demands succinctness—does not take the time to explain how his scientists and engineers should write reports. He has a simple acid test: any time a report comes to him with staples or paper clips holding it together he hands it to a subordinate and says, "Here, condense this for me." In other words, his idea of a good report is a one-page report. If the writer of a report for that division head does not provide his own opening component, he risks having a staff person interpret his report for him. The report writer will not communicate directly with his most important reader.

As our experiences make apparent, a report which addresses decision makers must be designed so that the primary audiences—the decision makers—can use the report without really reading it. Few managers read the discussion of a technical investigation, but most read the opening component. They are interested in the implications of a technical investigation for the organization, and they just do not have the time, the need, or even, perhaps, the technical knowledge to read through the discussion in a report. Thus the first component of a two-component report provides what the decision makers need. The second component provides what other audiences seek.

Reports which have decision makers as their primary audience are the most numerous. Other reports have primary audiences of persons who do not have administrative responsibility, but who must implement the technical recommendations and conclusions of the report. Usually, a decision has already been made, and it remains to implement those decisions. For example, in a report which explains specifically how to revise a procedure which some decision maker has already acknowledged to be faulty and already decided to change, the primary audience is not the decision maker. Here the primary audiences are those who must actually make the changes. They are the production, laboratory, and office personnel who take over after a decision has been made. In this sort of report, the first component addresses immediate audiences and important secondary audiences. Some of these audiences need to help expedite the report in the organization. Others need to know only generally what is happening. In this situation, the second component addresses the real users of the report.

The important point is that the two components of the report do not always have the same types of audiences. Figure 4-2, *Audiences addressed by the two components of a report aimed primarily at decision makers,* and Figure 4-3, *Audiences addressed by the two components of a report aimed primarily at persons without administrative responsibility,* illustrate two different kinds of reports. The figures show how the two components of the report are designed to address different audience needs. The emphasis in a report addressed primarily to decision makers is upon the opening component. In a report addressed primarily to people without administrative responsibility the emphasis is upon the second component. Notice that in each of these situations there are appropriately two components. It is only the relative importance of the components that shifts.

An example of effective design of a report with two basic components is the specimen report in Figure 4-4, *First page of a report addressed primarily to an audience of decision makers.* The report is a request by a government agency for proposals from industry. This request has the form of a seven-page cover document followed by a separately paginated, fourteen-page proposal. The "Background" segment presented in Figure 4-4 is immediately followed on page two by a segment entitled "Proposal Brief." The brief then is followed by discussions of proposal requirements and bidder qualifications. The first component of this report, therefore, presents the information the primary audience of decision makers want to know. The second component, the proposal itself, addresses secondary audiences who actually prepare proposals. The report is designed so that different audiences can use the material they need efficiently. It has two basic components.

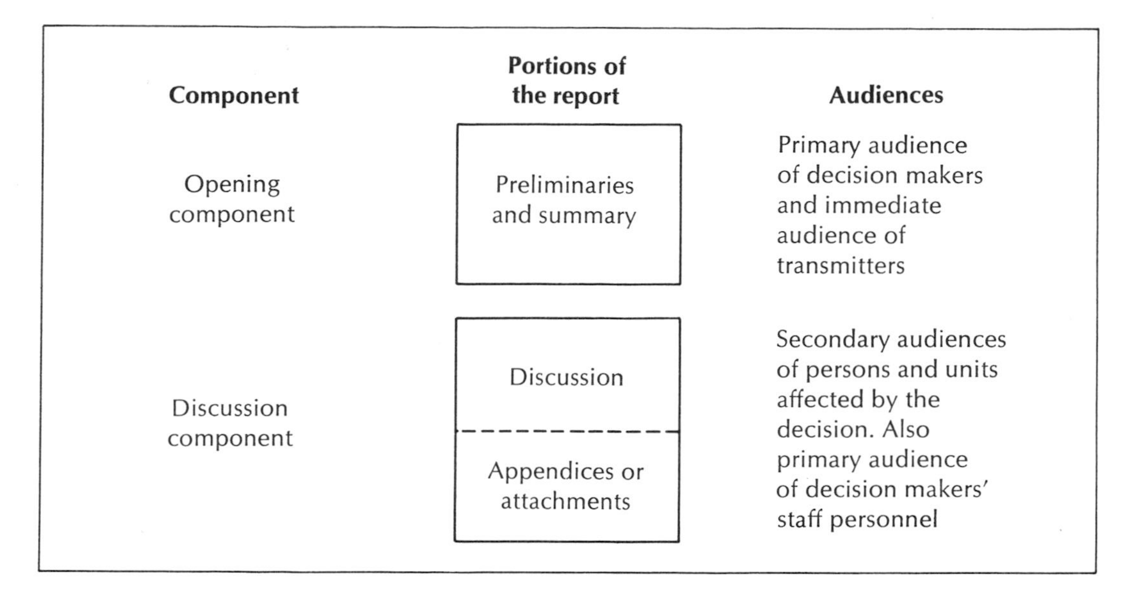

Figure 4–2. Audiences addressed by the two components of a report aimed primarily at decision makers.

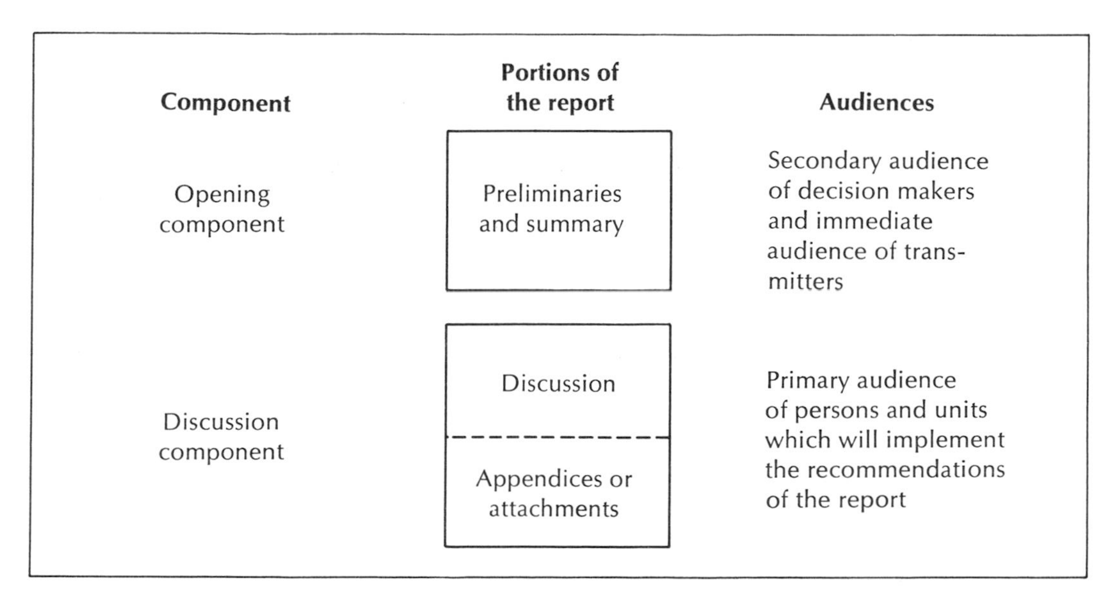

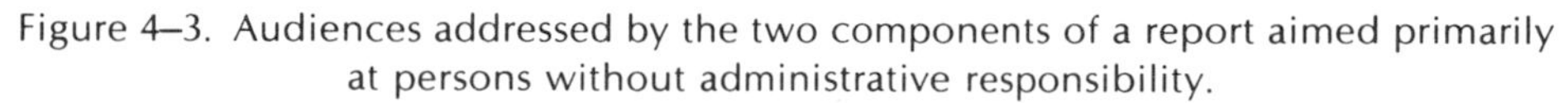
Figure 4–3. Audiences addressed by the two components of a report aimed primarily at persons without administrative responsibility.

A second example of a report with two basic components is the specimen report in Figure 4-5, *First page of a report addressed primarily to a person without administrative responsibility.* The writer, a transportation engineer who is being transferred to another division before he has completed an assignment, is explaining to his supervisor and to his replacement what remains to be done on his project. The opening component addresses the writer's supervisor in order to let him know generally what has happened and what remains to be done. The discussion component addresses the writer's primary audience, the replacement who will actually do what remains to be done. In this case the secondary audience, the supervisor, needs only to be assured that the project is under control. The primary audience, the engineer's replacement, needs very specific explanation of the technical investigation so far, especially the methodology to be followed. The two-component structure allows this writer to address his two types of audiences efficiently with one report.

No matter which type of report situation you face, our first suggestion is that you design a basic two-component structure. The analysis of your audience and your purpose determines which component of the report is instrumentally the most important.

Second suggestion: Design your report so that it moves from general to particular between the two components.

When an engineer begins to write a report, he sometimes fails to think in terms of general to particular. That is, he fails to adopt his reader's perspective. Instead, he still thinks in terms of his own perspective.

The engineer, and every professional who has just completed a technical or intellectual task, for the most part has worked from particular to general. He has worked from apparatus to data accumulation to interpretation of results to formulation of conclusions—from particulars to generalizations. Of course, this is a normal way of doing things. In addition, it is essentially the inductive process of scientific discovery. Thus, the unreflective report writer, especially one who has failed to put on his rhetorical hat, mistakenly designs his report in terms of what he has done, that is, from particular to general. Unfortunately, this tendency is reinforced in college laboratory and design courses, where students concentrate almost entirely on the particulars of their experiments and designs because those are the appropriate educational concerns of the professor. In such situations, the result is a report which has no instrumental purpose: the audience—the professor—makes no production decisions or budget commitments on the basis of a student's work. Consequently, due to the nature of his technical task and perhaps his college training, the unreflective report writer produces a report which is not effectively designed to be read by audiences who have solely an instrumental interest in what he writes.

Although the tendency of the writer is to move from particular to general, the tendency of audiences is to demand general to particular arrangement. Almost every reader needs to know immediately what the report asks of him. His very first response to this document that suddenly appears on his desk is, "What is this? Do I read it, route it, or skip it?" By arranging the report material to move from general to particular the writer answers those questions first.

From what we have seen in our discussion of audiences, any report has a number of readers who have no immediate need to know the particulars. The department head or supervisor may only transmit the report. In that case, the important question is, "Who should get this report?" If he can avoid it, he will not read any more of the report than necessary to get a very general idea of the report's

UNITED STATES DEPARTMENT OF COMMERCE
Maritime Administration
Washington, D.C. 20230

September 2, 1972

RFP-MA1-70:524

Gentlemen:

Subject: Request for Proposals—Phase I
Ship Design and Program Studies
for a U.S. Flag Merchant Fleet

Background

The Maritime Administration is seeking to establish a new maritime ship construction program of highly productive vessels at an annual rate above the present level. The proposed program will seek to provide increased efficiency in shipbuilding and ship operations to improve our competitive position in world shipping. As part of this program, Maritime seeks to develop ship designs for multi-year, multi-ship production which permit maximum economies of production and provide an incentive base for shipyard modernization. A three-phase program is contemplated. The first phase is concerned with the definition of requirements and preliminary designs of the proposed vessels. Phase II will encompass the engineering development of these designs and firm construction commitments. Phase III will be the construction of these vessels.

This proposal request is for Phase I which is to provide, through maximum industry participation, a design base in common program prerequisites for the follow-on long-term build program.

This Phase I study is to be in the areas of general cargo ships, dry bulk carriers, and liquid bulk carriers. The objective of this proposal request is to obtain from industry a series of standard ship preliminary designs suitable for low cost multi-year, multi-ship procurement with a minimum of government assistance.

Figure 4–4. First page of a report addressed primarily to an audience of decision makers (page 1 of 21).

CENTRAL NEW YORK RAILROAD
Northwest Division

DATE: June 14, 1974

TO: M. A. Holmes, Superintendent of Transportation
New Assistant Transportation Engineer

FROM: J. A. Sacher, Assistant Transportation Engineer

SUBJECT: Report to my successor on progress of developing new passenger train schedules.

FOREWORD

The Central New York RR is currently upgrading the mainline between Centralville and Smithville from 60 mph to 80 mph. The increase in speed will bring a need for new passenger schedules. I was assigned the task of developing these schedules by Mr. Holmes, the Superintendent of Transportation. However, effective June 15, 1974, I am being transferred to another division. My successor needs to know the state of the project if it is to be completed by July 31, 1974. This report addresses three specific areas:

1. method of doing the project
2. work done on the project
3. work remaining on the project

The purpose of this report is to inform my successor of the procedure used in solving the problem, to review the work already finished, and to explain the unfinished stages of the project.

SUMMARY

The method determining the new schedules has five specific stages:

1. collecting the initial data
2. formulating Speed-Distance curves
3. formulating Time-Distance curves
4. finding the expected speed performance
5. developing the timetables

Stages 1, 2, and 4 are finished. The results from the expected speed performance were good. The trains will take advantage of the increase in speed. The results are on file under project #53956. The remainder of the project, Time-Distance curves and timetables, should be finished by July 31, 1974. The Transportation Department requires an employee and a public timetable be . .

Figure 4–5. First page of a report addressed primarily to a person without administrative responsibility.

purpose and to answer that question. Other readers who have no need to know the particulars of the report are persons whose roles are indirectly affected by the report. Some managers and staff personnel, for example, need to know in general terms what decisions are being made, but need not concern themselves with details. Some production personnel and line management want to know only, "Are there future implications here for me?" To answer that question they need to know generally what the report is about. They need an overview. Decision makers need to know the problem and the solution. They want to know the total cost. They need to know whether a decision is feasible. Beyond that they have no need for the particulars and, as our study of management reading patterns mentioned above indicates, these readers seldom read more than the most general portions of the report.

Any report, of course, addresses some people who need to read into the particulars. However, even these readers need a general component first. As we indicated in the chapter on purpose, all readers need a beginning sense of overview. This allows them to read efficiently and selectively. First, the overview establishes the geography and tells the reader where he is going. This makes for efficient reading by allowing the reader to pace himself. It also allows him to skim by following the thread of the idea. Second, the overview amounts to an index and thus makes for selective reading. If there are only one or two specific segments of the report which are of interest to a reader, the beginning component tells him that and allows him to proceed to those segments. In fact, some formats call for the first component to be keyed specifically to segments in the second component so the reader can turn directly to the segment he is interested in. No need for the reader to grope along on his own and discover for himself how the report is put together.

As an example of the reader's need of an opening component despite his interest in the particulars, consider the report we just examined a moment ago. The transportation engineer's report primarily addresses his successor. It concentrates on detailed explanation of methodology. For his successor, this is the focus of the report. However, his successor needs to understand the project, why it was undertaken, and what its present status is. He must have some beginning generalization. If the report writer does not supply this generalization, the successor must recreate the framework of the project for himself. He cannot simply start applying the detailed methodology. The transportation engineer of course addresses a secondary audience also, in this case a supervisor. The supervisor needs the opening component to get an idea of how much remains to be done by the successor. He may well read selectively in the discussion, but he does not need detailed explanation of the methodology or the results to this point. In this case, then, the opening component provides overview which helps each of the two types of audiences addressed.

A good general-to-particular movement between the two components of a report is illustrated by the travel report in Figure 4-6, *Informal report arranged to move from general to particular.* The first component, the general component of the report, ends with item 4, "Action Required." Readers who do not need the particulars would read no further than "Action Required." Some readers, however, would use the opening component as an overview to guide them into the three-page detailed discussion, the particular component of the report. Most of them would read the details of the preliminary acceptance trials of the U.S.S. *Springfield*, while a few would search out discussion of the visit to the U.S.S. *Dewey*. This report thus has two basic components whose movement from general to particular serves the needs of quite diversified organizational readers.

If one thinks about it, the example above is a good illustration of the writer

NAVAL SHIP ENGINEERING CENTER
CENTER BUILDING
PRINCE GEORGE'S CENTER
HYATTSVILLE, MARYLAND 20782

IN REPLY REFER TO
CLG(TAL/9080)
Ser 420-212
9 August 1970

REPORT ON TRAVEL

Subj: Preliminary Acceptance Trials of USS SPRINGFIELD (CLG-7)

To: Mr. R. S. Johns, Head, Advanced Technology Branch
Mr. O. H. Oaks, Tech. Director, Ship Design Division

Persons Making Visit: Mr. T. S. Wick, SEC 6114, Mr. J. Smith, SEC 6114

Dates	Places Visited	Persons Consulted
25 July 1970	USS SPRINGFIELD (CLG-7)	INSURV Board Members Ships Officers and Men NAVSHIPYD BSN Officers Representatives of BUSHIPS, BUWEPS, NWP, ETC.
28 July 1970	USS DEWEY (DLG-14)	Ens. R. L. Tews, Asst. CIC Officer and JOD

Dist: 6410, 6420, 6421, 6236

1. Purpose. This trip was made to observe the Preliminary Acceptance Trials of the USS SPRINGFIELD (CLG-7) and familiarize ourselves with guided missile type ships. This report acquaints ship design groups with trial procedures and at sea conditions aboard these ships.

2. Background. The USS SPRINGFIELD (formerly CL-66) was converted to a guided missile light cruiser (CLG) with Fleet Flag facilities at Bethlehem Shipbuilding Corp., Quincy, Mass. It was delivered and commissioned at Boston Naval Shipyard. At the invitation of Code 521 for interested Design Division personnel to attend the Preliminary Acceptance Trials, we were selected to make the cruise.

3. Brief. The ship left Boston harbor and proceeded northward in the Atlantic Ocean, retracing its course several times while various tests were conducted for the scrutiny of the INSURV Board. These tests consisted of a full power run, "crash" astern run, high speed turning, anchor handling, electronic performance tests, control of aircraft and missile operating tests. At the conclusion of the various tests the ship returned to port and we availed ourselves the opportunity to visit the very recently commissioned guided missile frigate USS DEWEY (DLG-14)

4. Action Required. None

5. Detailed Discussion. We flew from Anacostia Naval Air Station in the INSURV Board plane with about thirty others who were members of the . . .

PEOPLE–PERFORMANCE PRIDE–PROFESSIONALISM

Figure 4–6. Informal report arranged to move from general to particular.

successfully overcoming his natural tendency to work from particular to general, about which we spoke earlier. For his audiences' benefit the writer has reversed the order and started with generalizations. Had he not been so aware of his audiences' needs, he would almost certainly have begun this report with what is now item 5. He would have started by saying, "We flew from Anacostia Naval Air Station . . . ." From a technical point of view that was the beginning of the investigation. From a rhetorical point of view, however, it is no place to begin the report. An ineffective opening which begins with the particulars of the technical investigation is, unfortunately, an extremely common pattern. In fact, the tendency to dive into the particulars immediately is probably the most prevalent fault in basic report design. It is a tendency which takes quite some hard work to overcome.

You should overcome this tendency even in short reports. The one-page memorandum report in Appendix A, for example, has general-to-particular structure. Or as another example, the following report outline suggested in the *Technical Report Manual* issued by the Technical Information Section of the Chrysler Corporation Engineering Office also stresses general-to-particular structure:

Introductory Section

Summation Section

Technical Section

Graphic Aids Section

Supplementary Section (Appendix)

But remember that this structure does not apply only to long reports; it applies to simple memoranda as well. The *Manual* says this "general organization is recommended for major, minor, and *one-page reports*" (our italics). Our second suggestion therefore applies to the basic structure of even routine one-page reports.

Third suggestion: Make the components self-sufficient and selectively redundant for any report of more than a few pages which addresses a complex audience.

Any reader has the time to read straight through a two-page report no matter how ineffectively it is designed, particularly if the material is not technically difficult. Of course, it is an inefficient use of his time, but for only a two-page report, the total block of time is negligible. A longer or technically difficult report, however, if ineffectively designed, may cause the reader to waste significant amounts of time. Our third suggestion tells you how to design reports which address complex audiences on technically difficult subjects or at length.

We have already established the premise that different audiences have different needs and therefore read selectively. The audience of decision makers as well as the others who need only a succinct overview will be satisfied by the opening component of a report. Audiences who need more particular information will read the discussion component. In long or technically difficult reports, the opening component cannot effectively meet the needs of audiences who need the particulars. One component cannot by itself clarify the problem, conclusions, and recommendations for all the report's users. If, on the one hand, the statement of the problem in the opening component is concise enough to address those who need a succinct overview, it cannot adequately explain the technical questions for audiences who need to comprehend the particulars in the discussion. Similarly, when the summary of conclusions and recommendations is concise enough for decision makers, it is inadequate for those who must implement the decisions. If, on the other hand, the statements of the problem, conclusions, and recommendations are developed adequately enough for the reader who needs the particulars, the opening component becomes too cumbersome for readers who need a succinct overview. The point is, in lengthy or technically difficult reports, neither a single concise statement nor a single fully developed statement of the problem, conclusions, and recommendations can serve both audiences equally well. The writer, therefore, must discuss the problem, conclusions, and recommendations in each component of the report. This redundancy permits the writer to adapt his material specifically to his audiences.

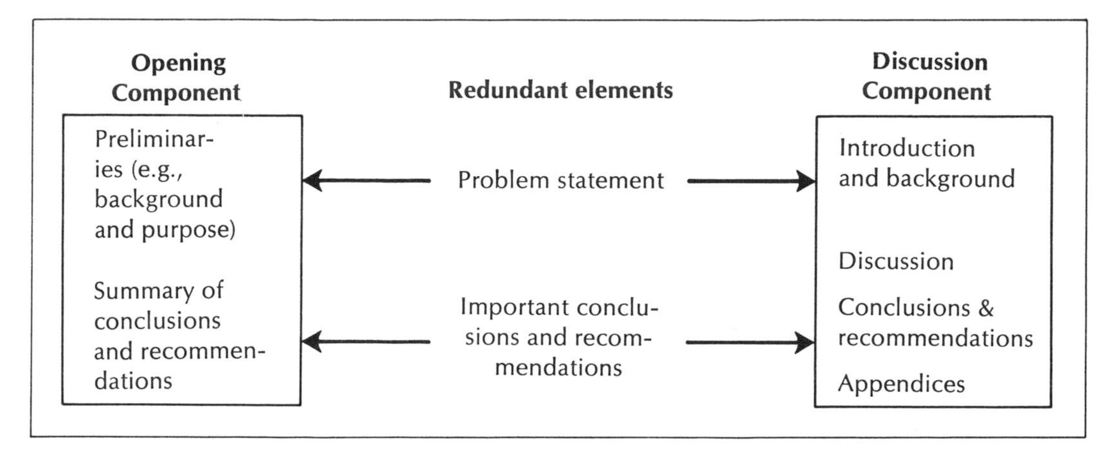

Figure 4–7. Redundant elements in reports addressed to complex audiences.

Figure 4-7, *Redundant elements in reports addressed to complex audiences,* identifies the redundant elements which appear in the two components. As you can see, each component contains some version of a problem statement and of important conclusions and recommendations. Notice that because of this redundancy each of these components is self-contained; that is, each can be read independently. It is important that the writer understand the need for redundancy. But it is crucial that the writer understand the difference between the two components. Although they are redundant, they are not simply repetitious. Each serves different needs, so each has its own focus. In the opening component, the problem statement focuses on the organizational problem and the rhetorical purpose of the report. It may not identify the specific technical questions investigated or identifies them in only very general terms. The problem statement in the discussion, however, focuses on the specific technical questions and must explain them in some detail so that the discussion of the technical investigation itself is clear to audiences who need particulars. It may de-emphasize the explanation of the organizational problem and even the rhetorical purpose of the report.

In the summary this difference in focus is achieved by selectivity. The summary contains important conclusions and recommendations selected from the discussion. In the foreword, this difference in focus is achieved either by condensation or by selectivity. That is, the foreword may condense the problem statement in the beginning of the discussion. Or the foreword may stress the organizational context while the beginning of the discussion stresses the technical problem and technical questions. These alternatives are shown in Figure 4-8, *Alternative tactics for distinguishing the foreword from the discussion problem statements.*

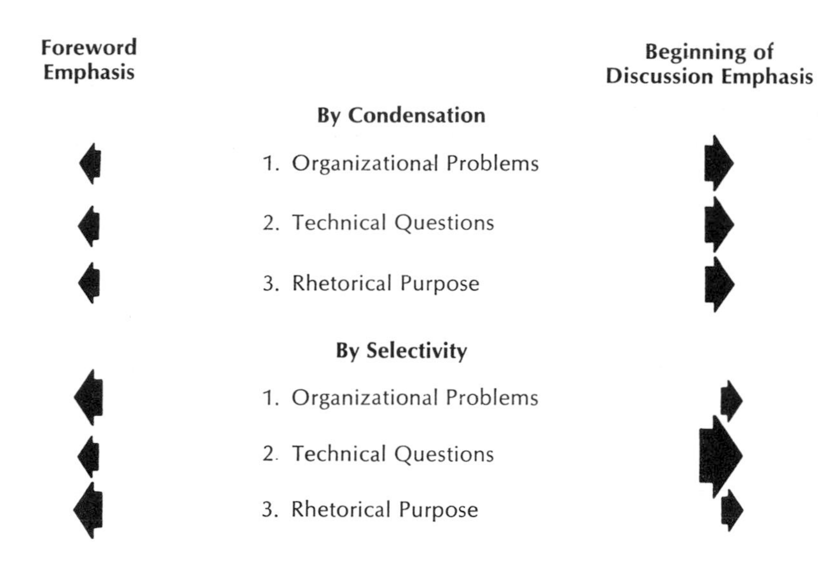

Figure 4–8. Alternative tactics for distinguishing the foreword from the discussion problem statements.

As an illustration of this difference in focus achieved by condensation, notice the differences between these two forms of a problem statement in a long report, one from the "Foreword," the other from the "Introduction" to the discussion. The two purpose statements come from the same report we examined earlier, the report which proposes a design for a lifting cradle and platform support structure:

Foreword

Propulsion tests conducted with ship models in the University Towing Tank often require more than one day to complete. However, the present hoist system is inadequate to lift ballasted models from the tank for temporary storage. Mr. Fred Miner, Director of Testing Operations, has requested that I design a means of temporarily storing large ship models out of the water in the towing tank. He has also requested that I study the possibility of a new supporting structure for the platform around the ballasting basin.

This report proposes a design for the construction of a new lifting cradle for model storage and a redesign of the platform supports.

. . .

Introduction

Propulsion tests conducted with ship models in the towing tank often require more than one working day to complete. However, the models cannot be left in the water overnight due to necessary leakage at the stern bearing. Therefore, the models are hoisted clear of the water for overnight storage.

When ballasted, many of the larger models are heavier than the capacity of the present hoist system. Deballasting to store the models consumes valuable working time. As an alternative to the deballasting procedure, the Director proposed a lifting cradle to supplement the present hoist system by lifting models clear of the water for temporary storage.

Associated with the cradle design is a redesign of the under-structure of the platform surrounding the ballasting basin. The present structure is inadequate for accepting the loads of the proposed cradle and is only marginally sound for its present use. The question is, can the present structure be reinforced or must it be replaced?

In the sections that follow, I present a design for, first, the construction of a lifting cradle, and second, a new supporting structure for the platform around the ballasting basin.

As an illustration of this difference in focus achieved by selectivity, notice the differences between these two forms of a problem statement from a relatively short report, in which there is no repetition of material at all:

foreword segment

On August 8, 1969, American President Lines authorized specification changes on five Seamasters through the issuance of Change Order No. 16 (ref. e). The Owner in making this economic decision relied upon a preliminary cost estimate given by Doe Shipbuilders, the Contractor. The Contractor now submits a final estimate exceeding the preliminary estimate by 600%. The Owner considers this final estimate unreasonable, and in accordance with standard contractual procedure asks Marad to establish a fair and reasonable cost. The Owner requests Marad to review and adjudicate this final estimate.

In examining the reasonableness of the final cost estimate submitted by the Contractor, the Owner thinks the following questions must be addressed:

1. Most important, do the charges relate specifically to the particular modifications required by the Change Order, or should the charges be considered as a development of the contract?
2. Are the charges realistic or inflated on the basis of typical construction procedures?
3. Are the charges provided for in the contract?

introduction to the discussion

The disparity between the Contractor's preliminary and final cost estimates suggests that the final estimate must be scrutinized.

To answer these questions, the modifications to the original contract specifications required by Change Order No. 16 must be noted. Then specific items in the Contractor's final estimate can be examined in light of the original contract and modifications. Other charges in the estimate can be examined separately.

The notions of a two-component structure and especially of redundant structures may be difficult to accept. The college student and the writer of scientific articles, and even the report writer who mistakenly assumes he has a homogeneous audience, tend to establish an essay structure requiring their readers to read straight through from beginning to end. Although such a structure can be arranged general-to-particular, it does not have two components with redundant characteristics. As we have shown, the report writer does not have a homogeneous audience unless he adopts the strategy of writing separate reports for each group of his audiences. However, this strategy will not suffice in most reporting situations. Consequently, the writer must design a basic report structure to meet complex audience needs. Our three suggestions in this chapter will help him to do this effectively.

CHAPTER 5

Designing the opening component

*When you have designed the basic structure of the whole report, you are ready to start drafting. You now must design individual segments. When you know the purpose of each segment in the opening component of a report, you can determine the content and structure of that segment.**

In Chapter four we suggested that reports are most effective for complex audiences when they are divided into two basic components. This design approach is effective because it derives from the uses those audiences have for the report. The first component introduces the report into the organizational system, and provides an overview of the problem and solution. The second component makes the report useful to the organization by providing the particular information necessary for audiences to understand and implement the solution. We have suggested the broad outline for a report; now we need to explain how to fill out the outline. In this chapter and the next, we will explain in some detail how to design the two basic components. To explain the design of the opening component, we divide it into three segments: the heading, the foreword, and the summary. Because these segments assume different forms in different reports, we then explain how to adapt our principles to alternative report formats.

Before we discuss the opening component segments, we should make an important point about report formats. Our analysis of the basic structure of a report identified two components. We believe these components should be present in all reports addressed to complex audiences. Unlike the authors of many texts on technical writing, we do not distinguish between the essential structures of the informal and formal report; thus the principles in this chapter are applicable to any report format. Of course, the formal report is superficially different from the informal report; it has a cover, title page, table of contents, and segments which start on separate pages. However, the basic structure lying behind the format of a

* These chapters are arranged in terms of general to particular; that is, we discuss the fundamental principles of report design, then move to particulars. When you write a report, however, you do not follow exactly the order of these chapters. Specifically, you should write the body of your report before you write the opening component discussed in this chapter. The opening component draws material from the discussion.

formal report and an informal report is the same. Too much emphasis has been placed on the superficial differences between report formats. We emphasize their basic similarity. This is illustrated in Figure 5-1, *Basic similarity of formal and informal report structures.*

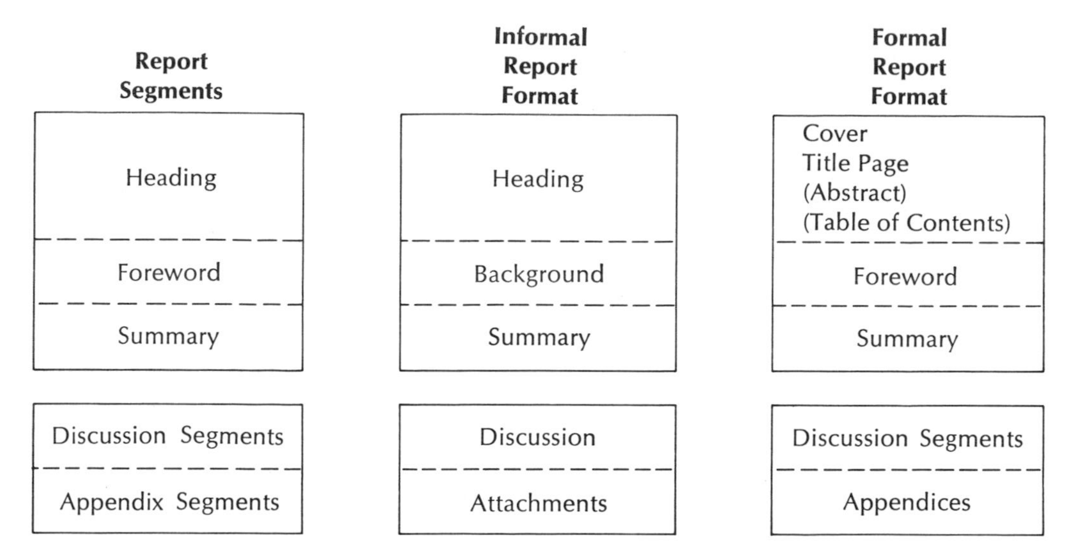

Figure 5–1. Basic similarity of formal and informal report structures.

The figure indicates that the only distinctive format item is the table of contents included with a formal report. Aside from that, any differences between these types of reports are primarily matters of layout and labeling. The summary of a formal report, for example, is on a separate page, while the summary segment of an informal report may be only an unlabeled paragraph. Even the label *summary* is arbitrary; instead the segment might be labeled "Conclusions and Recommendations," "Brief," or "Action Required." But these differences in layout and labeling are superficial. The summary performs identical structural functions no matter what it is called or how it is typed; it presents the solution to the problem established in the foreword. When you design a report, ignore the superficial format differences. Design the segments of the basic structural components in terms of their functions.

Another word of advice before we explain how to design opening component segments. The formal report format makes the formal report seem more important than the informal report. Even the term *formal report* suggests that somehow this sort of report is highly significant. Probably that is the reason for the undue attention the formal report receives in technical writing texts. However, from an organizational point of view, the informal report is at least as important as the formal report, if not more so. For one thing, informal reports are far more numerous in an organization than formal reports. But more importantly, in an organization most decisions are made on the basis of informal reports. Often the formal reports are material presented after the fact—for the record. They may even be more concerned with "PR" or "window dressing" than with instrumentally useful information. Thus the engineer must master informal report writing. He must not assume that the term *informal* suggests the report might be casual or unimportant.

The importance of the informal report means that the report writer must design the informal report as carefully as he designs the formal report. Both types of reports are designed to endure for some time within the organization, and often in other organizations and agencies as well. In both types of reports, for instrumental as well as information retrieval purposes, the elements of the purpose statement as well as of the technical investigation need to be developed in some detail. The opening component of a report, informal or formal, is especially important if the report is to be instrumentally useful over periods of time. The opening component should have three essential segments: the heading, the foreword, and the summary.

DESIGNING THE HEADING SEGMENT

The heading segment of a report introduces the report into the organizational system. To be able to use the report efficiently, the reader should be able at a glance to determine the functions of the report. His first question is, "Should I pay attention to this report?" The heading segment answers his question. To do so, the heading segment presents the following information, here arranged in the order of importance:

1. Subject of the report. The subject line, or title and subtitle, must state the topic and suggest the purpose of the report.
2. Source of the report. The writer and his role in the organizational unit must be identified.
3. Audiences of the report. Important audiences, their roles, and recipient organizations must be identified.
4. Reference information about the report. Relevant previous reports and communications are identified. Project and file numbers or retrieval codes are given. And the date of issuance is stated.

In formal reports, the heading segment includes an additional item:

5. The table of contents.

These items enable the reader to establish the communication routes of the report in the organization. They help establish the rhetorical purpose of the report, and start to introduce the problem. In the next few paragraphs we will explain specifically how to state and arrange this information.

1. Subject of the Report

Phrasing the subject line or title and subtitle is important because both the subject and the purpose of the report must be stated succinctly. It is not enough simply to announce the topic, that is, what the report is about. In an informal report, the subject line states both the subject and the purpose. In a formal report, the title states the subject, and the subtitle states the purpose. In other words, the title and subtitle in a formal report perform the same functions as the subject line in an informal report.

State the subject of the report with brevity and precision. If possible, the title should be ten words or fewer for quick comprehension and efficient retrieval. The first noun in the title must be significant and all nouns should be substantive. The title should not place nouns such as *study, analysis,* or *report* in initial positions. In fact, if possible, nouns such as these should not be used at all. Furthermore, the nouns in the title should be standard and comprehensible to any possible audience. In addition to substantive nouns, the title may also contain a verbal concept to clarify the nature of the investigation.

Examples of effective subject lines and titles which incorporate these suggestions are:

Subject: Clutch to Disengage Turbine Starting Pump
(*not* "Report on Turbine Starting Pump Clutch")

Subject: Correct Preparation of 1974 Program Profit Reports
(*not* "1974 Program Profit Reports")

TRANSPARENT METAL FILMS AS
PRECISION RESISTANCE THERMOMETERS

Feasibility Study and a Comparison
of Nickel Films and Gold Films

Notice the last example. It is a title and subtitle from a formal report. The subtitle specifies the subject of the report (the "metal films" are nickel and gold), the organizational function ("feasibility study"), and even the general methodology (a comparison, by implication to determine which film is best). The subtitle is the place for the descriptive and perhaps the technical terms many writers mistakenly put in the title itself, unless the format explicitly identifies the report type elsewhere. The subtitle, for example, identifies the report as a preliminary design, a theory, or production modification. In this case the subtitle identifies the report as a feasibility study. The subtitle thus clarifies the specific subject and the organizational function or purpose of the report.

2. Source of the Report

Identifying the source of the report is more than simply putting your name on it. Your role and unit must also be identified. Because organizations are complex and dynamic, distant audiences will recognize the role even though they do not recognize your name. Future audiences will recognize the role even though the name in that role may have changed. Role, not name, is organizationally important. Further, if your role is to be clear, it must be identified in terms of your specific department in the organization. It is not enough simply to say John Smith, Project Engineer. It is more helpful to say John Smith, Project Engineer, Coffee Product Development Department. On letterhead stationery the parent organization is identified. On plain stationery you must also supply that identification.

3. Audiences of the Report

Identifying the important audiences of a report is necessary to acknowledge and make explicit the communication routes to be followed by a report. Without that information the reader cannot determine the function of the report in the organizational system. Too often reports are issued with no audience information at all or with only minimal audience information. This reduces the report's instrumental usefulness considerably. Identification of important audiences means that the primary audiences and important secondary audiences are named; it is not enough to name an immediate audience who serves as a channel to these important audiences. The primary audience should be identified as the recipient of the report. Important secondary audiences can be identified on a distribution list on the title page. As above, when you identify the audience, use roles and organizational units in addition to or even in place of names.

AMERICAN PRESIDENT LINES
INTERNATIONAL BUILDING
601 CALIFORNIA STREET • SAN FRANCISCO, CALIFORNIA 94108 U. S. A.

Trans-Pacific • Round-the-World

August 16, 1972

File: Change Order No. 16

TO:	John Smith, Head Office of Ship Construction Maritime Administration U.S. Department of Commerce Washington, D.C. 20235	*names, roles, and departments identified*
FROM:	H. F. Monroe, Manager Engineering Department American President Lines	
SUBJECT:	Adjudication of final estimate for modifications: Take Home Motor Speed Control and Reversing	*subject of the report*
	APL Seamasters, MARAD DESIGN C4-5-69a CONTRACT NO. MA/H8B-46 DOE HULLS 489-493 Change Inquiry No. 8—Change Order No. 16	*project information and references*
DISTRIBUTION:	Doe Shipbuilding, Doeville Brown Naval Architects, New York Marad-Doeville	*important secondary audiences*

Dear Mr. Smith:

On August 8, 1969, American President Lines authorized specification changes on five Seamasters through the issuance of Change Order No. 16 (ref. e). The Owner in making this economic decision relied upon a preliminary cost estimate given by Doe Shipbuilding, the Contractor. The Contractor now submits a final estimate exceeding the preliminary estimate by 600%. The Owner considers this final estimate unreasonable, and in accordance with standard contractual procedure asks Marad to establish a fair and reasonable cost. The Owner requests Marad to review and adjudicate this final estimate.

Figure 5–2. Example of informal report heading incorporating letter format.

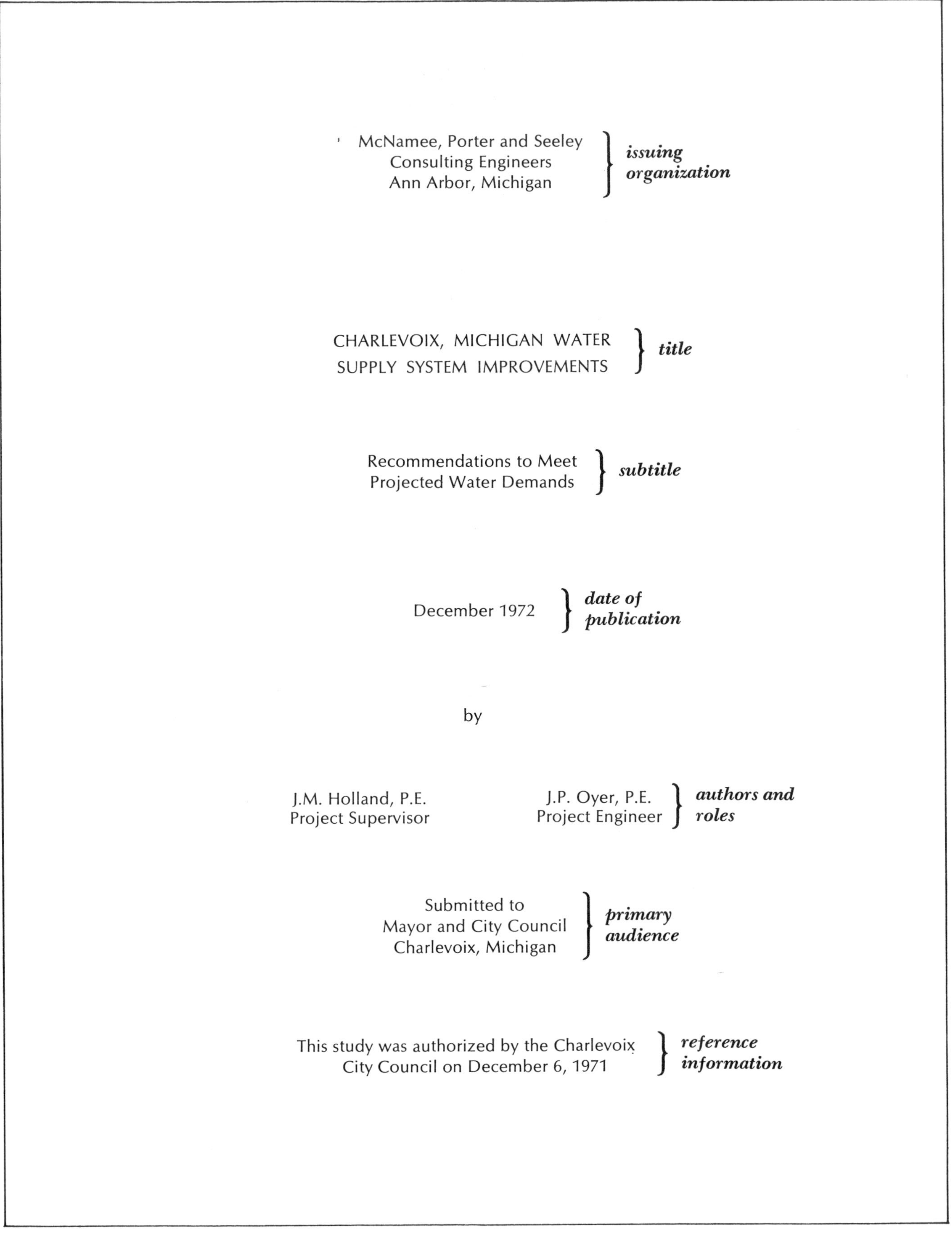

Figure 5–3. Example of formal report title page from a consultant firm report.

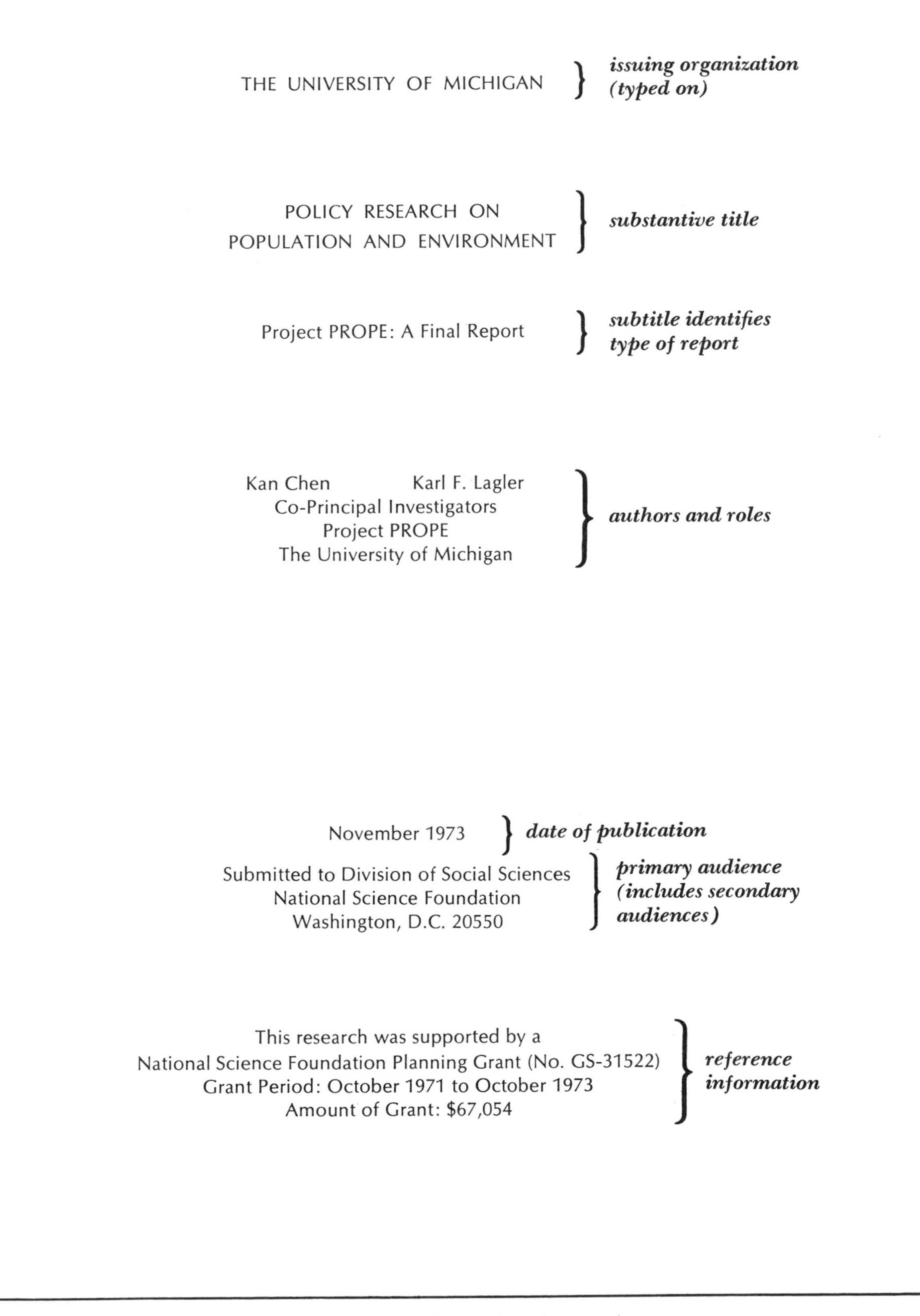
THE UNIVERSITY OF MICHIGAN

POLICY RESEARCH ON
POPULATION AND ENVIRONMENT

Project PROPE: A Final Report

Kan Chen   Karl F. Lagler
Co-Principal Investigators
Project PROPE
The University of Michigan

November 1973

Submitted to Division of Social Sciences
National Science Foundation
Washington, D.C. 20550

This research was supported by a
National Science Foundation Planning Grant (No. GS-31522)
Grant Period: October 1971 to October 1973
Amount of Grant: $67,054

Figure 5–4. Example of title page from a formal research report.

TECHNICAL REPORT STANDARD TITLE PAGE

1. Report No.	2. Government Accession No.	3. Recipient's Catalog No.
4. Title and Subtitle		5. Report Date
		6. Performing Organization Code
7. Author(s)		8. Performing Organization Report No.
9. Performing Organization Name and Address		10. Work Unit No.
		11. Contract or Grant No.
		13. Type of Report and Period Covered
12. Sponsoring Agency Name and Address ***primary audience***		
		14. Sponsoring Agency Code
15. Supplementary Notes		
16. Abstract ***Notice all the information this form calls for— all heading items we have mentioned, and more.***		
17. Key Words	18. Distribution Statement ***secondary audiences***	
19. Security Classif. (of this report)	20. Security Classif. (of this page)	21. No. of Pages / 22. Price

Figure 5–5. Standard title page form for report to government agency.

4. Reference Information About the Report

Providing reference information locates a report in time just as the source and audience information locates it in place. The date by itself usually is insufficient. A report does not exist independently of previous or future communications and actions. By means of references, file numbers, key-words, and retrieval codes, the reference information relates this report specifically to past communications and allows it to be accessible in the future.

Examples of effective heading segments which incorporate our suggestions are in Figure 5-2, *Example of informal report heading,* Figure 5-3, *Example of formal report title page,* Figure 5-4, *Example of title page,* and Figure 5-5, *Standard title page form.* To point up particular characteristics, we are annotating them directly rather than discussing them in the text. Our annotations illustrate the essential elements in an effective heading segment. Before we move on to our examination of the foreword segment, however, we want to talk about one more element which is a part of the heading segment in formal reports, the table of contents.

5. The Table of Contents

The function of a table of contents is to allow audiences to use the report efficiently. It provides an overview of the report structure and an index for selective reading of the report. It should list and give page references for the headings and major subheadings of the segments of the report, including the appendices. If there are sufficient items to warrant separate listing, lists of tables, figures, symbols, or glossaries of technical terms are sometimes included, each on a separate page immediately following the table of contents. These lists are indexed first in the table of contents.

The table of contents is the second page of most formal reports. (Variations occur when an abstract or letter of transmittal precedes the table of contents.) The table of contents page has the phrase *Table of Contents* centered at the top of the page. Headings and major subheadings of the segments of the report are listed and numbered exactly as they appear in the report.

Examples of effectively designed tables of contents are in Figures 5-6 and 5-7, *Example of table of contents. . . .* Again, they are annotated to point up particular features.

DESIGNING THE FOREWORD SEGMENT

The foreword provides the information all audiences need in order to understand the organizational aspects of the technical problems discussed in the report. It is literally a "fore-"word. It serves two specific functions: to introduce the organizational problem, and to present sufficient background so that the summary can be understood and used. If it is well designed, the foreword—along with the summary—allows many readers to avoid reading into the detailed discussion of the body.

In very short reports of one or two pages, the foreword segment serves all of the purpose statement functions discussed in Chapter three, "The problematic context: The purpose of the report." That is, very short reports lack the redundant purpose statements necessary in longer and more complex reports. A good example of the foreword of a very short report is seen in Figure 4-6, *Informal report arranged to*

PHYSIOLOGICAL RESPONSE TO STATIC GRIP
Preliminary Results

title stated

TABLE OF CONTENTS

lists indexed first when listed separately		LIST OF TABLES	iii	*page nos. included*
		FOREWORD	iv	
		SUMMARY	v	
major headings and sub-headings exactly as in text	1.	INTRODUCTION AND OVERVIEW	1	
	2.	PROBLEM DEFINITION AND METHODOLOGY	2	
		2.1 Problem Definition	2	
		2.2 Methodology	4	
	3.	PRESENTATION AND ANALYSIS OF RESULTS	7	
all-Arabic numbering system		3.1 Data Presentation	7	
		3.2 Data Analysis	11	
	4.	CONCLUSIONS AND RECOMMENDATIONS	16	
	Appendix A:	METHODOLOGY QUESTIONNAIRE	18	*appendices identified*
	Appendix B:	TABLES OF RESEARCH DATA	26	

Figure 5–6. Example of table of contents with all-arabic numbering system.

POLICY RESEARCH ON POPULATION AND ENVIRONMENT } *title stated*

TABLE OF CONTENTS

three levels of headings and sub-headings exactly as in text here identified to provide overview of structure

traditional numbering system

page nos.

Figure 5–7. Example of table of contents with traditional numbering.

move from general to particular. Notice particularly that the foreword is the only purpose statement in this short report; there is no introduction to the main body. Notice also that this foreword segment is not called "Foreword" but is broken into two units labeled "Purpose" and "Background."

In long reports, especially reports with complex audiences, the foreword segment presents information for the most part drawn from a subsequent introduction to the body. In this case, its essential function is to place the technical investigation clearly in its organizational context. This focus upon the organizational context of the report means that the foreword does not simply duplicate the introduction. It presents very specific organizational material, sometimes, for example, the name of the person making the technical assignment.

Whether the foreword segment is from a short or long report, it appropriately includes most of the following items:

1. The organizational problem: the conflict at issue in the organization.
2. The organizational context: the names of persons and departments involved in the issue.
3. The technical problem: the technical investigation needed to resolve the organizational issue.
4. The assignment: specifically what the writer of the report was asked to do.
5. The technical questions or tasks addressed and perhaps the hypothesis or solution. This item may be omitted when it would be too complicated for some audiences to understand or when the questions are not necessary for audiences to use the summary.
6. The rhetorical purpose: the instrumental purpose of the report. Be sure you state the *instrumental* purpose of the report, not what the report is about. State how the organization will use the report.

Grouping these six items into logical categories and arranging these categories yields the following outline for the foreword segment:

The Problem

the organizational problem
the organizational context
the technical problem

The Assignment

the assignment
the technical questions or tasks

The Rhetorical Purpose

The sequence of items in this outline is not inflexible, but at least in general terms the content is necessary for an effective foreword segment. Unless you see an especially good reason for varying the pattern, we suggest you simply adopt this outline. In following the outline, however, remember the essential function of the foreword and do not let the segment get unnecessarily long. It should be proportionate to the rest of the report. A one-page foreword for a three-page report suggests the writer has slavishly followed the outline but ignored both his readers and what he has to say. Figure 4-6 has a good example of a foreword for a short report. Examples of effective foreword segments from both short and long reports which incorporate our suggestions are in Figures 5-8, 5-9, 5-10, and 5-11, *Example*

FOREWORD

larger organizational problem — The National Aeronautics and Space Administration awarded a research contract to the Department of Mechanical Engineering at The University of Michigan to investigate the boiling problem common to liquid fuel tanks of spacecraft in outer space. Dr. H. Merte is conducting research into this phenom enon by means of laboratory simulation. An apparatus to simulate liquid fuel boiling was designed. A means of accurately measuring the inside surface temperature of the fuel tank was needed. — *organization context*

technical problem — Consequently, I was asked to investigate the feasibility of measuring the temperature of a transparent metal film used as a precision electrical resistance thermometer. — *assignment*

technical tasks — My investigation had two purposes:

1. To determine the feasibility of using a transparent metal film as a precision resistance thermometer while it is being used simultaneously as a heat source.
2. To determine whether transparent nickel film or transparent gold film would make the better resistance thermometer.

primary ıetorical purpose — The purpose of this report is to present the results of my investigation. The results will enable the apparatus to be built so that the team can proceed with experimental simulation.

The results of my investigation also will be used by Mr. A. Oker and Mr. S. Kodoli. They are doing doctoral research in the field of nucleate boiling with thin films in cryogenic liquids at zero gravity. — *secondary rhetorical purpose*

Paragraph 1	The Problem	
Paragraph 2 Paragraph 3	The Assignment	
Paragraph 4 Paragraph 5	The Rhetorical Purpose	(primary) (secondary)

outline by paragraphs

Figure 5–8. Example of an effective foreword segment from a research project report.

FOREWORD

The University of Michigan Committee on Human Flight (UMCHF) is presently involved in designing a Man-Powered Aircraft (MPA). The Committee realized that the first step in their design is to define the limiting factors of the project. The only limiting factor upon which all other design components are dependent is the power-to-weight (P/W) ratio. This factor is crucial because it determines whether or not a plane can gain enough lift for a take-off. — ***organizational problem and context***

Man can generally work harder for a shorter period of time than for a long one, but there are few systematic studies of the exact way in which power output diminishes as the duration of the exercise increases. Extensive physiological studies have been made on running, but in this form of exercise little external work is being done. Therefore, previous studies and the results obtained are of only indirect use for our present purpose—the design of a Man-Powered Aircraft. The need exists for further investigation to determine the work that man can generate. — ***technical problem***

This reason alone brought about the request from Professor Anderson, UMCHF project supervisor, to find the maximum work output that can be generated by man. An attempt is made to examine the properties of man in considering him purely as a source of mechanical power. We will be concerned here with only the production of work by the human arm and leg muscles. The report answers the following questions posed by the Committee on Human Flight: — ***assignment***

technical questions —

1. What is the maximum power output that an average man can generate pedaling a bicycle transmission, and how long can he maintain this rate?
2. Will hand-cranking affect the above figure significantly?
3. Should supplying additional oxygen be considered in increasing the power output?
4. Should we consider obtaining a professional cyclist in powering our MPA so that a higher power output can be realized?

The purpose of this report is to state the conclusions obtained in my investigation. Recommendations are made so that decisions can be made by the Committee on Human Flight in the design of the MPA. The knowledge gained from these further studies will also be of considerable interest in the design of all types of man-driven machines. — ***rhetorical purpose***

Figure 5–9. Example of an effective foreword segment from a student design project report.

UNITED STATES DEPARTMENT OF COMMERCE
Maritime Administration
Washington, D.C. 20230

September 2, 1972

RFP-MA1-70:524

Gentlemen:

Subject: Request for Proposals—Phase I Ship Design and Program Studies for a U.S. Flag Merchant Fleet } ***rhetorical purpose***

Background

organizational problems { The Maritime Administration is seeking to establish a new maritime ship construction program of highly productive vessels at an annual rate above the present level. The proposed program will seek to provide increased efficiency in shipbuilding and ship operations to improve our competitive position in world shipping. As part of this program, Maritime seeks to develop ship designs for multi-year, multi-ship production which permit maximum economies of production and provide an incentive base for shipyard modernization. A three-phase program is contemplated. The first phase is concerned with the definition of requirements and preliminary designs of the proposed vessels. Phase II will encompass the engineering development of these designs and firm construction commitments. Phase III will be the construction of these vessels. } ***technical problem***

rhetorical purpose { This proposal request is for Phase I which is to provide, through maximum industry participation, a design base in common program prerequisites for the follow-on long-term build program. } ***the assignment***

This Phase I study is to be in the areas of general cargo ships, dry bulk carriers, and liquid bulk carriers. The objective of this proposal request is to obtain from industry a series of standard ship preliminary designs suitable for low cost multi-year, multi-ship procurement with a minimum of government assistance. } ***specific technical tasks***

AMERICAN REVOLUTION BICENTENNIAL 1776-1976

Figure 5–10. Example of an effective foreword segment from a request for proposals.

DEPARTMENT OF CIVIL ENGINEERING

COLLEGE OF ENGINEERING
The University of Michigan
Ann Arbor, Michigan, 48104
(313) 764-8495

DATE: March 27, 1975

TO: Professor R. B. Harris

FROM: G. R. Madison, Research Assistant

SUBJECT: Packing Assignment Algorithm Manual: A procedure for solving CPM network leveling problems

REF: Status of Harris-Leubecker Procedure for Leveling, R. B. Harris, December 27, 1972

DIST: G. Ponce-Campos, CE-536 Professor
CE-536 Students
CE-430 Packing Algorithm File

The Need for a Packing Assignment Algorithm Manual

In Civil Engineering 536, the present Critical Path Method (CPM) network leveling procedures used by students produce satisfactory results in terms of an optimal solution. However, these procedures require excessive computational time. Solution by hand is particularly tedious and awkward. Therefore, I have recently been assisting Professor R. B. Harris in developing a Packing Assignment Algorithm to generate solutions more quickly and with better results than the procedures presently used. Faculty members would like to use this algorithm in CE 536 for classroom instruction in CPM leveling procedures, but no suitable guide exists.

organization problem and technical tasks

Consequently, as part of my independent research for CE 430, Professor Harris asked me to prepare a manual for potential users of the algorithm. This report transmits and summarizes the attached manual of procedures for using the Packing Assignment Algorithm in hand solutions to CPM network leveling problems. The manual will supplement the CE 536 lectures on the basic concepts involved in solving leveling problems. This report and the manual are being submitted in partial fulfillment of my CE 430 course requirements.

the assignment

primary rhetorical purpose

secondary rhetorical purpose

The Basic Steps for Use of the Algorithm

Detailed knowledge of the algorithm and of matrix leveling procedures is

Figure 5–11. Example of an effective foreword segment from an informal report with user's manual attached.

of an effective foreword segment. . . . Again, to point up particular characteristics, we annotate them.

DESIGNING THE SUMMARY SEGMENT

The summary provides a condensed statement of the conclusions and recommendations developed in the report. It presents the solution to the problem posed in the foreword segment. It allows some audiences to act, and provides a useful overview for all audiences. Because the focus of the summary segment is upon conclusions and recommendations, the term *summary* is somewhat misleading. The segment does not really summarize; it condenses selected material from the body in a very disproportionate manner. Therefore a more accurate label for this segment might be *Summary of Conclusions and Recommendations*—and this term is sometimes used. However, because we have to repeat the term so often in this book, and because the shortened term is so widely used, we use the term *summary.*

The summary appropriately presents two types of information. First, it presents information to help explain the conclusions and recommendations. Second, it states what the reader is supposed to learn from the report. In concise form, it includes most of the following information (which should be taken entirely from the discussion):

1. A brief statement of the objective or hypothesis of the investigation.
2. The methodology or experimental procedures used.
3. The results of the investigation.
4. The conclusions of the investigation.
5. The recommendations for organizational action.
6. Subsequent action or investigation, if any is called for.
7. Costs and benefits.

Grouping these seven items into logical categories and arranging these categories yields the following outline for the summary segment:

Objectives and Background

- objective or hypothesis
- methodology
- results

Conclusions

Recommendations

Implications for the Organization

- subsequent action
- costs and benefits

Again, the sequence of items in this outline is not inflexible. For example, you might want to group results with the conclusions or recommendations. In general terms, however, the outline provides a guide for the design of an effective summary segment. In some instances the content of the summary segment may be identified by headings corresponding to those in this outline. Sometimes the headings might be omitted and a simple term such as *actions required* might be used. Whatever the case, the essential content of the segment would be the same.

In following this guide, keep these two points in mind. First, do not bury the conclusions and recommendations by disproportionately discussing the objective, methodology, and results of the investigation. A few sentences will do. Second, keep the summary segment proportionate to the foreword as well as to the rest of the report. Annotated examples of effective summary segments from both short and long reports which incorporate our suggestions are in Figures 5-12, 5-13, 5-14, and 5-15, *Example of an effective summary*. . . .

ALTERNATIVE PATTERNS

In our discussion of the opening component of a report, we have divided it into three sections: the heading segment, foreword segment, and summary segment. This is a typical division of the opening component but not necessarily the only way to segment the material. When the material is segmented in other ways, as conventions or situations sometimes demand, the writer must not lose sight of the functions the segments in the opening component must serve.

A good example of a slightly different alternative format is outlined in Figure 5-16, *Format guide for research institute reports.* Because this format is required of reports addressed to the Department of Defense, many technical reports must conform to it. Despite different segment arrangement, the opening component in this format serves essentially the same functions we have discussed. The Department of Defense format places the summary before the preface, and then places the contents between this opening component and the discussion (introduction, main text, conclusions, and recommendations). Of these differences, placing the summary before the preface may seem an important change. In practice, however, reports conforming to this format include some of the purpose statement information (foreword material) at the beginning of the summary because this information is necessary for a reader to understand the summary. The remaining foreword information is the substance of the preface in a Department of Defense report. Thus, in this format the summary segment becomes the important opening component segment, and an outline for it would be devised by combining the outlines we suggest for the foreword and summary. The other noticeable difference in this Department of Defense format—placing the contents between the opening component and the discussion—is a device to put the summary at the very beginning of the report. This arrangement accentuates the two-component structure of the report.

Other formats may, at first glance, also seem quite different from the format we have discussed. But again, these usually serve the same functions. To help a writer keep his attention focused upon these functions, we will conclude this chapter by looking at some of these alternative patterns. Specifically, we will talk about letter reports, letters of transmittal, and abstracts. Of course, these are not the only alternative formats, but our discussion of these should indicate how the principles of a good opening component should operate in whatever format you use.

Letter Reports

Many reports take the form of letters because they are sent outside the organization or to distant divisions of the organization. However, letter reports are

SUMMARY

hypothesis question

The problem is: is it feasible to use a thin metal film as a precision resistance thermometer while it is being used simultaneously as a heat source?

Results of Investigation

Calibration tests were performed on a nickel test film and on a gold test film to determine if linear relationships did exist between their resistances and temperatures. Both films were found to possess a linear relationship between resistance and temperature. However, nickel film undergoes an aging effect. Gold film does not undergo an aging effect.

methodology and results

Conclusions and Recommendations

1st conclusion and recommendation

1. The calibration tests established that a linear relationship does exist between the resistance of a transparent metal film and its temperature. Therefore, I recommend that a transparent metal film be used to simulate the metal shell of a fuel tank.

2nd conclusion

2. Gold film is superior to nickel film for precision resistance thermometers. Comparison of the tests established that the linear relationship for the nickel film varies because nickel ages. The linear relationship for the gold film does not vary because gold does not age. In addition, the linear relationship for the nickel film is much more sensitive to measurement limitations than that of the gold film.

Therefore, I recommend that a transparent gold film be used in the experimental apparatus. Gold films provide a more stable and more accurate means of determining temperature from resistance readings.

2nd recommendation

The transparent gold film can be used simultaneously as a heat source and as a precision resistance thermometer in the experimental apparatus.

answer to hypothesis question

Figure 5–12. Example of an effective summary from a research project report.

SUMMARY

The problem this report is concerned with is to find the maximum horsepower that man can generate with a standard bicycle transmission.

objective

Results of Investigation

Ten volunteers pedaled a Sturmey-Archer bicycle transmission. The results show that pacing efforts, individual training, and simultaneous hand-cranking and cycling all increase the horsepower that can be generated. Utilizing these methods together can give a maximum of 1.5 horsepower for take-off procedures and 0.5 horsepower for cruising speed.

summary of results

Conclusions and Recommendations

results and conclusions

1. There is a steady oxidative energy production of 0.5 horsepower falling to 0.475 horsepower after 25 minutes as a result of long term fatigue. Therefore this horsepower must be sufficient to maintain a cruising speed for the MPA. *#1*
2. Simultaneous cycling and hand-cranking procedures yield 50 per cent more power than cyling alone, but can be maintained for only short periods of time. The maximum horsepower available utilizing this procedure is 1.5 and can be maintained for less than fifteen seconds without substantial fatigue. Therefore this horsepower must be suffcient for an MPA to overcome drag and gain enough lift for a take-off. *#2*
3. The effects of breathing pure oxygen during prolonged periods of strenuous cycling were found to be extremely dangerous. Therefore this technique does not suffice as a possible method for maximizing power output. *#3*
4. Training the individuals to pace themselves increased their power output 20 to 30 per cent due to the increased efficiency of maintaining a desired speed on the bicycle ergometer. Therefore it is not necessary to obtain a professional cyclist for needs since an amateur can be trained to generate a greater output. *#4*

On the basis of these findings I recommend that the cyclist/pilot controlling the MPA hyperventilate prior to take-off. This is to maximize his oxygen debt and provide the most oxygen available to his lungs and muscles. Also the cyclist should pedal at the optimum speed of 60 revolutions per minute. This rate can and should be increased to 120 RPM so that efficiency is sacrificed for work output to gain lift for a take-off. Finally, the cyclist should be trained to pace himself to prevent confusion and maximize his transfer of muscle power to mechanical work.

recommendation

Figure 5–13. Example of an effective summary from a student design project report.

Dear Mr. Smith:

(foreword)

On August 8, 1969, American President Lines authorized specification changes on five Seamasters through the issuance of Change Order No. 16 (ref. e). The Owner in making this economic decision relied upon a preliminary cost estimate given by Doe Shipbuilding, the Contractor. The Contractor now submits a final estimate exceeding the preliminary estimate by 600%. The Owner considers this final estimate unreasonable, and in accordance with standard contractual procedure asks Marad to establish a fair and reasonable cost. The Owner requests Marad to review and adjudicate this final estimate.

summary — *results* / *conclusions* / *recommendation*

Modifications to the motor, controllers, and resistor bank drew a preliminary cost estimate by the Contractor of $42,221 for the five ships (ref. c). The final estimate is $279,866 (ref. g). The Owner feels that costs related to a change in gear ratio, structural changes to bulkheads, modification of gears and coupling, and related material costs are not chargeable to the Change Order and should be considered as a development of the Contract. The Owner furthermore questions the costs due to disruption in man-hours (678). The Owner also observes an escalation in costs for successive vessels not provided for in the contract. The final estimate by the Contractor, therefore, is excessive, and should be made consistent with the preliminary estimate and with the specific changes requested in Change Order No. 16.

(introduction to discussion)

In examining the reasonableness of the final cost estimate submitted by the Contractor, the Owner thinks the following questions must be addressed:

1. Most important, do the charges relate specifically to the particular modifications required by the Change Order, or should the charges be considered as a development of the contract?
2. Are the charges realistic or inflated on the basis of typical construction procedures?
3. Are the charges provided for in the contract?

The disparity between the Contractor's preliminary and final cost estimates suggests that the final estimate must be scrutinized.

To answer these questions, the modifications to the original contract specifications required by Change Order No. 16 must be noted. Then specific items in the Contractor's final estimate can be examined in light of the original contract and modifications. Other charges in the estimate can be examined separately.

Figure 5–14. Example of an effective summary from an informal report in letter format.

THE DOW CHEMICAL COMPANY

BAY CITY PLANTS & HYDROCARBONS

August 7, 1974

TO: E. R. Wegner, Manager, Petrochemicals Section

FROM: E.C. Williams, Superintendent, Butadiene Plant
D.H. Smith, Superintendent, Ethylene Plant

SUBJECT: PROGRESS ACHIEVED BY THE PETRO CHEMICAL DEPARTMENT DURING FIRST HALF 1974

ATTACHMENTS: Maintenance statements by foremen R. Jenkins and H. Chase

At your request we are submitting a semi-annual progress report for the Petrochemical Department. This report covers production, safety, and the status of the Expanded Day (5th Shift) Program for the first six months of 1974. Since there have been extensive job restructuring and maintenance preparation for the new 5th Shift, a report on the status of the department, though routine, is especially important now.

Brief

The maintenance and job restructuring required for the 5th Shift program have not harmed either productivity or safety in the department. Productivity during the first quarter exceeded prior records. Despite a slight decrease in the second quarter, we project a record year for production. Safety also has been good. We had no disabling injuries in the department during the reporting period.

The Expanded Day Program development continues to progress as scheduled. Necessary maintenance has been done because the men have adapted well despite the obstacles. The first step of job restructuring has already been accomplished. To continue this progress we now need to begin hiring, to provide equipment, and to provide an adequate building. The Unifier needs a minor maintenance building of 400-500 square feet.

summary segment

Productivity

The 1974 productivity of the department has been excellent. During the first quarter, production exceeded previous production records by 7%. During the second quarter

Figure 5–15. Example of an effective summary from a progress report.

ERIM-168

ASSEMBLY SHEET
(DoD Format per MIL-STD 847-A)

In the left column, supply page number for each item being included.
Add explanatory details to each line as necessary.
CROSS OUT all items not included.
Asterisked (*) items are mandatory.

Report No. ____________

Sheet ______ of ______

Page	Item	Annotation
	*Cover	
	*Inside Front Cover: Notices (disclaimers, disposition instructions, etc.); Review & approval	
	*DD Form 1473 - Report Documentation Page (includes Abstract)	
	Back of DD 1473	
		(in this format the summary includes the problem statement)
	Summary	opening component
	Back of Summary	opening component
	Preface (includes any Acknowledgments)	opening component
	Back of Preface: Blank	opening component
	Contents: Right-hand page	opening component
	Illustrations, List	opening component
	Tables, List	opening component
	Introduction: Right-hand page	main body; discussion component
		main body; discussion component
	* Main Text	main body; discussion component
		main body; discussion component
	Conclusions	main body; discussion component
	Recommendations	main body; discussion component
		discussion component
	References	appendices; discussion component
	Bibliography	appendices; discussion component
	Appendixes: Right-hand page (use letters, not numerals)	appendices; discussion component
		appendices; discussion component
	Glossary of Terms	appendices; discussion component
	Abbreviations & Acronyms; Symbols	appendices; discussion component
	Index	discussion component
	Distribution List	discussion component
	Inside Back Cover	
	Outside Back Cover	

Figure 5–16. Format guide for research institute reports.

essentially the same as informal reports within the organization. In many instances they may have precisely the same form, as for example in Figure 5-2, *Example of informal report heading incorporating letter format.* In other instances they are in purely letter form. However, the letter form is only superficially different from the informal report form. The heading segment information might be distributed at the beginning and the ending. The reader might have to turn to the end of the letter to find out who the sender is. Nevertheless, the source and the audiences are identified; the subject is announced; and the date is given. Similarly, the foreword and summary segments in most letter reports are not identified by heading but instead are conventionally paragraphed. Nevertheless, these paragraphs serve the same purposes we have discussed. The foreword and summary information are still presented essentially as we suggest above. Figure 5-17 is an annotated example of an effectively designed letter report.

Letters of Transmittal

Many reports are accompanied by letters of transmittal. These serve either to introduce the report to its readers or to function as the opening component for the report. The writer must decide which of these two functions the letter of transmittal is to perform. If the letter functions simply to introduce the report, the writer must make sure the report itself has the basic two-component structure. If the letter functions as the opening component for the report, the writer must design the letter to bear that burden. He must make the letter itself serve the functions which heading, foreword, and summary segments ordinarily serve. A word of warning: letters of transmittal not bound into the text of the report (they are often bound preceding the title page) are easily separated from the report. For that reason, reports with unbound letters of transmittal should have a basic two component design no matter what is in the letter. Figures 5-18 and 5-19 are annotated examples of each of the basic types of letters of transmittal.

Abstracts

Many reports and articles are accompanied by an abstract. This is an extremely compressed overview used separately from the report as a locating device and an aid to potential readers who must determine if they need to read the report or article. Abstracts vary in length from perhaps one hundred words to two hundred or two hundred and fifty words. However, since they will be published or circulated separately from the report as well as filed in indexes and information retrieval systems, they must be very compressed. Generally the entire abstract should be capable of being typed on a three-by-five-inch file card. Often, in fact, several copies of the abstract, printed on perforated file cards, are bound into the report so they can be conveniently torn out and filed or circulated. In formal reports, abstracts usually appear on a separate page between the title page and the table of contents page.

The first thing the writer must realize is that the abstract is not a substitute for the opening component of the report. It is a supplement which in no way should affect the basic two-component design of the report. Second, the writer must realize that the abstract is not a miniature version of the opening component. Unlike the summary, it does not focus upon conclusions and recommendations and is not intended, by itself, to help readers act—except to decide if they need to read or obtain the report. It does not convey sufficient information to be instrumentally useful without the report. Instead, the abstract is a miniature version of the second component of the report, the discussion, and it is intended solely to help readers locate and screen potentially useful reports.

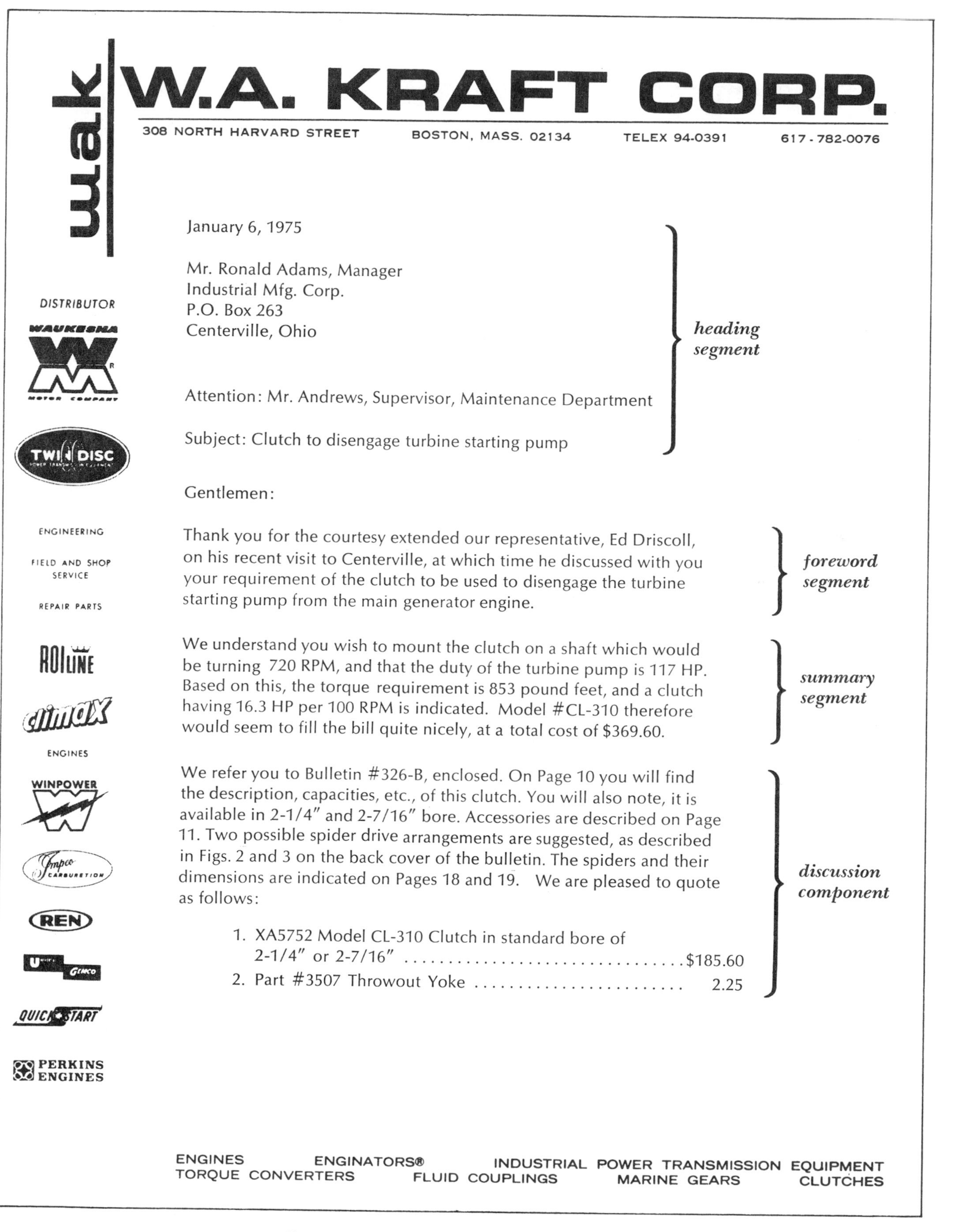

wak | **W.A. KRAFT CORP.**

308 NORTH HARVARD STREET BOSTON, MASS. 02134 TELEX 94-0391 617-782-0076

January 6, 1975

Mr. Ronald Adams, Manager
Industrial Mfg. Corp.
P.O. Box 263
Centerville, Ohio

Attention: Mr. Andrews, Supervisor, Maintenance Department

Subject: Clutch to disengage turbine starting pump

Gentlemen:

Thank you for the courtesy extended our representative, Ed Driscoll, on his recent visit to Centerville, at which time he discussed with you your requirement of the clutch to be used to disengage the turbine starting pump from the main generator engine.

We understand you wish to mount the clutch on a shaft which would be turning 720 RPM, and that the duty of the turbine pump is 117 HP. Based on this, the torque requirement is 853 pound feet, and a clutch having 16.3 HP per 100 RPM is indicated. Model #CL-310 therefore would seem to fill the bill quite nicely, at a total cost of $369.60.

We refer you to Bulletin #326-B, enclosed. On Page 10 you will find the description, capacities, etc., of this clutch. You will also note, it is available in 2-1/4″ and 2-7/16″ bore. Accessories are described on Page 11. Two possible spider drive arrangements are suggested, as described in Figs. 2 and 3 on the back cover of the bulletin. The spiders and their dimensions are indicated on Pages 18 and 19. We are pleased to quote as follows:

1. XA5752 Model CL-310 Clutch in standard bore of 2-1/4″ or 2-7/16″ $185.60
2. Part #3507 Throwout Yoke 2.25

Figure 5–17. Effectively designed letter report.

w.a.k **W.A. KRAFT CORP.** 308 NORTH HARVARD STREET BOSTON, MASS. 02134 617/782-0076

Mr. Ronald Adams, Manager
Industrial Mfg. Corp.
Mr. Andrews, Maintenance Dept.
January 6, 1975

} ***heading information restated***

3. Part #3039 Hand Lever ....$ 6.70
4. Part #1144-E Operating Shaft 7.95
5. X8152-3-B Spider (maximum bore 2.439/2.441") 167.10

Please note, the X8152-3-B Spider quoted is the through shaft spider described on Page 18, which, you will note, has a hub outside diameter of 4-1/4". On this hub, you would mount the driving sheave. In short, the driving sheave would have to be of such construction that it could be bored out 4-1/4". If this were not possible, then the alternate arrangement (Fig. 2 on back cover) would be applicable, in which case we would recommend Spider Flange A-5932 at $78.65 in lieu of the above $167.10.

Thank you for your opportunity to quote. We look forward to being favored with your order.

} ***discussion component cont'd***

Very truly yours,

W. A. KRAFT CORP.

Leo A. Fitzpatrick
Sales Engineer

} ***heading information***

LAF/jmf
ENCLOSURE

} ***appendix material***

AYRES, LEWIS, NORRIS & MAY, INC.

3983 RESEARCH PARK DRIVE, ANN ARBOR, MICHIGAN 48104 (313) 761-1010

January 19, 1974

the content of this letter essentially goes beyond that of the report itself

Berrien County Board of Public Works
County Building
Saint Joseph, Michigan 49085

Attention: Mr. Thomas Sinn, Director

Explains how to read the report. Note reference to the two-component structure in the report itself

Gentlemen:

The attached report of water supply and sanitary sewerage needs for Northern Berrien County has been completed in accordance with your requests. Your attention is particularly directed to the report summary given on pages 4 through 9, which has been referenced to key pages in the study. Because of the many demands on the time of public officials, the summary and references will permit a quick review of the report without reading the details presented .

We believe that the report presents a sound approach to orderly utility planning and development in Northern Berrien County. However, to be of value, the plan has to be implemented at an early date.

introduction of recommendations with emphasis on subsequent action needed

Our recommendations for project implementation are described in detail in the final section of the report dealing with financing. We further recommend that bonding counsel be retained at this time to assist in developing contracts and financing. We also recommend that meetings be scheduled to include representatives of the communities involved, the Department of Public Works, bonding counsel and our firm to work out details of the program. These meetings might logically be divided into two groups: one dealing with the Twin Cities' system to include the communities there involved, and another dealing with the Paw Paw Lake area to include those communities.

We hope that our study will fulfill your needs in developing the public utilities for Northern Berrien County. We have appreciated the opportunity to work with you and your representatives in developing this important public improvement and the courtesies and complete cooperation given us by all communities in the area.

We stand ready to provide any further assistance you need in bringing water and sanitary sewerage systems to the communities of Northern Berrien County.

Very truly yours,

AYRES, LEWIS, NORRIS & MAY

By [signature]

Raymond J. Smit

RJS:pr
enclosure

Figure 5–18. Letter of transmittal introducing report.

McNamee, Porter and Seeley

2223 PACKARD ROAD · ANN ARBOR, MICH. 48104 · AREA CODE 313 769-9220

Consulting Engineers

ACTIVE PARTNERS

J. C. SEELEY — R. M. BACHTEAL
J. M. HOLLAND — D. H. NOLAND, JR.
M. R. VAN EYCK — S. C. WARTINBEE

RETIRED PARTNERS

R. L. McNAMEE — W. S. HERBERT

December 27, 1972

Honorable Mayor
Councilmen and
Mr. Thomas G. Hanna, City Administrator
Charlevoix, Michigan 49720

Gentlemen:

foreword

On December 6, 1971, the City Council of Charlevoix, Michigan, authorized McNamee, Porter and Seeley to study the Charlevoix Water Supply System and make recommendations for improvements to meet future needs. Since the installation of the present system in the early 1920's, the population of Charlevoix has doubled. Recent projections indicate that the population will double again by the year 2000.

Our study to review the adequacy of the present system began in May 1972 and has just recently been completed. This report, being submitted to you for your approval, makes recommendations for improvements to the water supply system to meet the year 2000 projected water demands. The report includes detailed cost estimates for these improvements. ***purpose statement with rhetorical purpose***

summary

The study used population projections to compare the capacity of the present water supply system with future water supply requirements. Charlevoix should anticipate a significant increase in population by the year 2000. The permanent population should increase from 3519 to 4800, and seasonal population from 7000 to 10000.

results

Population to be served	Permanent	Seasonal
Present (Based on 1970 Census)	3519	7000
1980	3940	8000
1990	4370	9000
2000	4800	10000

conclusions

The present water supply system cannot now provide the theoretical maximum day flow or maximum potential fire flow demand during summer population peaks. The maximum safe capacity of the present water system is 1.89 million gallons per day. Population projections indicate that in the year 2000, maximum day flow can be expected to be 3.5 million gallons per day. Without improvements, the present water system would have a deficiency of 1.61 million gallons per day.

Figure 5–19. Letter of transmittal functioning as opening component.

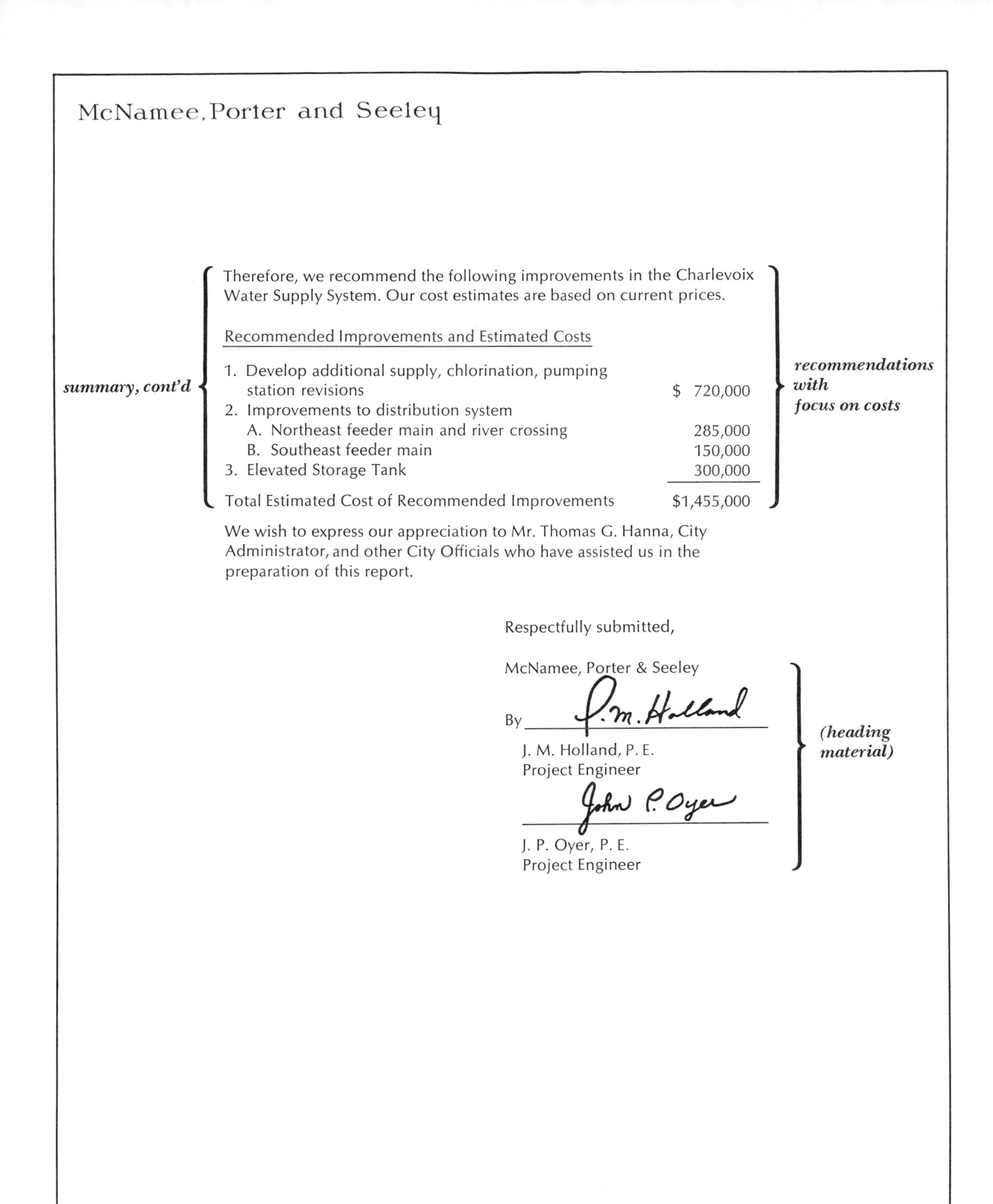

McNamee, Porter and Seeley

summary, cont'd

Therefore, we recommend the following improvements in the Charlevoix Water Supply System. Our cost estimates are based on current prices.

Recommended Improvements and Estimated Costs

1. Develop additional supply, chlorination, pumping station revisions	$ 720,000
2. Improvements to distribution system	
A. Northeast feeder main and river crossing	285,000
B. Southeast feeder main	150,000
3. Elevated Storage Tank	300,000
Total Estimated Cost of Recommended Improvements	$1,455,000

recommendations with focus on costs

We wish to express our appreciation to Mr. Thomas G. Hanna, City Administrator, and other City Officials who have assisted us in the preparation of this report.

Respectfully submitted,

McNamee, Porter & Seeley

By J. M. Holland

J. M. Holland, P. E.
Project Engineer

John P. Oyer

J. P. Oyer, P. E.
Project Engineer

(heading material)

To perform its functions, an abstract necessarily has two important features. The first consists of accessing information which, as card catalogue numbers on library books, allows readers to identify and locate reports or articles of potential interest. Depending upon where and how the abstracts are stored, the accessing information on abstracts may be either very simple or quite elaborate. It may consist of only the author's name, a title, and facts of publication, as in the following example:

> **Dunmire, Daniel E., and Englerth, George H.**
> **1967. Development of a computer method for predicting lumber cutting yields. U.S. Forest Serv. Res. Pap. NC-15, 7 pp., illus. N. Cent. Forest Exp. Sta., St. Paul, Minn.**
>
> **A system of locating defects in a board by intersecting coordinate points was developed and a computer program devised that used these points to locate all possible clear areas in the board. The computer determined the yields by placing any given size or sizes of cuttings in these clear areas, and furthermore stated the type, location, and number of saw cuts. The computer improved on the manual act of extracting cuttings from a board. This was demonstrated conclusively in the yields obtained from a representative sample of 4/4 hard maple lumber in the FAS, Select, and No. 1 Common lumber grades. Thus, a plant manager can obtain an accurate estimate of cutting yields from any grades of hard maple lumber.**

Figure 5–20. Abstract with card catalogue accessing information.

Or it may consist of a complex of information retrieval code numbers and key words, such as in this example:

> CROSS-GRAIN KNIFE PLANING IMPROVES SURFACE QUALITY AND UTILIZATION OF ASPEN. H. STEVENS.
> Research Note NC-127.
>
> ABSTRACT.—Aspen at 6 percent moisture content was planed parallel to the grain and across the grain on a cabinet planer with a 25° rake angle, 1/16-, and 1/32-inch depth of cut, and 20 knife marks per inch. Aspen was also cross-grain knife planed with a 45° rake angle, 1/32-, 1/16-, and 1/8-inch depths of cut, and 20, 10, 5, and 2.5 knife marks per inch. Cross-grain knife planing at all machine settings produced a better surface than finish knife planing parallel to the grain.
>
> ---
>
> OXFORD: 823.1:832.18:176.1 Populus spp. KEY WORDS: particleboard, machining.

Figure 5–21. Abstract with information retrieval system accessing information.

If the accessing information is as uncomplex as in the first example, the abstract will probably be stored in a simple, manually accessed index file. If it is as complex as in the second example, the abstract will probably be stored in an automated information retrieval system. If the abstract is to be stored in an information retrieval system, there are necessarily strict rules governing the nature and arrangement of accessing information. It is beyond the scope of our purpose here to describe these rules. Suffice it to say, however, that any good technical library includes a variety of both general and highly specialized pamphlets on abstracting. For most technical reports, the following guides will be useful:

Guide for Source Indexing and Abstracting of Engineering Literature, published by the Engineer's Joint Council; this presents a standardized approach for abstracting engineering information published in journals, publications of technical societies, private companies, and government departments.

Thesaurus of Engineering Terms, published by the Engineer's Joint Council; this provides a standard vocabulary of engineering and technical terms for use as key words, et cetera.

Directions for Abstractors, published by the Chemical Abstracts Service; this is a standard guide for abstracting scientific research articles and reports.

Guidelines for Cataloguing and Abstracting, published by the Defense Documentation Center; this is the guide for abstracts to be stored in the federal government's centralized information retrieval system.

If the writer needs to write abstracts which will be stored in information retrieval systems and needs therefore to know specifically what indexing codes, key word vocabularies, and patterns of arrangement apply in his area, he should consult with his company or college librarian.

The second part of an abstract consists of an informative condensation of the original report or article. It should not simply describe or *indicate* what is in the report; it should *present* what is in the report. Although you will see both indicative and informative abstracts, informative abstracts are generally much more useful to readers and should, therefore, be the sort of abstract you routinely write. As illustrations of the difference between indicative abstracts and informative abstracts, consider the following two examples. The first, Figure 5-22, is indicative, that is, it just describes what is in the report.

The second, Figure 5-23, is informative because it condenses the substance of the investigation.

As you write abstracts, resist the temptation to describe your report. Instead, present the most economical statement possible of your report's contents. For example, do not say: "In this report two proposed plans for nitric acid supply are compared as responses to Environmental Protection Agency regulations." Instead say: "The Environmental Protection Agency requires reduction of emissions of nitric acid fumes in our Arnold, Texas plant. We can either modify our production facility or we can stop production of nitric acid and instead depend upon external suppliers."

Do not say: "A comparison of the two plans' costs is presented." Instead say: "Modifying the production facility would cost approximately $1,125,000 in capital outlay. Depending entirely upon external suppliers would cost approximately $227,000 per year more than it presently costs us to supply ourselves."

New Farm Fencing Declines—Use of Treated Posts Up

ABSTRACT.—Describes the amount and type of farm fencing installed on commercial farms in the Central and Appalachian States during the years 1963-1966.

OXFORD: 831.51:764(74/77)

Figure 5–22. Indicative abstract.

DUNMIRE, DANIEL E.

1966. Effect of initial moisture content on performance of hardwood pallets. N. Cent. Forest Exp. Sta., St. Paul, Minn., 12 pp., illus. (U.S. Forest Serv. Res. Paper NC-4.)

Four years of service-testing 90 red oak pallets showed that those made with predrilled air-dry deckboards and green stringers gave better service and cost less to maintain than pallets made entirely from either green or air-dry lumber.

Figure 5–23. Informative abstract.

Do not say: "A recommendation is presented." Instead say: "Although modification of our facility would cost $1,125,000, over five years we would regain that initial cost. For that reason we recommend facility modification."

Figure 5-24 is an example of an effectively designed informative abstract. In 187 words, the writer enables the reader to decide whether or not he needs to obtain the report.

We said earlier that the principles of a good opening component should figure in whatever format you use. Although we have been discussing alternative patterns

CAUSES OF LOWER END DEFECTS ON ENGINES RETURNED TO PLANT #33, 1 JAN-1 JUNE 1973.
(Fred W. Shippy, 23 Sept. 1973)

Between 1 Jan 1973 and 1 July 1973 ninety-five 350 cu. in. engines were returned to plant #33 because of lower end defects. (The lower end includes all moving parts below the camshaft except the oil pump). These engines were disassembled and inspected for indications of the causes of failure. Inspection included examination of torque on all bolts and visual examination of all engines for evidences of defective parts and either missing parts or presence of foreign matter. Failures seem attributable to three major causes: fatigue in connecting rod (41 failures); rod bearing failures (29); connecting rod breakage (9). Other failures were caused by oil pan/crankshaft interference (5); main bearing failure (2); incorrect rod cap (3); stock in oil pan (6). By far the most consistent causative factor was low torque on rod bolts. (Low torque is less than 50 ft.-lbs of torque.) 83% of the engines exhibited this fault. Conclusions to be drawn are that we must improve methods to control rod bolt torque. Assembly procedures were clearly lax. Further, we must improve quality control inspection methods to verify proper rod bolt torque.

Figure 5–24. Example of effectively designed informative abstract.

for handling the opening component, do not let our discussion of alternatives divert your attention from the basic principles. Your opening, whatever its particular format, should address the organization, state the problem, and summarize important conclusions and recommendations. Remember, after the opening component, many of your readers should have all they need from your report; the others should be prepared to read on into the discussion.

CHAPTER 6

Designing the discussion component

The discussion of the report is a structural component which attracts remarkably little concern in most books on technical writing and about which company style sheets have virtually nothing to say. These sources perhaps lay down suggestions on the general format for a whole report, but when it comes to designing the discussion component of a report, writers may find themselves on their own, no help in sight. Yet careful organization of the discussion component is essential, for this component forms the center around which all the other report elements cluster. The purpose of this chapter is to help the writer find ways of designing a genuinely clear structure for the discussion component of the report.

In Chapter five we could establish a fairly precise set of directives for writing the opening component of a report. We could identify the segments, list what goes into each, and give them an order. We could set up a kind of formula for patterning the opening of your report, a formula based upon logic and convention. When we move to the discussion component, however, we are dealing with a much more slippery topic. If there are reasonably rigid patterns for ordering the opening of the report, a variety of patterns is possible for ordering the discussion. Accordingly in this chapter we present some guidelines for establishing an intelligent structure. Take them as they are intended, and they should make a difference in the effectiveness of your discussion.

The premise of our suggestions is that the discussion of a report should be a complete structural entity. It should be a complete but selective presentation of information necessary to accomplish your rhetorical objectives. It should be capable of standing clearly on its own without depending excessively on preliminaries or appendices.

Perhaps one of the most common failings we see in technical reports is that it is nearly impossible to read through the discussion without having constantly to flip forwards and backwards to piece together a whole explanation. The purpose statement is in the foreword; the recommendations are in the summary; the conclusions are spread throughout the discussion; and all the figures are in the appendices. And there the discussion of the report stands like a nearly completed do-it-yourself kit, parts scattered all over the place, waiting for someone to come along and assemble it. Of course that someone is thc hapless reader. This disconcerting situation tends naturally enough to destroy any continuity or unity in the report. There is no center to hold it all together. To unify the report and meet the needs of your important audiences, the discussion needs to be self-contained.

Of course, we are not arguing that the discussion of the report should be denuded of references to the appendices or that the discussion must contain everything that is worthwhile in the report as a whole. Not at all. The other parts of the report serve perfectly legitimate functions which should not be preempted by the discussion. The appendices especially serve to rid the discussion of data, calculations, and detailed figures that would blur the rhetorical focus if they were left unselectively in the discussion. But the discussion of any report will accomplish your rhetorical purpose much more effectively if the discussion can be read as one coherent and structurally complete whole.

Obviously there are a number of important implications in the idea that a discussion of a report should be able to stand essentially on its own. But structurally what do we mean when we say a report's discussion should be capable of standing clearly on its own? We mean the report's discussion must be whole. That is, it must have a *beginning,* a *middle,* and an *end.* It should not, as so many reports do, depend for the beginning and ending functions upon the preliminaries. If it does, it becomes a formless middle which cannot be read by itself. We will start by looking at the beginnings and endings. Then we will look at middles. Finally, we will briefly examine the relationships between the discussion and the appendices.

BEGINNING THE DISCUSSION

In our chapter on the purpose of the report, we said it is always necessary for the introductory section of the discussion to contain a fully developed problem statement. The statement includes:

1. A statement of the problem, i.e., the conflict at issue in the organization;
2. The posing of specific technical questions or tasks arising out of that problem, and addressed by the technical investigation;
3. The statement of the rhetorical purpose, i.e., a statement of what the report is designed to do in relation to the organizational conflict and the consequent technical questions and tasks.

As we also made clear in our chapter on opening component segments, the problem statement in the beginning of the discussion particularly develops the specific technical questions addressed by the investigation. This problem statement provides the essential content of the beginning. However, in many reports additional information is necessary. There are at least three types of additional information.

First, a forecast of the structure of the discussion frequently follows the problem statement. A forecast is particularly necessary in long or complicated reports. It serves to divide the discussion into manageable parts and to predict their sequence. Here is an example of such a forecast:

> An electrocardiogram (ECG) records electric signals from the heart muscle as it beats. A normal ECG is taken while the patient is lying perfectly still because electric signals from the motion of other muscles can interfere with the signals from the heart. However, doctors have found that the shape of the ECG sometimes changes if it is taken while the patient is exercising because the heart activity increases. By

comparing the ECG taken at rest with the one taken during exercise, doctors can obtain additional information about the condition of the patient's heart.

The problem is that the muscle activity during exercise produces electric signals which interfere with the electric signals from the heart. The muscle activity makes the exercise ECG difficult to read. To improve the exercise ECG signal, we investigated methods of eliminating the interference from muscle activity. The method had to be conducive to further research.

This report examines the four questions addressed in the technical investigation.

1. What is the nature of the ECG signal, and why is it useful to study the electrocardiogram during exercise?
2. Why was filtering, and not signal averaging, chosen as a means of eliminating the interference from the exercise ECG?
3. Why is filtering not the solution to eliminating the interference?
4. What are some alternative solutions?

With this forecast as a guide, the reader knows what the segments of the discussion are and what their sequence will be.

Second, amplification of some elements of the problem statement may be needed. For example, the issue of organizational concern giving rise to the assignment might require you to explain the background or context in which the work exists. The technical problem needing solution might require you to describe previous experimental work. The specific technical questions might require you to set forth the specifications or instructions which served as the basis for the present work. Here is an example which explains the theoretical background of the problem being investigated:

Corona is generated when a partial electrical discharge occurs across a gas-filled gap. This discharge may require rather high voltages depending on the size of the gap and the type of gas (usually air). Each time a discharge occurs some damage is caused to the insulation by erosion of the material, and radio frequency interference (RFI) is created. If corona is prevented from occurring, these problems are avoided and the insulation life will theoretically be infinite. The insulation life decreases significantly as the voltage reaches a level high enough to create corona. (A curve illustrating the life versus applied voltage relationship is shown in Fig. 1). Thus, for infinite life, the wire must be designed so there will be no corona.

In most applications, a shorter-than-infinite life is satisfactory. However, RFI is still produced by corona and is usually a significant problem. Corona frequently renders the associated circuitry useless.

Our testing was mainly concerned with determining the corona initiation voltage (CIV) and the corona extinction voltage (CEV) levels for a particular wire construction. The next section will discuss the equipment, method, and samples we used for our evaluation.

Third, potentially unfamiliar concepts may need explanation. This is particularly true in difficult reports which involve recent technology. Sometimes this involves a glossary of terms, sometimes an extended definition. Here is an example of the latter:

1.2 Volume Control *vs.* Weight Control

Modern destroyers are "volume-controlled," so the significant design features are those relating to utilization of volume. Formerly, destroyers were "weight-controlled," and the naval architectural characteristics such as buoyancy and stability were those chosen to carry the weight of hull, machinery, ship systems, people, and payload. The resulting enclosed space was ample to contain the total volume demanded by these elements. Modern destroyers are volume-controlled in that the volume required by these elements is greater than that associated with an equivalent weight-controlled design. Thus the hull must be larger and heavier and this produces secondary increases in machinery, fuel, and system weights, all of which result in a total displacement increase above a weight-controlled design.

This segment briefly explains the new concept of volume-controlled design. Without the explanation, the subsequent discussion might be unclear to many readers.

The first of the three additional types of information we have identified—the forecast—is usually included in the introduction following the problem statement. The second and third additional types of information—the amplified problem statement and the explanation of concepts—usually come next. If not much space is required, they also can be placed in the introduction. Ordinarily, however, they form a segment immediately following the introduction. This segment is characteristically labeled "Background" or "Theory." This segment appropriately concludes the beginning of the discussion.

ENDING THE DISCUSSION

Too many report writers, apparently feeling some misdirected urge to "let the facts speak for themselves," or feeling that the summary in the opening component is conclusion enough, omit any kind of conclusion from the discussion. They just stop. Yet, if the discussion is to stand on its own it must be ended. To leave out the report's ending is to obscure the detailed conclusions some readers seek. You are expecting them to add up the discussion for themselves, and to do it correctly. If you are lucky, your readers may be both willing to do that and capable of doing it well. More likely, however, you will find they resent having to do it and that they are not able to do it as well as you would like. The solution, therefore, is obvious; to complete the discussion and complement the purpose stated in the beginning, you must always include an ending.

What goes into an effective ending? Again there are no invariably effective formulae. The subject, the audience, or the purpose for any report will suggest how the task might best be handled. We can, however, identify some ending strategies to consider. Endings can:

1. Reintroduce the technical problem and summarize the problem-solution process; or
2. Reintroduce the rhetorical purpose and address the organization.

If you reintroduce the technical problem, your objective is to clarify the intellectual problem-solving process you went through. The logic of the process may become obscured by the particulars of the technical investigation or by the arrangement of the discussion. Depending on how the middle is organized, you might want to do one or several of the following:

Restate the conclusions you have reached in the course of the work.
Restate the objectives of the investigation and the problem or cause that gave rise to the report in the first place.
Summarize the main points or arguments.

Your rhetorical purpose will indicate if reintroducing the technical problem should be the focus of the ending. Here is an example which restates the purpose and summarizes the main points:

V. Conclusions and Recommendations

My objective in this project was to find a method of eliminating the interference from the exercise electrocardiogram signal and thereby enable Dr. Santinga to conduct further research.

Signal averaging does produce an interference-free ECG. However, signal averaging tends to eliminate any changes that occur in the ECG from beat to beat. Because these changes might be important, I felt a better way to study the exercise electrocardiogram would be on a beat to beat basis.

Filtering was examined as a method of removing the interference from the exercise ECG on a beat to beat basis. However, filtering proved to be unsatisfactory because the level of filtering necessary to significantly reduce the interference distorted the ECG signal. Filtering removes too much information from the ECG signal to make it useful to further research.

There may, however, be alternatives to filtering which will remove the interference on a beat to beat basis.

Professor Edmonson has suggested a computer integration of the ECG as one alternative. Analysis of the area under the exercise ECG curve should yield the same kind of information that measurements of ST slope and depression do. Computer integration should also tend to eliminate the effect of the interference.

Another alternative might be a computer program which would fit a curve to the average variations of the interference. Such a program should selectively filter the interference, but not disturb the actual ECG signal.

Notice that in addition to stating the purpose and summarizing the main points, this example also addresses the organization.

If you reintroduce the rhetorical purpose, the second ending strategy identified above, you may find it appropriate to address the organization. You can:

Recommend specific subsequent actions on the part of the reader.
Pose specific questions for subsequent investigation.

Your investigation yielded conclusions which form the basis for recommendations to resolve the issue of organizational concern. Many reports end with a recommendations section because the recommendations are the final step of the problem-solving process, and because they reintroduce the rhetorical purpose. These recommendations do more than just restate those presented in the opening component. Here they are for different audiences and therefore must be more detailed. They are what many reports are all about.

The problem statement and consequent rhetorical purpose might mean that the ending should pose specific questions for subsequent investigation. This will be the case if your investigation was part of a larger investigation, if your solution leads to a new problem, or if your investigation determines the nature of subsequent company actions. Here is an example from a report on an experimental device used to measure the effects of mercury poisoning on the central nervous system.

RECOMMENDATIONS FOR FUTURE INVESTIGATION

There are four major recommendations for future research based upon this study.

1. It appears as if the reliability of the measurement system is strongly related to the S/N ratio and possibly to learning effects. Therefore, added effort should be

devoted to attempting to find better ways to condition the input signal. An increase in the S/N ratio may serve to make the system more reliable.

Further data should be taken to ascertain how the learning process affects the system. In particular, one is interested in how it occurs and when it levels off. It is possible that a warm-up should be made part of the operating procedure. This warm-up would give the operator a few tremor inputs before beginning the actual analysis.

2. Some sort of skill in which people differ greatly may be involved in performing the tracking task with the designed machine. This hypothesis should be tested. If the reliability does vary significantly from person to person, an analysis should be made to determine what this skill is so that operators may be picked accordingly.
3. The effect of better quality components should be investigated. This has a cost trade-off and should be considered in relation to its effect on the entire system's worth.
4. The variability of the mode in a test-retest situation must be evaluated to determine:

   a. the reliability of the test-retest scheme as a detection device; and
   b. what is an acceptable deviation at the man-machine level.

This process testing should be run on a number of patients using both the computer analysis system and the device designed for this project. A comparison against blood samples might be desirable.

You should take care to end the report rather than to let the investigation "speak for itself." In the body you are addressing numerous audiences who are not directly enough involved with the problem to be able to interpret the investigation for themselves.

STRUCTURING THE MIDDLE

The first thing to realize is that you must do three kinds of things in order to design an effective middle for your report. First, you must *segment* the mass of information generated during the technical investigation, partitioning it into coherent steps, parts, pieces. Second, you must *select* from among those segments those which—in terms of purpose and audiences—genuinely belong in the middle of the discussion component. Some will not. Third, you must *arrange* those segments; that is, you must put them into an order which is logical, appropriate to your purposes, and apparent to your readers. Graphically this process is illustrated in Figure 6-1, *The structuring process.*

Notice that we introduce a feedback loop at the end; however, at any given point you may be forced back to reexamine your prior choices. For example, arranging the material may force you to reexamine information you have earlier thrown out. In short, the process is dynamic.

To the writer faced with these three tasks—segmenting, selecting, and arranging his material—the report's middle may seem an unmanageable, amorphous mass of information. What is needed, obviously, is some intelligent method by which he can proceed. The beginning establishes a "general to particular" pattern in the report's discussion, but the particulars in the discussion still need to be segmented and ordered. Here is a method of approach that will help. It derives from ideas we introduced in previous chapters.

We originally said that the communication situation always involves a writer, a message, and an audience. If you analyze the implications of each of these, you can establish three methods for segmenting, selecting, and arranging material:

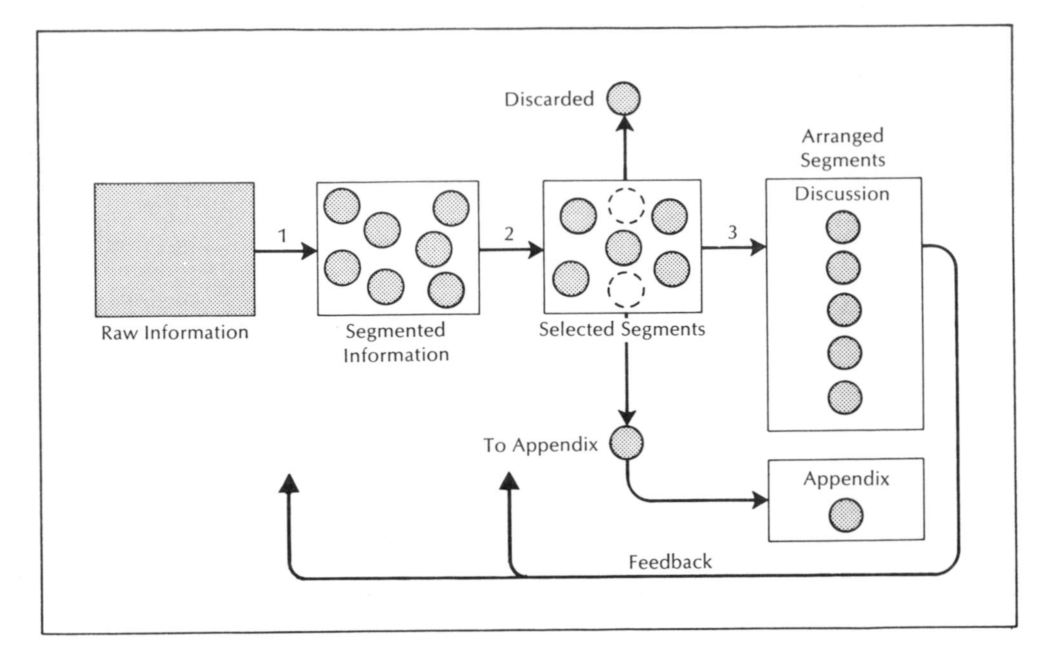

Figure 6–1. The structuring process.

1. The implications of audience suggest a structure based upon rhetorical purpose.
2. The implications of the writer's role suggest a structure based upon an intellectual problem-solving process.
3. The implications of message suggest a structure based upon the subject matter of the technical investigation.

Structuring by Rhetorical Purpose

First, think of structuring in terms of rhetorical purpose, because that is what the report is all about. If you still are wearing your investigator's hat, you may structure your report in terms of the problem-solving process or subject matter when structuring in terms of rhetorical purpose would be more effective.

What does it mean to say, structure your report according to rhetorical purpose? Essentially it means that the report is organized in terms of what the writer seeks to accomplish by writing the report rather than in terms of his method of investigation or of his subject matter. As we have shown, reports serve instrumental purposes in an organization, that is, reports help persons to make decisions and to act. To do that reports often need not stress details of the investigation or subject matter. Reports which are structured in terms of rhetorical purpose directly address audience needs.

As an example, assume you are writing a report on a satellite design, and the purpose of the report is to convince your audiences that the satellite will accomplish its mission. If you do not think in terms of rhetorical purpose, you might structure your report in terms of the subsystems within the satellite, that is, in terms of the subject matter. Yet enumeration of the subsystems, no matter how methodical, does not allow you to focus on the distinctive and thus convincing features of the satellite design. It would be far more effective to structure your report as a series of supports for your assertion that the satellite will be successful.

If you structure in terms of rhetorical purpose, you can subordinate, relegate to appendices, or even discard segments of particular but routine information, such as information about the launch vehicle.

Specific reports provide illustrations of the concept. One writer, writing a progress report, structured it as an answer to the question, "What progress have I made?" His report was structured as a series of steps toward organizational objectives. The segments were not units of time or units of subject matter, they were units of evidence defined by the relevant concept of "progress." As another example, a writer writing an economic justification report structured it as a series of evidences of cost comparisons. His problem was, "Can it be demonstrated with reasonable certainty that the added investment will produce a return of at least 10 per cent per annum after taxes over the life of the investment?" Here the report was structured to make the case convincing to management. The segments were not units of subject matter, but units of the comparison chosen and arranged to sell. As a final example, another writer was making a recommendation to purchase a particular switch from among those on the market. He structured his report as a series of demonstrations of advantage. Again, he did not structure according to subject matter, that is, types of switches; he structured according to the criterion of advantage (cost being just one type of advantage).

When you structure by rhetorical purpose—as you can see from the examples—your primary concern usually is to support an hypothesis or make an assertion. Sometimes, of course, the hypothesis is routine and even unstated. (A progress report usually is expected to support the hypothesis, "I've made progress"; a travel report usually is expected to assert that your travel time and the company's money were well spent.) The persuasive element in such reports yields a possible structural outline, consisting of four elements. The beginning of the discussion establishes the problem and provides the solution. Next comes a definition of the criteria to be used to evaluate the evidence. Then come the segments of support. And finally comes a restatement of the conclusions and recommendations.

The first element of a report structured according to rhetorical purpose establishes the problem and provides the solution. These are the tasks of the beginning section of the discussion. (Since we have talked at length about this already, we will not elaborate further here.)

The second element of a report structured according to rhetorical purpose establishes the criteria to be used to evaluate the evidence. Depending on the complexity of this task, this is either an amplification segment in the beginning or the first segment of the middle. If the criteria are readily apparent and easily stated, they can be put into the beginning; if they have to be argued themselves, the criteria segment becomes the first segment of the middle. For example, the assertion, "this investment will return in excess of 10 per cent per annum over its lifetime," is a statement of criterion from the beginning of a report requesting government subsidy for design modification. In this case the criterion is easily stated since it is explicit and clear cut. In other cases, criteria must themselves be established. Need, feasibility, reliability, practicability, fairness, safety, significance, and value are all relative concepts which must be defined before they are applied. In such cases a separate segment of the report defines the concept; it is an extended definition of the criterion.

The third element of a report structured according to rhetorical purpose—usually several segments—provides the support for the assertion. These are the supports for the conclusion. They are arranged in descending sequence of significance and frequently move from positive to negative. That is, they support the assertion first with arguments in favor. Then they support the assertion by showing that arguments against it and for other alternatives are not decisive. As an example, a report which proposes purchasing a particular pump would begin by demonstrating that the recommended pump met the criteria established. Then it might move to demonstrate that alternative pumps which meet some of the criteria are less desirable than the pump being recommended. There are two groups of support here—one positive, one negative. Thus there are several segments in this third element which themselves need to be arranged according to rhetorical purpose.

The final element of a report structured according to rhetorical purpose restates the assertion which the report has supported. (Again, since we have already discussed variations of ending a report, we will not discuss this element further.)

Detailed examples might be helpful to clarify the concept of structuring according to rhetorical purpose. Here are outlines of three particular reports structured according to this method. The first illustrates the difference between structuring according to rhetorical purpose and structuring according to another method (in this case, the problem-solving process). The second illustrates the four elements of the basic outline of the rhetorically structured report. The third illustrates a positively and negatively supported assertion.

The first example is already partly familiar. It is an outline of the progress report by the transportation engineer who was being transferred to another division before he finished revising train schedules. His report is structured in terms of what has been done and what remains to be done, units of progress. The outline is as follows:

I. Stages of Investigation Completed
   A. Formulation of Speed-Distance Curves
   B. Determination of Expected Speed Performance
II. Stages of Investigation Remaining
   A. Formulation of Time-Distance Curves
   B. Preparation of Time Tables

The units of progress, Roman numerals I and II, provide the basic structural segmentation and arrangement. By using this method of structuring the report, the writer makes the status of the investigation immediately clear to his audiences. Had he structured the report according to the problem-solving process, the status would not have been immediately clear. Such a structure, as you can see from the following outline, would have been rhetorically illogical. Stages one, two, and four are completed; stages three and five remain to be done:

I. Collecting the Initial Data
II. Formulating Speed-Distance Curves
III. Formulating Time-Distance Curves
IV. Finding the Expected Speed Performance
V. Proposing the Time Tables

The first outline is much more directly a reflection of the writer's purpose; it makes his progress immediately apparent to his readers. Notice also that structuring according to rhetorical purpose enabled the writer to relegate to the appendix less important information, in this case information on collecting the initial data. In other words, analysis of his rhetorical purpose enabled the writer to segment, select, and arrange his material in this report.

The second example is from a report in which a ship company is seeking subsidy from the Maritime Administration to help pay for design modifications of cargo hatches on ships already under subsidized construction. The outline is of a straightforward, positively supported assertion that the subsidy is justified. The four major divisions correspond exactly to the four elements of rhetorical purpose structure:

Subject: Design Modification for Construction Differential Subsidy.

I. Purpose of Design Modification Authorization

II. Construction Differential Subsidy Policy

III. Economic Justification for Subsidy

  A. Basis I. Same Cubic Volume of Cargo in Fewer Containers

  B. Basis II. Increased Cubic Volume of Cargo in Same Number of Containers

  C. The Competitive Advantage of Increased Flexibility

IV. Application for Construction Differential Subsidy

The first element establishes the problem; the second the criterion for making the judgment. (These two elements form the beginning of the discussion.) The third element provides the support for the assertion that the subsidy is justified. The fourth restates the conclusion and recommendation. Although it is not apparent from the information in the outline, the writer presents his three arguments in support of the subsidy in a descending sequence of importance. This is a fundamental principle of arrangement according to rhetorical purpose. Furthermore, the writer has been very selective in structuring his material, also not apparent in the outline. He has omitted technical information on modification of the hatches (subject matter information). He has also omitted material on the need to enlarge the hatches (problem-solving material). In this particular report, the issue is not whether the modification is needed (that has been established by an earlier report); the issue is, who will pay the bill?

The third example is from a report we have not mentioned before. The report proposes a system of evaluating parts suppliers, that is, vendors, to an automobile assembly plant. The old system had been acknowledged to be inadequate because it did not enable the company to monitor suppliers whose parts were excessively prone to defects. The writer proposes an alternative system. The first segment states the problem; the second establishes criteria for evaluating the solution; the third, fourth, and fifth support the proposal, and the sixth presents the recommendations.

Vendor Quality Index to Identify Problem Vendors

1. The Need for a New Vendor Rating System

2. The Objectives of the Vendor Rating System

3. The Advantages of the Proposed Vendor Quality Index

  3.1 The Vendor Quality Index

    3.1.1. Defect Severity Classification System

    3.1.2. Computation of Vendor Quality Index Rating

3.2 Use of the Vendor Quality Index

3.3 Effectiveness of Vendor Quality Index

3.3.1. Impact of Defect Severity on Vendor Rating

3.3.2. Efficiency of Defect Information Retrieval

3.3.3. Additional Cost of the Index

4. Disadvantages of the Monthly Vendor Summary

4.1 Current Ratings by Monthly Vendor Summary

4.2 Failure to Rate Vendors Effectively

4.2.1. Severity of Defect Unweighted

4.2.2. Inaccuracy of Vendor Ratings

5. Comparison of Vendor Quality Index Ratings with Monthly Vendor Summary Ratings

6. Implementation of the Vendor Quality Index Rating System

This outline is a particularly good example of supporting an assertion with both positive and negative arguments. The third segment presents positive supports. The fourth segment presents the negative; it shows the disadvantages of the current system. The fifth segment synthesizes by comparing the old and new systems to show that the new system is more desirable than the old.

These three detailed examples illustrate how your rhetorical purpose enables you to segment, select, and arrange your material appropriately. They even yield a rough outline for any report which makes some sort of assertion, implicitly or explicitly. This outline will provide at least a starting place for organizing according to rhetorical purpose.

I. The Problem Statement and Solution

II. Criteria to Evaluate Evidence

III. The Support (internally ordered either in a descending sequence of importance or as positive and negative evidence)

IV. Restatement of Conclusions and Recommendations

Faced with the problem of structuring a report, you should first consider this method. Before deciding to use this method, however, consider the alternatives to be sure that structuring by rhetorical purpose is most appropriate.

Structuring by the Problem-Solving Process

This method requires the writer to discover segmentation and order in the *intellectual process* which includes the technical task he has just performed. Buried within the intellectual process is an order the writer can use to help his readers understand what has happened. The trick is to look at your technical activities as part of an intellectual process of problem solving.

We ordinarily and mistakenly tend to think of technical activity as physical activity. By looking at the *thing* rather than the *process* we tend to ignore the preliminary intellectual stages. However, the problem-solving process does not begin when you set up your equipment or make your first measurements. Nor does it end when you have added up the results and drawn your conclusions. Thus an adequate description of the process must be seen as an account of an intellectual activity beginning well before you start measuring or weighing things and extending long after your equipment is put away.

The intellectual problem-solving process can be divided logically into nine stages. (Here we are simply amplifying our explanation of the engineer's role, as illustrated in Figure 3-2, p. 30):

Stage One: The organization encounters a problem or discovers the need for technical investigation.

Stage Two: The technical task is assigned to you or is undertaken voluntarily by you.

Stage Three: You define the specific technical questions to be investigated or tasks to be performed.

Stage Four: You determine how to answer the questions or perform the tasks, that is, what equipment and procedure (means and methodology) to use.

Stage Five: You actually perform the technical investigation, probably in a carefully planned series of steps.

Stage Six: You collect and tabulate the results.

Stage Seven: You analyze the results and draw conclusions.

Stage Eight: You formulate recommendations based upon your conclusions.

Stage Nine: You write the report to create the organizational responses necessary to solve the problem.

This nine-stage description of the problem-solving process gets at the underlying intellectual order behind a technical investigation. It may not describe the order in which the technical investigation actually is performed. The technical investigation is often a blur of intellectual activity, full of methodological assumptions, intuitive leaps, short cuts, and waste motion. The hypothesis which governs the inquiry might occur at any stage in the problem-solving process. Some of the conclusions can precede and even guide the collection and tabulation of further results. Our description ignores the actual sequence of events of a technical investigation and gets at the fundamental order of the intellectual process. It shows how your mind works as it meets with a problem, analyzes it, formulates a hypothesis, tests it, and draws conclusions—but without the waste motion. Looked at carefully, this description of the intellectual process can provide a segmentation and ordering of your material that will enable your reader to encounter the process and come to understand it as you have.

Here is an outline based upon the intellectual problem-solving process of a particular investigation; we have also identified the stages of the problem-solving process.

I. **Introduction:** *Stages One* and *Two* in our description of the intellectual problem-solving process

   A. **The problem:** An explanation of the organizational problem that gave rise to the report

   B. **The objective:** A statement of the assignment

   C. **The method of this report:** A forecast of the structure

II. **Background:** *Stage Three*

  A. **Previous work:** An explanation of what already has been done about the problem

  B. **Specifications:** An explanation of detailed instructions or specifications that served as the basis for the present work

III. **Experimental procedure:** *Stages Four* and *Five*

  A. **Materials:** A description of equipment used

  B. **Methods:** A step-by-step description of the procedure followed [Warning —this is where writers find the chronological order they mistake for the intellectual order.]

IV. **Discussion of results:** An extended explanation of precisely what was learned—*Stages Six* and *Seven*

V. **Synthesis:** *Stage Eight*

  A. **Problem restated:** A restatement of the report's objective and of the problem that gave rise to the report

  B. **Summary:** Reviewing the main points

  C. **Recommendations:** Explaining subsequent action or posing specific questions for investigation

Perhaps if we strip away some of the explanations and the additional beginning and ending ingredients in this outline you can better see the structure of the intellectual problem-solving process.

I. **The problem:** *Stages One* and *Two*

II. **Background:** *Stage Three*

  A. **Previous work**

  B. **Specifications**

III. **Experimental procedure:** *Stages Four* and *Five*

  A. **Materials**

  B. **Methods**

IV. **Discussion of results:** *Stages Six* and *Seven*

V. **Synthesis:** *Stage Eight*

The important point to remember is this: Segmenting and arranging by the intellectual problem-solving process does not mean that you present a chronological step-by-step description of your actual technical work. In fact, because the actual performance of the technical investigation (*Stage Five* of our description) is the least important stage intellectually, it certainly should not yield the structure of the report. Unfortunately it often does. Your intellectual process begins before you enter the laboratory and ends after you have your results. Almost no report reader (other than professors) wants a detailed description of what you actually did.

A good example of how the intellectual problem-solving process (rather than the actual process of the investigation) should determine the structure of the report was given to us by an engineer working for the National Highway Traffic Safety Administration. He gave us the following description of how his study was reported—as opposed to how it was actually done. The study was "A Human Factor Analysis of Most Responsible Drivers in Fatal Accidents." Here is his explanation:

> The study was originally designed as an exploratory investigation of the human factors present in fatal motor vehicle accidents. We decided to key upon the driver most responsible for the accident and gather a standard set of data on him via an interviewing technique. After data collection, we were going to try to relate some of the findings to the causes of the accident and see if these drivers differed in any respects from other types of drivers. The objectives were general and exploratory in nature.
>
> We then designed the data forms, gathered the data, and began analyzing the data.
>
> We created certain groupings and categories on a logical basis. For example, we separated drivers who were killed (Type I) from drivers who survived (Type II) and from drivers who killed pedestrians (Type III). When analyses showed differences between these groups, this aroused our curiosity. We were finding something we did not expect. Type II drivers were differing significantly from the other two groups in very important respects. Therefore, we formulated the null hypothesis and then went on to show that it could be rejected with respect to these variables. Thus the basic thesis of our report was formed.
>
> In summary, our problem-solving process went this way:
>
> (1) We formulated some general objectives due to a need for information.
> (2) We designed the methodology for data collection.
> (3) We collected data, and began tabulation.
> (4) We began to group the data in logical categories.
> (5) We did exploratory factor analysis on the data groups.
> (6) We discovered something that we did not expect.
> (7) We formulated the null hypothesis knowing that we would reject it.
> (8) We narrowed the study to this objective.
> (9) We completed the study.
> (10) We wrote the paper.
>
> These ten stages were the sequence of the investigation. However, we wrote the report as if the investigation had been conducted this way:
>
> Data need
>
> Hypothesis
>
> Design
>
> Method
>
> Results
>
> Discussion/Recommendations

To show how his report structure conforms to the stages of the intellectual problem-solving process structure as we describe it, here is his outline with our stages of the process indicated:

> Data need (*Stage One*)
>
> Hypothesis (*Stage Two*)
>
> Design (*Stage Three*)
>
> Method (*Stage Four*)
>
> Results (*Stages Five* and *Six*)
>
> Discussion/Recommendations (*Stages Seven* and *Eight*)

This structure clarifies the problem-solving process which a chronological narrative of the actual investigation would have obscured.

In the example above, the headings clearly correspond to the stages of the problem-solving process. In the example that follows, the headings are substantive but the intellectual process still determines the structure of the report. The report on filtering electrocardiogram signals is already familiar to you from the beginning and ending segments. As above, we indicate the stages.

TABLE OF CONTENTS

The three examples we have presented should be sufficient to show that the stages of the intellectual problem-solving process can yield an effective report structure. For some kinds of reports, particularly research reports, you may find this method for structuring the most appropriate.

For reports other than research reports you also may find a problem-solving structure appropriate. Reports issued by consultant firms often use this structure. A consultant firm might be contracted to suggest expansion of a water treatment plant to meet the demands of a city's projected population growth. In such a situation the firm often would design its report as follows:

Projected water consumption trends.

Current water treatment capacity.

Necessary improvements to meet the needs.

This three-part structure follows the simple problem-solving formula:

Needs − Capacity = Necessary Improvements

This simple problem-solving formula is analogous to the standard deductive pattern used for solving some problems:

Major Premise

Minor Premise

Conclusion

An engineer frequently finds the problem-solving structure appropriate because that is what an engineer is hired to be—a problem-solver.

Structuring by Subject Matter

If neither rhetorical purpose nor intellectual process enables you to structure your material, you may have to fall back upon the last of the three structural methods we suggest. A workable alternative, but less desirable than either of the other two, this method requires the writer to look carefully at his subject matter. Often he can segment and arrange the material according to natural and obvious "joints" in the subject matter itself. Particularly in reports dealing with complex systems, the system itself yields a convenient way to segment the material.

As an example of a report structured according to subject matter, here is the Table of Contents of a report by an aerospace engineer.

1. Introduction
    1.1. The Mission Explained
    1.2. The Problem
    1.3. Procedure
2. Launch Vehicle Compatibility Check
    2.1. Payload Envelope
    2.2. Separation System
3. Satellite Compatibility Check
    3.1. Sensors
    3.2. Attitude Control System
    3.3. Power Supply System
    3.4. Communication System
    3.5. Thermal Protection System
4. Compatibility Conclusions

Here the subject of the writer's technical task is apparent: after design modifications, he had to check a complex subsystem for compatibility with another intricate subsystem. In order to structure his material he has simply taken advantage of the hierarchy of subsystems within the complex launch vehicle and satellite system. One at a time he ticks off the subsystems and checks to see if their component subsystems are still within the compatibility parameters established by the preliminary design. The system itself becomes his organizer. That is, his segments are the subsystems and component subsystems of the launch vehicle-satellite system; his structure is a systemic structure. He has, in short, simply let the subject matter suggest its own structure.

Here is another example of a report structured according to subject matter. Although from a different field, the report has the same structure as that of the first example.

I. The Need for a S/3—Modem Interface

II. Communicating with the Modem
    A. Serial Data Transmission
    B. Connections between the I/F and Modem

III. Communicating with the S/3

  A. Data Flow between I/F and S/3
  B. Effect of I/O Instructions on the I/F

IV. Basic Form of the S/3—Modem Interface

  A. Multiplexing Unit
  B. Modem—S/3 Unit
  C. Operation of the S/3—Modem Interface

V. Conclusion

The subject of this report yields a three-part division: the System 3 computer, the Modem data transmission device, and the Interface coupling the two units. The purpose of the report is to present a design proposal for the Interface. To do that, however, the writer decided first to describe the two units to be connected. Then he presents the design itself, again according to the logical three-part segmentation inherent in the subject matter.

A final example should suffice to show that subject matter can sometimes yield a workable structure for a report. This one comes from a technology assessment report:

I. Dial-A-Ride and the Merchant

II. Background: The Dial-A-Ride Assessment Project

III. Impact of Dial-A-Ride on Ann Arbor

  A. Technological Considerations
  B. Economic Considerations
  C. Environmental Considerations
  D. Political Considerations
  E. Social Considerations

IV. Specific Benefits and Costs to Merchants

  A. Benefits to the Merchants
  B. Costs to the Merchants

V. Future Developments

The assessment yields two basic types of impacts, categories of impacts on the community in general (item III on the outline) and the benefit/cost impact for the merchants (item IV). Within the discussion of each type of impact, the structure again is determined entirely by subject matter.

The three methods of structuring the material in the discussion component provide you with alternative strategies for segmenting, selecting, and arranging the material. But let us not leave you with the impression that these are equally desirable alternatives. As we said, we have discussed them in a decreasing order of rhetorical effectiveness. Structuring by rhetorical purpose is the most effective because it provides you with a means of segmenting and selecting as well as arranging the material. The strength of this method is that it forces you to be selective. Structuring by the intellectual problem-solving process is also effective in many report writing situations because it provides you with a means of segmenting and arranging your material clearly. The strength of this method is that it is especially helpful for arranging material, although it might not lead you to be selective. Structuring by subject matter at least helps you to segment your material. Subject matter by itself provides no hints for selectivity and few for arrangement.

The strengths of structuring by rhetorical purpose and by the problem-solving process make them the more desirable alternatives. They are also desirable in another respect. The beginning and ending segments of the discussion involve both rhetorical purpose and the problem-solving process. As our outlines for structuring by these methods indicated, the beginning and ending segments are intrinsically related to the structuring of the middle. For example, *Stages One* and *Two* of the problem-solving process form the beginning segments of the discussion. However, in a report structured by subject matter neither the beginning nor the ending segments are found in the subject matter itself. For that reason, structuring by rhetorical purpose and structuring by problem-solving process are usually more effective than structuring by subject matter.

USING THE APPENDICES

We finish our suggestions for designing the discussion component with some advice on how to use appendices. We include appendices in the discussion component because the appendices are an integral part of the report despite the way they are treated by many report writers. Some writers depend upon appendices excessively, and consequently their reports are all uninterpreted data. Others ignore appendices, and consequently they bury their rhetorical objectives under a mountain of details. To avoid either of these two extremes, the report writer can apply the principle of selectivity and, to a lesser extent, of proportion.

The principle of selectivity requires the writer to distinguish between material which is immediately necessary for him to accomplish his rhetorical objectives and material which is necessary to document the technical investigation fully. The report must contain test results, site visit accounts, memoranda, printouts, tables of measurements, and so on if it is to be useful to numerous audiences for any length of time. However, if the report is to be useful to important audiences, this information must not clutter up the discussion of the report. The discussion must be a straightforward development of the purpose. If some secondary audiences need very particular information, the appendices provide it.

As illustrations of the principle of selectivity, let us review some examples we have already seen. In the report which proposes a clutch to disengage a starting pump (Figure 5-17, p. 83) the writer put all the technical information about the clutch itself into the appendices. He uses the discussion to assert that the clutch meets the need and to explain its cost. In other words, his rhetorical objective was to sell the clutch; a technical description would have interfered with that function. Thus he avoided the temptation, to which many engineers succumb, to be excessively concerned with the hardware.

In the report which asks for adjudication of a cost overrun by a shipbuilding company (Figure 5-14, p. 79) the writer uses attachments to supply memoranda which would not be easily accessible to most readers. To quote extensively from these memoranda or to include them in the report body would have interfered with the argument. They are necessary information for documentary purposes; however, the documentation should not be mixed in with the argument.

In the report which demonstrates the feasibility of using thin metal films as precision resistance thermometers (Figure 5-12, p. 77), the writer uses the appendices to supply the calibration curves and his mathematical computations. These are the specific evidences for his conclusions; however, the discussion summarizes and interprets this evidence for the readers who neither need nor want to see the raw evidence.

In all three of these examples the writers apply the principle of selectivity to maintain focus on their rhetorical purposes. They relegate whole segments of material to the appendices. They also use the appendices, as we suggest, to rid discussion segments of excessive detail—calculations, raw data, and detailed figures—that would blur the focus if left in. In the discussion they interpret information; in the appendices they supply raw information.

Although the principle of selectivity is the most important, the writer can also use the appendices to keep segments of the discussion proportionate to each other. If he has, for example, five segments in his discussion, his readers will expect them to be presented with roughly equal development unless there are clear reasons for not doing so. The readers would not expect one segment, especially an intermediate or less important segment, to balloon up to several pages while another is presented in a few sentences. The writer should use the appendices to maintain the relative importance of the discussion segments to each other.

But do not be misled. Proportion is less important than selectivity and effective explanation. Do not overwrite to achieve proportion at the expense of focus on purpose. And do not ignore differences in complexity of subject matter which require you to develop one topic at greater length than others. Nevertheless, use the appendices when appropriate to maintain proportion.

A writer should not fear using appendices. Some reports will require very little appendix material. But other reports may consist almost entirely of appendix materials. The effective report writer uses appendices as much as is necessary to accomplish his rhetorical purpose. If you are really wearing your report writer's hat, you could possibly end up with a ten-page discussion followed by ninety pages of appendices. Learn to accept that possibility.

THREE
Writing and Editing the Report

ARRANGING REPORT SEGMENTS AND UNITS

EDITING SENTENCES

ADDITIONAL DESIGN FEATURES:
LAYOUT AND VISUAL AIDS

REPORT DESIGN: GUIDE AND CHECKLIST

CHAPTER 7

Arranging report segments and units

After the writer has determined the basic design of his report, he comes to the actual writing of individual segments. Here he must apply basic structural strategies and then develop specific tactics for arranging major segments of the report, units within those segments, and even individual paragraphs within those units. If he is aware of alternative patterns of arrangement by which to accomplish his rhetorical purposes, he should be able to carry clear structure right down to the paragraph level.

It is time to take stock of where we are and where we are going.

To this point in the book we have been concerned with basic design strategies. These are basic in the sense that they operate—for the most part at least—in the prewriting stage of report writing. That is, these strategies must be devised before the writer begins drafting a report. Of course some of them involve drafting as well as preliminary planning. Formulating a purpose statement, for example, is partly a matter of actual writing, partly a matter of making fundamental plans before writing. However, our emphasis to this point, except for our discussion of the opening component, has been upon strategies of design which operate before the writer ever picks up his pencil.

Now we turn to the actual writing of a report. The writing begins with individual segments and units of the discussion component. (As you recall, the opening component of a report should be written after the discussion component because it draws material from the discussion.) The purpose of this chapter is to show how basic strategies should operate as the writer begins to draft his report, and to suggest particular tactics for applying those strategies. Specifically, we offer six methods of developing segments, units, and paragraphs.

As we talk in this chapter about segments, units, and paragraphs we refer to the divisions of the discussion illustrated by the three-level outline, Figure 7-1, *Segments, units, and paragraphs identified on a three-level outline.* On the outline, for example, our annotation identifies segments, units, and paragraphs as we use the terms.

As you can see from our annotation, the term *segment* as we use it encompasses not only the basic structural divisions of the report, but also the major subdivisions

Segment	Unit	Subunit	Outline heading	Paragraphs
segment			1. The Need for a New Vendor Rating System	*two paragraphs*
segment			2. The Objectives of the Vendor Rating System	*two paragraphs*
segment			3. The Advantages of the Proposed Vendor Quality Index	*one paragraph (overview)*
	unit		3.1. The Vendor Quality Index	*one paragraph (overview)*
		unit	3.1.1. Defect Severity Classification System	*two paragraphs*
		unit	3.1.2. Computation of Vendor Quality Index Rating	*four paragraphs*
	unit		3.2. Use of the Vendor Quality Index	*three paragraphs*
	unit		3.3. Effectiveness of Vendor Quality Index	*one paragraph (overview)*
		unit	3.3.1. Effect of Defect Severity on Vendor Rating	*two paragraphs*
		unit	3.3.2. Efficiency of Defect Information Retrieval	*three paragraphs*
		unit	3.3.3. Additional Cost of the Index	*one paragraph*
segment			4. Disadvantages of the Monthly Vendor Summary	*one paragraph (overview)*
	unit		4.1. Current Ratings by Monthly Vendor Summary	*two paragraphs*
	unit		4.2. Failure to Rate Vendors Effectively	*one paragraph (overview)*
		unit	4.2.1. Severity of Defect Unweighted	*two paragraphs*
		unit	4.2.2. Inaccuracy of Vendor Ratings	*two paragraphs*
segment			5. Comparison of Vendor Quality Index Ratings With Monthly Vendor Summary Ratings	*four paragraphs*
segment			6. Implementation of the Vendor Quality Index Rating System	*three paragraphs*

Figure 7–1. Segments, units, and paragraphs identified on a three-level outline.

within them. Both types of segments are usually identified in the text by major heading and numbering. The term *units* as we use it describes clusters of related paragraphs which are combined to make up these segments. Often they too are identified in the text by subordinate headings and numbering. (In short reports, of course, there is no distinction between segments and units.) The term *paragraph* as we use it describes a cluster of related sentences. Several clusters combine to make up a unit. These are usually identified only by indentation or spacing; however, numbering without headings is possible too.[1]

Although we need to use the terms *segment, unit,* and *paragraph* to indicate the different levels of complexity, degrees of generality, and even length of the divisions of the discussion, we believe that all three can be developed according to the same fundamental strategies and tactics. Each states a controlling idea and develops it fully in an orderly and coherent fashion. There may be differences in the degree of amplification; however, structurally they are the same thing in different sizes and shapes.

BASIC STRATEGIES

We have already talked at length about two basic strategies for structuring the whole report and its two components. These same strategies also apply to writing report segments. Learn to use them automatically.

General to Particular

You should arrange your segments, units, and paragraphs in the same general-to-particular order we have recommended for the whole report and for each of its two basic components. The dominant movement within segments should follow the pattern of the whole: overview to details, generalizations to particulars.

Especially at the level of particularity represented by individual segments of a report, you should give readers an overview before giving them the details. Readers need the overview to follow the details. Most importantly, however, readers want generalizations more than they want particulars. Report readers want results before they hear about the process by which the results were gotten. They always say, in effect, "Tell me whether this segment is worth reading. Give me the 'bottom line' first; then—if I am interested—I'll look for the details. If not, I'll take your word for it." Whether it is a segment such as "Costs" or "Experimental Procedure," a unit of several related paragraphs, or even a single paragraph, the principle is the same: generalization first, particulars next.

In both a unit and a segment, this strategy means put the *core paragraph* first. In a paragraph it means put the *core sentence* first. This paragraph or this sentence controls the structure of the whole or sums up the details developed in each of the subsequent paragraphs. In short, this strategy means that you should always start with the largest level of generalization presented in the segment.

---

[1] In this chapter attention is focused on structures beyond the paragraph. For a discussion of conventional paragraph structures, see Sheridan Baker and Dwight W. Stevenson, *Problems in Exposition* (New York: Thomas Y. Crowell, 1972), chaps. 6, 7.

Descending Significance

You should arrange the particulars in each segment, unit, and paragraph in a descending sequence of significance. This will not always be possible with some types of material, but when possible you should routinely do it.

This strategy is an obvious extension of the first. Your objective is to get significant information to the reader as quickly, economically, and efficiently as possible. Thus, having begun the segment with the conclusion or overview (the largest level of generalization), you should move next to support that conclusion by presenting your most telling and important evidence.

If there are, for example, three justifications for the assertion made in the core paragraph or the topic sentence, you should arrange them in the order in which they would tend logically to persuade a reader: strongest support first, weakest last. Or, as another example, if you are discussing possible consequences of a particular act, probability determines the sequence: the more likely the consequence, the closer it moves to the top of the list. When a reader encounters the segment, he is confronted with the most certain information first; the less certain comes last. Or, as a final example, if you are describing the various uses of a piece of equipment, the significance of the functions determines the sequence. You simply construct a hierarchy of functions: the most important comes first, the lesser ones after.

Find and use the hierarchy of significance in the supporting details that will make up most of any segment. Sometimes you will find the key in the subject itself. Other times you will have to consider the specific interests of your readers or your rhetorical purposes before you can discover the hierarchy of significance of your details. Your objective, as before, is to help the reader as much as possible by giving him what he seeks: significant information presented quickly and clearly.

SPECIFIC TACTICS

The basic strategies will help you to "rough out" the structure of your report segments. But it is not enough to say, "put the conclusion first and arrange the details in a descending sequence." After all, purposes of segments vary. Some seek to describe, some to predict, some to persuade, and so on. They are like tools designed to accomplish distinctly different tasks. If you are to design segments well, you must devise tactics appropriate to the specific purposes of the segments. These specific tactics, as we shall see, can be treated as simple variations on our basic general-to-particular approach. Each tactic provides distinctive internal general-to-particular design. In the pages which follow we will discuss six patterns for internal design of segments. The six patterns we discuss are:

- Persuasive Segments
- Descriptive Segments
- Process Segments
- Cause/Effect Segments
- Question/Answer Segments
- Comparative Segments

For each pattern we will explain the purpose, illustrate effective arrangements, and provide a typical outline.

Persuasive Segments

This order is employed when your primary purpose is to defend a judgment or to present a conclusion in such a way as to make it credible to your readers. It involves your saying, in effect, "I believe this is true, and I think you will agree when you see my reasons. They are . . . ."

Of course, in technical discourse there are always unstated persuasive elements and assumptions—that the equipment used was the best choice, that the procedure was intelligent, that the results really are what they appear to be. It is only in the appendices of a report that a writer is likely to present information without judgment. There, his tables and graphs and printouts present "raw" data. But we are concerned here with more direct passages of persuasive writing, with segments of the report in which the writer proposes explicitly to persuade us to accept his view.

Begin with a succinct statement of your conclusion or hypothesis. This statement controls all that follows and is its whole reason for being. It is the simplest, most direct statement of your point. In segments and units this statement would be a short core paragraph. In paragraphs it would be a sentence.

Next, enumerate your reasons for holding the conclusion to be true. These reasons, of course, should be arranged in the descending order of significance we recommended earlier. For a simple conclusion not requiring much explanation, perhaps a single paragraph will do. For most conclusions each of the reasons becomes a paragraph of its own. In other words, make a cluster of related paragraphs: conclusion plus supports.

Finally, look at your conclusion from the negative perspective. Are there arguments against it? If so, why have you not accepted them? Are there alternatives? Then why have you rejected them? In short, anticipate and meet the objections that might be raised to your conclusion. Again, this might take a few sentences, a separate paragraph, or even a cluster of paragraphs.

The pattern—conclusion, positive support, anticipated objections—is an appropriate one to use whenever you have an explicitly persuasive purpose. Too often, however, report writers are reluctant to be this direct and open. Trying to maintain the facade of objectivity, they obscure the fact that they do hold opinions and are making a judgment. This facade of objectivity may lead readers to read inefficiently, to misinterpret, or to be unconvinced. In some measure all of these failures are likely in the following example from a memorandum in which the writer states his position on an issue faced by his company:

> There are a number of us that do not want to put container cranes on the V ships unless we absolutely have to; and there are a number of persons who consider the need for cranes absolutely obvious. Without question cranes are costly to install, costly to maintain, have adverse effect on crew size, and reduce the overall container lift capacity. On the other hand, the arguments in favor of cranes may be summarized as follows:
>
> A. Approximately 20% of the projected revenues come from ports that have no cranes.
>
> B. A shipboard crane will supplement port cranes when traffic is heavy and hence reduce the lost time in port. Floating cranes and "jerry-rigged" facilities may work, but not with the efficiency of a shipboard crane.
>
> Our conclusion is that we have given realistic estimates for installing and maintaining the cranes, and that either of the two reasons above appears economically to justify installing the cranes. If they help generate revenue in a tough, competitive situation or reduce port time, the costs of installation and maintenance are more than offset.

negative example

Here the writer advances his conclusion and his reasons for believing it the best alternative. He hopes to swing his readers to his view. The passage is an explicitly persuasive segment. But the arrangement of the segment is backward. The reader cannot tell from the first paragraph where the writer stands on the issue. In fact, the reader might assume quite wrongly that the writer opposes putting cranes on the ships. When the writer says, ". . . a number of us don't want to put cranes on the ships unless we absolutely have to . . ." a reader might easily assume the "us" includes the writer. This misinterpretation becomes all the more likely when the writer uses the impersonal term *persons* to represent his own position. Thus instead of beginning with a clear assertion, the writer begins by merely stating the issue: there is a case for, and against, the cranes.

Next the writer lists the arguments against the cranes: they are costly to install, costly to maintain, and so on. No hint that he rejects these arguments, or that he is in any way sceptical; he just lists them. Then he goes on to enumerate the arguments in favor of cranes: they increase potential customers and reduce lost time in ports. Again, no hint of where he stands. He carefully preserves the facade of objectivity.

Finally, in the next to last sentence, he arrives at his conclusion: he thinks cranes should be installed. But even here he is tentative, saying "either of the two reasons *appears* (our italics) economically to justify installing the cranes." Translation: "I'm in favor of installing cranes."

The arrangement in the example is ineffective for several reasons. To discover the conclusion, the reader must read all of the segment. If he does not, or if he skims, he risks being misled. There is also the chance that the reader may be convinced by the arguments opposed to the conclusion before he gets to the arguments in favor. (This is especially true because the arguments against the conclusion are based on solid evidence of cost while the arguments for the conclusion are based on the uncertain evidence of potential market advantage.)

To accomplish his purpose the writer should have reversed the arrangement of his persuasive segment. He should have begun with the conclusion—and clearly signaled that it was a conclusion. (None of the "appears to justify" business.) Then he should have offered the reasons for holding his conclusion. And only then should he have brought up the opposing arguments. That arrangement would have allowed the reader to know ahead of time precisely what sort of segment it was. No surprises at the end. And it would have allowed the reader to skim without missing the point of the segment or, worse still, drawing the wrong conclusion.

As an example of an effectively arranged persuasive segment, consider the following passage. Notice how much more direct and clear it is than the previous example:

8. Lack of evidence of rock salt deposits

Our study indicates that there is no salt present in the Silurian Salina Formation which underlies the Enrico Fermi site. This conclusion is supported by visual inspection of cores taken from the upper part of the Salina Formation at the site, by published reports on the Salina in Michigan and Ontario, and by well-logs and drillers' reports from the area.

8.1. First, on the basis of a visual inspection of the core from the several deep borings made for Detroit Edison both at the Fermi site and the Monroe site I can report that no salt was encountered in that portion of the Salina penetrated (approximately 210 feet at Fermi and 150 feet at Monroe). As the Salina is approximately 500 feet thick in the Monroe area I can therefore say that the upper 30 to 40 per cent of the Salina is free of salt.

8.2. Second, the published literature on the Salina in Michigan and Ontario indicates that no salt is present at the Fermi site. Statements from four of the most recent such reports should verify this finding.

8.2.1. K. K. Landes, in the text accompanying the U.S. Geological Survey Oil and Gas Investigations Preliminary Map 40 (1945, The Salina and Bass Islands Rocks in the Michigan Basin) says of the Salina salt beds, "The F and lower salts disappear in southern Wayne County . . . . The D salt is not present in the Macomb County well, and it disappears with the higher and lower salts in southern Wayne County . . . . The B salt, like the D and F salt, ends abruptly in southern Wayne County."

8.2.2. C. S. Evans, in a paper entitled, "Underground Hunting in the Silurian of Southwestern Ontario" (1950, Proc. Geol. Assoc. of Canada, Vol. 3, pp. 55–85) includes a map showing the southern limit of salt a mile or so south of Amherstburg, a town on the east bank of the Detroit River seven miles north of the Fermi site.

8.2.3. B. V. Sanford (1965, Salina salt beds, Southwestern Ontario, Geol. Surv. Can.: Paper 65–9, 7 pp.) states that the B salt has the widest distribution of any of the saline units. Both his figure showing the distribution of the B salt and the figure showing the total salt thickness within the Salina and its distribution show that there is no salt within the Salina Formation in Ontario south of a point one mile south of Amherstburg.

8.2.4. In an American Chemical Society Monograph (1960, Monograph 145, edited by D. W. Kaufman) K. K. Landes discusses the salt deposits of the United States (Chap. 5). In the section on Michigan (pages 71–74) he states that the bedded salt deposits of the state are in either the Salina (Silurian) or Lucas (Detroit River Group-Devonian) formations. The Devonian salt occurs only in the northern part of the Southern Peninsula. All salt mining is done from the Salina, and in the Detroit-Windsor area. His Figure 1, on page 72, showing the aggregate thickness of Silurian salt deposits in northeastern United States and Ontario shows the southern limit of Silurian salt to be north of the Wayne-Monroe County line; in fact, according to the figure there is no salt present anywhere in the section beneath any of Monroe County. The natural brines (distinct from the bedded salt deposits discussed above that are either mined or artificially brined) utilized within Michigan are from the Parma, Marshall, Berea, and Dundee formations, all of which occur only to the north and northwest of the Fermi site, and are higher stratigraphically than the rocks that underlie the Fermi site.

8.3. Third, well-logs and drillers' reports do not indicate that salt may be present at the Fermi site. A careful reading of over 100 well-logs for the Monroe County area turned up only two in which salt was even mentioned, and in neither case is the evidence persuasive that salt might be present at the Fermi site. The one report mentioned salt at Milan, where one might expect salt to be present. The other mentioned salt near Lambertville; however, the latter report, not backed up by samples reviewed by a geologist, is perhaps questionable.

8.3.1. The first mention of salt in the logs reviewed was at Milan, Michigan, first reported on by A. C. Lane in the Michigan Geological Survey Annual Reports for 1901 and 1903. The log for this well, based on samples, showed 5 feet of rock salt near the base of the Monroe group (which includes the Salina) at depths between 1540 and 1545 feet. The Milan area is about 30 miles west-northwest of the Fermi site; the possibility of salt occurring in the section should increase in that direction, which is towards the center of the Michigan Basin. Although twelve detailed logs (some of which are also based on samples) from the townships between Milan and the Fermi site were carefully looked at, no further mention of salt could be found.

8.3.2. The other mention of salt noted in the many logs reviewed was in that of a well located some two and a half miles north of Lambertville, about 24 miles southwest of the Fermi site. The well-log in question was based simply on the driller's log, rather than on samples evaluated by a geologist; it noted "dolomite and salt" in a 15 foot interval near the base of the Salina, between depths of 525 and 540 feet. Mounted samples from a well some two miles away and logs of eleven other wells between the one near Lambertville and the Fermi site again give no further indication that salt is present in the section.

8.4. In sum, the evidence indicates no salt underlies the Fermi site. The evidence consists of close visual inspection of the cores taken from the upper part of the Salina for Detroit Edison, four recent reports based on a detailed study of carefully selected well-logs and cuttings, and all the older published reports each based on only a few rough drillers' logs.

The arrangement of the passage above[2] exactly follows the pattern we have proposed: conclusion, positive support, rejection of alternative views. The first paragraph (8.) states the conclusion that no rock salt is present in the formations beneath a proposed nuclear power plant site. This paragraph also summarizes the evidence which leads to that conclusion and thereby forecasts the structure of the rest of the segment.

The second paragraph (8.1.) presents the strongest support for the conclusion, the evidence of borings made at the site.

The third paragraph (8.2.) presents an overview for a cluster of five paragraphs which together offer the second support for the conclusion, the evidence of literature which describes the formations in the general area of the site. Paragraphs 8.2.1., 8.2.2., 8.2.3., and 8.2.4. each supply one particular support for the generalization in the overview paragraph, 8.2.

This same pattern—an overview paragraph with two particular supporting paragraphs—is followed in paragraphs 8.3., 8.3.1., and 8.3.2. In this case the evidence presented is negative evidence. That is, the writer rejects the apparently contradictory evidence of two drillers' logs that salt may be present in the site area.

The final paragraph (8.4.) sums up the evidence and restates the conclusion.

Now before ending our discussion of persuasive segments, let us introduce an additional point. We have said you should routinely arrange persuasive segments to move from general to particular. In certain situations, however, the reverse order may be more effective. Assume, for example, that passions have been running quite high on the opposing sides of a hotly debated issue—and nobody wants to hear any position stated but his own. In such a situation you might find it effective to start by reducing the sense of threat and hostility your opposition will feel when you state your conclusion. Start tactfully by showing that you take their arguments seriously, that while you have not been convinced by them, you nonetheless respect them. Or give a neutral summary of both sides. Give the reader as much time to "settle down" as you can. Then, after the threat of raising the issue has moderated, move on to the support of your conclusion, but this time start with the lesser of your supports and build up to the stronger. Finally state your conclusion. This is an effective order; however, save it for only those instances in which tact is an overriding concern. To use it at other times will be to obscure and even to undermine your conclusion.

The persuasive segment, then, is a simple three-step segment. When your primary purpose is to defend a judgment or present a conclusion, the segment may be effectively arranged according to this outline:

---

[2] Here, as in most of the other examples in this chapter, the specimen segment comes from a formal report. The arrangement of an entire informal report is often the same as the arrangement of a single segment from a long report.

1. Statement of conclusion or hypothesis.
2. Support for the conclusion, arranged in descending sequence of significance.
3. Anticipation and refutation of objections. (Of course, sometimes the third step might be appropriately omitted).

In long segments a fourth step might be added: a restatement of your conclusions.

Descriptive Segments

This order is employed when your primary purpose is to describe a thing, that is, to create for your reader both a visual and functional understanding of a physical object. It involves your saying, in effect, "here is what it looks like and how it works."

Technical description differs from conventional description in that its emphasis is primarily functional.

As an example, suppose you want to describe a particular piece of test equipment. You want your reader to understand it so that he will accept your results. Primarily, of course, the reader should have a functional understanding of the equipment. However, in order to accomplish that, you will probably have to create for him a visual image as well. First he has to "see" the equipment in his mind's eye; then he has to understand the functional interrelationships among the various parts. How are the parts arranged? How do they work? Your explanation is part blueprint and part working instructions. It is both visual and functional.

As you can see, there are essentially two types of explanation going on in a descriptive segment. One is spatial description; the other functional explanation. Spatial signals such as the phrase, "mounted on the heat extractor end of the cryogenic cooler," help the reader to orient himself, to construct a reliable visual image of the object being described. Operational signals such as, "to collect the precipitates, an Anderson sampler," help him to construct a functional understanding of the object being described. Clear technical description must take advantage of these two types of explanation and should make them both distinct and well signaled. An effective pattern for description emerges directly from the need for two types of explanation.

The descriptive segment should begin with an overview of the object's function. In segments and units this would be a short core paragraph. In a paragraph it would be the first sentence. This is your most economical explanation of the object being described.

Next, present a physical description of the object. Often, of course, the weight of this description should be carried by figures in the text. (Not buried in the appendices, for the chances are that many of your readers will not bother to hunt for Figure X.) If you use a figure, briefly introduce it by identifying its major features. After the figure you may explain further how to interpret the particular details. You should use illustrative figures when possible, but if you cannot—and thus need to rely on verbal explanation—follow the same procedure. Identify the object's main features; then describe its parts in a systematic and logical sequence.

Whether you use a figure or rely upon verbal description, however, you are moving into particular information. The physical description you provide prepares the reader to understand the most important particulars of your description, the functional particulars. As the reader looks at or visualizes the object he begins to understand how it functions as you have already said it does. For some of your readers, of course, that is explanation enough.

Finally, present the functional description of the object. Here you divide the object's functions into meaningful stages and take them up in logical sequence. This sequence will be either of two types: a causal sequence—part A moves part B which moves part C and so on, or a hierarchical sequence based upon the movement from most significant to least significant parts of the system. (In a hierarchy, the parts are interrelated but not necessarily in a causal sequence. A satellite, for example, has a thermal protection system and a communications system. In a functional description of this satellite, you should describe the communication system first because it is the functionally more important of the two systems.)

In either of these two types of explanation, focus on the significant stages of the object being described and do not bury yourself and the reader in excessive details. Here is an example of a descriptive segment in which the writer buries the reader in excessive details.

negative example

The following are physical characteristics of the Northern Seahorse class, of which the Maritime Venture is the first to be built:

Length Overall	250′–0″	
Beam	50′–0″	
Depth to Main Deck	22′–6″	
Designed Draft	18′–0″	
Displacement at Design Draft		3565 Tons
Cargo Deadweights:		
Deck Cargo		1070 Tons
Liquid Cargo		0450 Tons
Dry Cement		0250 Tons
	TOTAL	1770 Tons

Crew: 6 Officers, 9 Seamen

Passengers: 6

Propulsion: Four medium-speed diesels, each pair driving a 10-foot diameter ducted, controllable-pitch propeller. Total continuous shaft horsepower, 14,000. Maximum bollard pull, 120 tons.

Service Speed: 16.5 Knots

Maximum Range at Service Speed: 5,500 Nautical Miles

These new vessels are to be built by North Sea shipyards for Caladonian Exploration to meet the added demands of near-Arctic offshore drilling service. They will have the ability to remain with the rig for extended periods of time and will be able to singlehandedly tow the rig from one drillsite to the next as well as handle anchoring duties at the new location.

The Northern Seahorses are notable in several respects. They are to date the largest offshore tug/supply vessels in the world. Their increased dimensions will enable them to carry over 700 tons more cargo than the typical 190 footer in service today. They are also among the first tug/supply vessels to be built to American Bureau of Shipping Ice Class 1-A standards, which will give them virtually an unlimited latitude of operation.

These new vessels have many automated features. The machinery has been designed for a one-man engine room watch. Bridge control of engine speed and computer correlated propeller pitch settings have been provided. Critical pressures and temperatures are automatically recorded and can be monitored from the control station console. Bilge alarms and fire retardation systems are all automatic.

Ship's power is provided by four shaft-driven generators, each clutched off an engine, giving a variety of power capabilities to meet the varying needs of anchor and tow winch operations at the drillsite. An independent generator will be used for ship's services during standby at the rig or at dockside.

The cargo deck is equipped with tandem hydraulic anchor handling and towing winches mounted on the centerline. Each winch has a static load of 250 tons, and can be operated from either the aft command station in the pilot house or from controls on the supply deck. The winches are located as far forward as possible to allow maximum utilization of the supply deck. The forward half of the supply deck is equipped with a rolling deck platform which enables the forward cargo to be moved aft previous to being unloaded from the ship, which is anchored in a stern-to position during this operation. This feature enables the vessel to maintain a greater distance, thus reducing the chance of a collision.

Dry cement pressure tanks are located low and forward in the ship. Ballast tanks are interspersed through the length of the ship to facilitate proper trim at different load conditions. Fuel oil, which is transferred by hose for use at the rig, is located in centerline deep tanks to afford maximum protection in the event of a collision.

Spacious air-conditioned accommodations are furnished for all crew members and passengers. Officers are provided with private staterooms, and the remaining crew and passengers will be berthed in double cabins. All staterooms will be attractively wood paneled and fully carpeted to afford the most in luxury and comfort.

The Northern Seahorses are the finest example of offshore service vessel design and construction. This versatile class will combine unmatched cargo capacity, towing ability, and range with the maximum in safety and comfort.

In this example, by launching into particular details of his description, the writer fails to provide the reader with any helpful overview of the ship's function. The reader has no means of determining which are the significant details, and what the purpose of the description is. The reader is taken through paragraphs and units of minute details on physical characteristics, power plant, automated features, cargo facilities, and accommodations, yet he has no way of knowing which are important and what the point of the whole segment is. The writer has not provided a functional description.

As an example of an effectively arranged descriptive segment, page 124 is a passage in which the writer uses a figure and clear causal sequence to explain an object's function. Evident in this example is a three-step outline for descriptive segments:

1. Overview of the object's function.
2. Physical description of the object, with figure when appropriate.
3. Functional description of the object.

You should use this outline whenever your primary purpose is to describe an object.

Process Segments

This order is employed when your primary concern is to create an understanding of a process, set of operations, or event that has duration in time. You say, in effect, "Here is how it is done," or, "Here is how it occurred." In technical discourse the distinction between description and narration often is a matter of focus. Things perform processes, and processes are performed with things. In technical description you focus on the object, although you usually explain how it operates—the process by which it does what it does. In technical narration you focus on the process, and describe the object as necessary to make the process clear to the reader. You must decide whether the primary purpose of a segment is to describe an object or explain a process. It is the emphasis on explaining a process with which we are concerned here.

Chronology provides the basic means by which to organize a process segment.

The Separation System

The separation system provides for the separation of the spent fourth stage of the launch vehicle from the satellite. This is done by firing explosive bolts to sever the steel bands which hold the satellite to the fourth stage. The explosion severs the bands and allows springs to push the satellite and the fourth stage apart with sufficient force so that the two cannot collide.

Here is a cross-sectional view of the system:

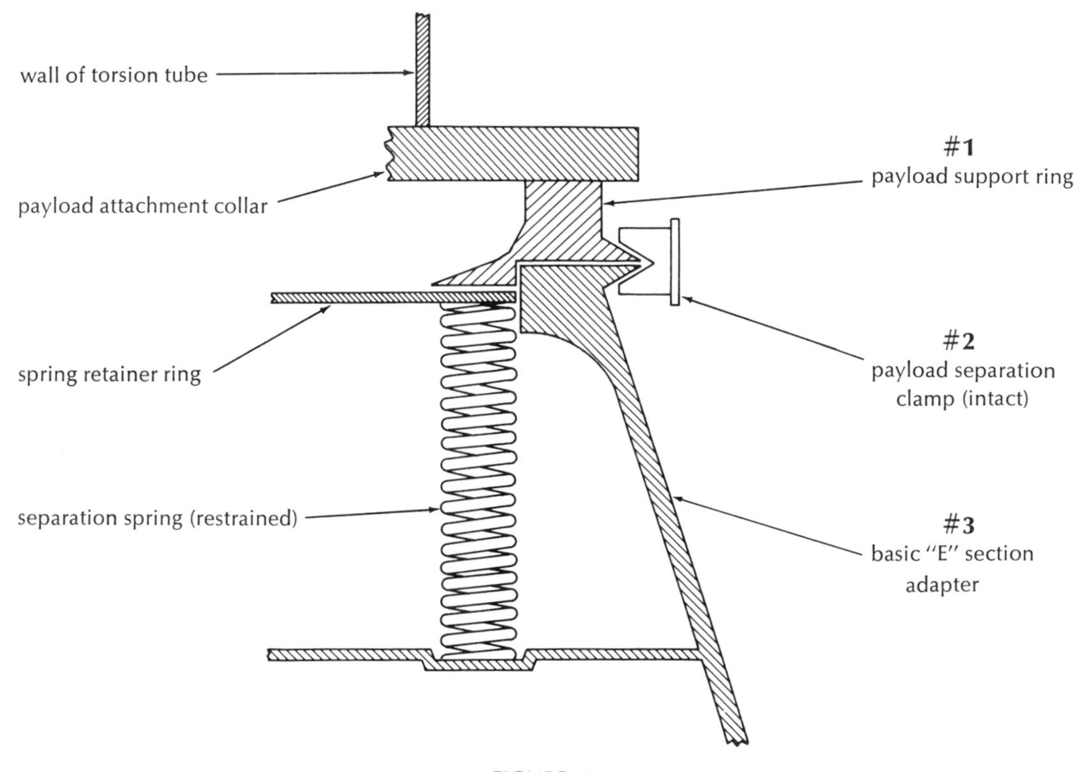

FIGURE 4
Three Basic Parts of Separation System

The three basic parts of the *"E"* payload separation system chosen for Project SCOPE are:

1. The *payload support ring* which mates the top of the fourth stage to the bottom of the satellite.
2. The *payload separation clamp,* steel bands with V-block which hold the section adapter (3) and the payload support ring (1) together prior to separation.
3. The *"E" section adapter,* a conical magnesium structure in which are mounted the springs which supply the force for the separation.

Once the payload has reached the desired orbit, the explosive bolts attached to the steel bands of the payload separation clamp are fired. This severs the bands and allows them to drift harmlessly off into space. As soon as the bands are severed, the previously restrained springs in the section adapter force the satellite and the fourth stage apart. Once the two are sufficiently separated, the springs are also completely free and drift off into space. The separation is completed.

The severed parts of the bands, the springs, and the fourth stage should be quickly clear of the satellite. When they are, the solar paddles and antennas can be opened. . . .

However, since the process is invariably performed for a reason and with consequence, you should begin with an overview of the objectives of the process. This overview might also briefly state the method by which these objectives are accomplished. Notice how this core paragraph provides an appropriate process overview:

> The extraction process uses methylene chloride as a solvent to recover the dimethylformamide (DMF) from the demineralized water used to scrub the synthetic fibers. The recovered DMF must be pure enough to be recycled; the water stream must be pure enough to send to waste treatment. The process uses a three-stage method—an extractor column, a steam stripper, and a distillation column—to purify the DMF and the waste water.

Next, if necessary, give the reader an explanation of any hardware or unfamiliar physical objects you have to introduce to explain the process. You want to avoid interrupting your subsequent explanation of the process itself. As in descriptive segments, in process segments you can often use a visual aid to provide most of this explanation.

Finally, explain the particulars of the process or event, being careful to signal each stage as you go. You have already alerted the reader to expect a certain number of stages and given him at least a preliminary sense of what they are. Thus he has some understanding of the process already. Now, however, you must "flesh out" the stages so that your reader gets an adequately detailed understanding of each without losing his sense of the whole process. It is not difficult to establish an order for these particulars; chronology does that for you. Nevertheless, some writers have difficulty segmenting their chronological explanations meaningfully. The sequence of the stages may be clear to them because they have "lived through" the process, but the significant stages of the process may not be. The reader is interested in the general stages of the process; therefore, do not let your narrative of particulars obscure them.

The procedure we describe is a simple one. Judging from the evidence of example, however, many writers explain processes ineffectively. They provide inadequately detailed explanation; they fail to impose meaningful segmentation; and they enumerate details in a completely linear fashion without overview. These problems are illustrated in the following examples. The first is on Geneva wheel indexing:

> The indexing action of an external Geneva wheel results from the interaction of the two rotating components, the wheel and the driver. The driver consists of a pin attached at the end of a crank arm that is secured to a shaft that is rotating at a constant speed. The wheel is a disc with equally spaced radial slots open at the outer diameter and is attached to a shaft. As the driver pin approaches the wheel, the slot is in a position to receive the pin. The pin enters the slot and rotates the disc until the pin and slot disengage. The next slot should be in a position to receive the pin on the next revolution of the driver.

negative example

This writer does not give us enough information about the action he is trying to explain. That is, there is not enough development here unless you already know about Geneva wheel indexing. It is a case of underdevelopment. In addition to that, he fails to give any sense of overview or objective. He assumes you already know.

The second example is on the operation of a Time Domain Reflectometer:

> With TDR a pulse (voltage step) is generated and transmitted down the transmission line. When it encounters a discontinuity, part of the pulse is reflected. The

negative example

unreflected part continues along and perhaps suffers a second or third reflection. By studying the nature of the reflected pulse, in comparison to the incident pulse, one is able to deduce the size and type of discontinuity. By measuring the time between the transmitted and reflected pulses, one can determine the distance to each of the discontinuities. The TDR is usually used with an oscilloscope on which the reflected voltage appears algebraically added to the incident pulse. The TDR operating bandwidth (frequency range) is proportional to the reciprocal of the rise time of the pulse.

This writer fails to give any sense of overview, but in addition to that, he fails to impose any meaningful segmentation on the process. He just buries us in a chaos of particulars which could be sorted out efficiently only by extremely well-informed audiences with time on their hands. Upon first reading, in fact, the paragraph seems completely ill-organized.

The last example is on detecting mercury contamination in fish.

negative example

Weighed fish samples are placed in individually sealed quartz tubes. Tissue from the same specimens plus an exactly known amount of $HgNO_3$ as an internal standard is also individually sealed in a quartz tube. The quartz ampules containing both the unknown sample and the internal standard are simultaneously irradiated in the nuclear reactor for 36 hours. After a delay of two to three days to allow the high level of radioactivity of the samples to decay to a level where they can be safely handled, each ampule is broken open, the contents having been first frozen in liquid nitrogen to retard the decay rate.

The samples are covered with a watch glass and digested with just sufficient 1:1 concentrated $HNO_3H_2SO_4$ to bring the samples into solution. After cooling, the solutions are diluted to 25 ml with 1 M HCl. This solution is passed through an ion-exchange column. The flask is rinsed several times with 1 M HCl and the rinsings are also passed through the column. The resin is removed from the column, placed in a glass vial, and its activity analyzed with a Ge(Li) detector and a Multichannel Analyzer. The resulting gamma ray differential pulse height spectra obtained is shown in Fig. 1.

This writer not only fails to provide an overview, he replaces it with a rigorously linear account of the process in which every action is as important as every other action. The reader could, of course, extract some understanding of the process from this, but the reader would have to supply his own meaningful segmentation of the process since the writer supplies none, and most readers could not.

Although the negative examples help clarify the nature of the problems you might have with process segments, positive examples will provide guidelines for you to follow when writing them. Here are two examples of effectively designed process segments. The first is a simple passage from a report on an investigation of a fire aboard a cargo ship. It clearly and efficiently explains the purpose and sequence of the process.

Watch standers were trained to respond to a boiler room fire with this five-step sequence of actions designed to initiate fire-fighting action on the ship:

1. Close the quick-closing fuel supply valve located at the boiler room control panel. This would shut off the supply of fuel from the storage tanks to the piping system.
2. Using the remote control located at the boiler room control panel, turn off the fuel oil pump. This would stop fuel already in the piping system from being pumped on into the burner.
3. Using the switches on the main switchboard, turn off the boiler room ventilation

system. This would help to smother the fire by denying it oxygen and help to confine the smoke and flame to the engine room.

4. Using the telephone at the main switchboard, inform the Officer of the Day so that he can organize a fire-fighting party.
5. Using the portable fire extinguishers beside the main switchboard, attempt to contain the fire until the fire-fighting party arrives.

The itemization clearly segments the details so that the reader (and the watch standers) does not become lost in the particulars. The overview sentence generalizes about the stages of the process and the objectives of the process. Each item then particularizes one stage and its complementary objective.

The following segment is an explanation of the experimental methodology used in a project to develop an instrument to detect exposure to mercury fumes in industrial situations. Clearly, explanation of the methodology used to test the device is necessary in order to convince the reader that the instrument does indeed do what it is purported to do: measure the body's nervous system signals to distinguish between employees exposed to mercury poisoning and those not exposed.

3.5. The Experimental Test Process.

To test how reliable use of the instrument by an operator is, two questions must be answered:

(1) Given a well-defined peak in the output curve, will different operators of the instrument perceive the same mode?
(2) Will the perceived differences in the exposed and nonexposed tremor modes be significant enough to allow an operator to differentiate between these two groups of employees?

To answer these questions a two-stage experimental test process was devised. The process involved four test subjects as potential operators of the instrument. These subjects read and interpreted data supplied to the instrument by a magnetic tape recording. Before providing the data input, the tape recording supplied the subject-operator with instructions for operating the instrument. The entire experimental test process took about one hour with each subject-operator.

3.5.1. Testing the Operation of the Instrument.

The magnetic tape recording first provided operating instructions and then supplied seventeen sets of data to the instrument. These seventeen sets of data were of two types: oscillator signals and employee tremor signals. These signal types formed the two-stage experimental process needed to answer the two questions.

The tape recording first explained the purpose of the experiment and told the subject how to operate the instrument. After this introduction, the tape was stopped to allow the subject to ask questions to make certain he knew exactly what to do. Then the tape presented a sample set of data for the subject to interpret on the instrument meter. If the subject used the instrument to interpret the data correctly, the tape then proceeded to supply the seventeen sets of data for interpretation. (If the subject had difficulty interpreting the data correctly, the tape was rerun as often as necessary for the subject to learn how to use the instrument correctly.)

The tape then supplied seventeen sets of data for the subject to interpret from the instrument meter. Each data set contained ninety seconds of signal input followed by forty-five seconds of rest period to reduce the chance of mental fatigue in the subject-operator. The subject-operator had to analyze each data set as follows:

State the signal frequency he judged to cause the maximum deflection on the meter;

State the degree of certainty (1% to 100%) he felt about his judgment;

Comment briefly on any distinctive characteristics of the data set.

The time required for the subjects to state judgments with a high degree of certainty was monitored to determine the amount of data required for an operator to test an employee for exposure to mercury poisoning.

3.5.2. Stage One of the Experiment.

This stage addressed the question: Given a well-defined peak in the output curve, will different operators of the instrument perceive the same mode? This stage involved the first six and the seventeenth sets of data. These sets were oscillator-produced signals which allowed direct comparison of the subjects tested.

The first six data sets were put on tape from an oscillator. These oscillator signals were used to determine how reliable and reproducible the output of this instrument was when interpreted by different operators. The signal-to-noise ratio (S/N ratio) was varied from 100% signal to 50% signal-50% background noise. These S/N ratios were used because they span the range of S/N ratios on actual tremor recordings. The performance of the operators was monitored as the S/N ratio increased.

The seventeenth data set also came from the oscillator. This set was used to determine the increase in skill of the subject-operator after he had analyzed sixteen sets of data.

3.5.3. Stage Two of the Experiment.

This stage addressed the question: Will the perceived differences in the exposed and nonexposed tremor modes be significant enough to allow an operator to differentiate between these two groups of employees? This stage involved the seventh through sixteenth sets of data. These sets were employee tremor signals.

These ten data sets formed a random sequence of tremor responses from five nonexposed employees and from five mercury-exposed employees (the subject-operators, of course, did not know ahead of time which data sets were which). These sets of signals had been previously obtained in the original Institute study of mercury poisoning. They were from employees matched with respect to such characteristics as age and weight. The ability of the operators to distinguish between signals from exposed employees and from nonexposed employees was evaluated.

To help further evaluate the four subject-operators' analyses of these data sets in stage two of the experiment, their judgments on differences in signal characteristics were compared to differences in signal characteristics as determined by computer analysis of the original signals.

The purpose of this process segment was to explain the experimental procedure. (Analysis of results was in the following segments.) In the first paragraph of unit 3.5. the writer states the objectives (to answer two questions), and in the second he provides an overview of the entire process. However, notice that there are actually *three* related processes involved—the process involving the subject from beginning to end, the process of the experiment proper (the two stages), and the process of designing the experiment and subsequently evaluating the results. Because three different types of actions were occurring, at times simultaneously, the writer had to segment his material very carefully. His segmentation involved distinguishing between the three related actions that occurred, and subordinating the sequential steps within each of the various stages of the entire segment (testing, stage one of the experiment, stage two of the experiment, and evaluation). In addition, he provided core overview paragraphs for each of the subsequent units (3.5.1., 3.5.2., and 3.5.3.). Most important, he did not allow the many particulars, some of them background details antecedent to the process (for instance, the source of the employee signals), to obscure the stages and steps of the processes. Although sequence and chronology provide means of ordering various units of the segment, the basic ordering derives from the segmentation. In this example, the stage-by-stage explanation of the particulars derives from the segmentation of the entire process, not from the chronology of the actions involved. It is a good example of a writer's segmenting his chronological explanations meaningfully.

If you examine the outline of the previous example, you will see that in general it has three steps: 3.5., 3.5.1., and 3.5.2. and 3.5.3. This is a basic pattern in effectively designed process segments. To design process segments, start with the following three-step outline:

1. Overview of the objectives of the process. (Usually supplemented by introduction of the method of the process and number of stages).
2. Explanation of physical objects necessary to explain the process.
3. A stage-by-stage discussion of the particulars of the process or event.

Cause/Effect Segments

This order is employed when your primary purpose is to predict the likely consequences of an observed fact or to trace back to the causes lying behind the fact. It involves the writer's saying, "Here is how we got where we are" or "Here is what to expect next."

A cause/effect segment is different from a process segment which presents a series of causally related events. Some processes consist of cause-effect chains. In such a causal process, the focus is upon the sequence because it is a known sequence; thus the pattern is a process pattern. However, a cause/effect segment investigates unknown causes and effects which do not form a sequence. For example, excessive noise levels in a foundry were discovered to result from the operations of a prepared sand feeder, of a casting oscillator, and of molding machines. Each of these contributed independently to the noise problem in the foundry; the molding machines made noise even when the sand feeder was not running. Because there was no causal connection between the sounds produced by these machines, a segment identifying the causes of the excessive noise could not be patterned as a process segment; it had to be patterned as a cause/effect segment. In cause/effect segments the focus is upon identifying and demonstrating the provable relationships between multiple causes of an effect or multiple effects of a cause.

Either of two distinct emphases will characterize cause/effect segments. One emphasizes the past—identifying causes; the other the future—predicting consequences. Yet if there are totally different emphases in these segments, the basic design is precisely the same. In fact, it is a simple pattern which differs only slightly from the pattern we saw earlier when we discussed persuasive segments: conclusion, support, anticipated objections. Thus to understand the cause/effect pattern you have only to learn its distinctive features. These occur in the opening.

First, both cause and effect segments originate in a question posed about a present fact: What caused it? What will result from it? For this reason, begin with a statement which clearly signals the present fact in question and the nature of the question. Is it a question of cause or of effect? The temptation, of course, will be to skip that explicit statement of the fact and to jump right into the particulars, that is, the causes or consequences. If you should, a careless reader might misinterpret the fact in question and consequently misunderstand the causes or consequences being demonstrated.

In addition to the statement of the fact and indication of the nature of the question, the opening of a cause/effect segment should provide a forecast of the answer to the question that has been posed. At the very least, you can indicate that there are to be four causes identified or that three possible consequences are to be demonstrated. Thus you can alert the reader to the design for the rest of the unit. But most of the time you can easily go beyond that in your opening statement: you can identify the causes or effects to be demonstrated. A simple list will predict what is to follow.

Second, in a cause/effect segment as in persuasive segments, you should support your conclusion. This consists of demonstrating that the causes or effects you have identified are convincing. Begin with the most significant or certain. In the case of causes, begin with *direct causes,* that is, causes without which the effect could not have happened. From there move to *contributory causes,* that is, causes which assisted in bringing about the present fact. In the case of effects, begin with the most probable; that is, of the likely consequences of a cause, which are most certain to come about. (Because you are predicting the future, it is imperative that you present adequate support for your assertions of probability.) In both cause and effect segments the second element presents positive support for your conclusion.

Finally, in a cause/effect segment, as in a persuasive segment, look at your conclusion from the negative perspective; that is, anticipate objections. Since you are dealing with a subject in which it is impossible to demonstrate your conclusions with absolute certainty, you must explain why you reject other causes or effects which some persons might consider plausible. In short, anticipate and meet the objections that might be raised.

The three elements in this outline for cause/effect segments are clearly like those in the outline for persuasive segments. The only significant difference occurs in the first step of the outline. To see just how important careful development of this first element is, consider the following opening element of a report segment. Consisting of a single line, it does not state the fact under discussion, indicate the nature of the question, or forecast the structure of the answer. The writer just begins this way: "The Board of Investigation determined the following facts about the engine room fire." Now that seems clear enough, doesn't it? Or does it?

This sentence explains neither the fact nor the nature of the investigation. Was the Board seeking the causes of the fire? If so, there is no clear signal of that. The writer says only that the Board determined "facts about" a fire. Causes? Consequences? He does not say. Yet when the reader goes on to read this segment of the report it is clear that the Board was not really concerned with the causes of the fire at all; the cause was conceded to be a ruptured pipe. Rather, they were concerned with the causes of the unusually extensive damage done to the ship by this particular fire. This segment of the report seeks to answer a specific question, but the writer's opening line does not make the investigation clear; the reader learns only that there will follow a list of "facts about" an engine room fire.

A rewritten version of the opening clarifies the fact, the nature of the investigation, and the arrangement of the segment:

> The Board of Investigation seeking the causes of the unusually severe damage done by the engine room fire identified three separate factors:
> 1. The watch standers failed to shut off the supply of fuel to the engine room;
> 2. The watch standers failed to shut off the engine room ventilation system;
> 3. The fire-fighting party could not fight the fire effectively from outside the engine room nor could they enter the room.

This version of the opening sets up the pattern of the rest of the segment. Each of the causes, arranged in descending order of importance, will be the subject of a unit within the segment.

Here is an example of a brief cause/effect segment in which opening statement, support, and refutation of objections are all effectively handled:

> Failure Analysis of Impeller Shaft P/N 53127X
>
> An investigation was conducted to determine the cause of failure of the steel impeller shaft (P/N 53127X) in the 13PV Turbine Pump after only one month of service. Examination of the failed shaft and analysis of the stress situation both indicate that the shaft failed in fatigue. An inspection of the fracture surfaces reveals the beach marks and final rupture area that are characteristic of fatigue failures. Analysis also indicates that the fatigue strength of the shaft is marginal, with the situation worsened by the apparent use of a chipped cutting tool in the cutting of the threads where the failure occurred.
>
> 1. The appearance of beach marks on a smooth fatigue zone (see Fig. 1) and a coarse, crystalline final rupture zone clearly indicate that this is a fatigue failure. The fatigue crack initiated near point 'I' (Fig. 1) and progressed across the shaft to final rupture at area 'R'. The changing angular orientation of the beach marks as they progress to the final failure zone indicate that there was a rotating bending load causing the failure. This is consistent with the loading known to be acting on the shaft.
> 2. An analysis of the fatigue strength of the steel impeller shaft with respect to its nominal loading shows that its endurance strength is only 3% above the expected minimum that would be required for infinite life. This is a very marginal value, especially when viewed in the light of the high variability seen in the fatigue strengths of steel parts, even when tested under rigidly controlled conditions.
> 3. An inspection of the threads where the failure occurred reveals that they had been machine-cut (as opposed to rolled which would have improved the stress situation by approximately 25%). Furthermore, microexamination seems to indicate that the cutting tool used to cut the threads was chipped, thereby decreasing the thread root radius and increasing the stress raiser effect of the threads.
>
> A calculation of the total life of the shaft shows that the shaft failed after 20 million cycles, which is on the high side of the distribution of fatigue cycles at failure in steel parts. However, considering the marginal endurance strength and the possibility of occasional overloads beyond the nominal loading level, this fact is not inconsistent with fatigue failure experience. Therefore it is my conclusion that the 13PV Turbine Pump impeller shaft (P/N 53127X) failed in fatigue because of an insufficient endurance strength which was due in part to the stress raiser effect of an improperly manufactured threaded section.

This segment comes from the body of a memorandum report attached to a seventy page report on Line Shaft Design. Except for the figures and the detailed calculations, the segment makes up most of the memorandum report. We include it to illustrate that the principles of segment design which we have been discussing are applicable to segments from reports of any degree of complexity. We are not limiting ourselves to segments from the discussion of long formal reports.

This example, although simple, illustrates the basic cause/effect pattern. An outline of the three steps in a well designed cause/effect segment is as follows:

1. Statement of the fact in question and the nature of the question. In addition, the structure of the answer should be forecast. The answers may also be stated.

2. Support, that is, demonstration of causes arranged from most significant to least, from direct to contributory, or effects arranged from most probable to least.
3. Anticipation of objections and refutation of alternatives.

Question/Answer Segments

This order (frequently called problem/solution) is employed when your primary purpose is to respond to a question which has been, or might be, raised by a reader. It involves saying, in effect, "Here is a problem that needs solving—and here is how I would handle it."

There are two likely uses for question/answer segments in the middle of a report. We stress this because you are likely to see only one of the two uses and thus to ignore a good deal of the segment's potential. First, question/answer segments can be used as a convenient way of responding to a specific technical question posed as part of the assignment. That is the obvious use. Second, question/answer segments can be used as a way of anticipating and responding to questions which might be—or should be—posed by a perceptive reader. You are second-guessing your readers, anticipating the confusions, objections, questions they might raise as they read the report. This use is less obvious but necessary because some of your audiences are not familiar with the assignment. For them you must introduce considerations that go beyond those called for by the assignment. If your audience analysis has been productive, you should be able to anticipate these necessary additional considerations.

The pattern for a question/answer segment is quite similar to the cause/effect pattern we just explained. Begin with an explanation of the question or issue. That is, what puzzle, confusion, or objection does it address? For your perceptive readers, of course, you are merely voicing what they may already feel; i.e., something is odd or puzzling. For your less perceptive readers you are pointing up a problem they perhaps should be thinking about. In either case you are trying to stand outside your role as writer and to voice the confusion, doubt, or scepticism a reader might feel at this particular point in the report. You may do this by posing a specific hypothesis question (answerable by "yes" or "no") or by identifying unknowns (how? when? with what?) which need investigation.

Next, you state your hypothesis or your answer. You may also introduce explanation of the method which supports the hypothesis or of the criteria which identify the unknown.

Finally, as in the cause/effect pattern, you present the evidence which supports the hypothesis. The evidence supplies detailed information necessary to resolve the puzzle that causes a question to be raised in the first place. The evidence is arranged in a descending sequence of significance.

Here is an example of a question/answer segment in which the writer anticipates the reader's question:

6. Grants.

The total direct cost to the public of implementing the recommendations for water supply and sewerage improvements proposed in this report would be $21,575,000. We propose $10,245,000 in bonds for water supply and $11,330,000 in bonds for sewerage improvements. These bonds would be issued by the DPW and repaid by contracts with local communities, the revenues coming from utility rates, special assessments, and local taxation.

Financing the water supply and sewerage improvements entirely with locally generated funds is, according to our study, entirely feasible. We have nevertheless reviewed available grant programs to see if federal funds might materially reduce the financial requirements at the local level. In recent years programs of the federal government have played a major role in financing improvements of the sort you propose. For that reason we have investigated the three government programs for which Berrien County could conceivably qualify. Our conclusion is that the outlook for government funds is dim.

6.1. The first program for which Berrien County might qualify is a program administered by the Department of the Interior. It provides for grants to certain communities for construction of waste treatment plants and certain intercepting sewers. The program provides funds to qualifying communities for 30% of costs of constructing waste treatment facilities. Under this program, the Village of Stevensville has been offered a grant to cover 30% of treatment costs currently proposed for construction. The program authorized $18 million for Michigan in the current fiscal year but the federal government has curtailed this program so that only $1.23 million is available. Although you might qualify for this source of funds, the small amount of money available does not suggest much likelihood of your receiving a grant.

6.2. A second grant program falls under the Farmers Home Administration. Under act of 1965, loans and grants were to have been available to small farm communities (under 5500 population) for planning and constructing water and sewer facilities in rural areas. Grants of up to 50% of project cost were contemplated for these projects; however, appropriations have so limited the program that only eight of 200 applying Michigan communities have benefited since 1965. Current federal spending levels leave little hope of this program providing funds to the area.

6.3. The third program for which the county might qualify is administered by the Department of Housing and Urban Development. Under act of 1965, the Department could grant up to 50% of the construction cost of basic water or public sewer facilities, excepting sewage treatment works. The program as originally authorized by Congress provided for $800 million dollars to be appropriated over four years. Due to national financial problems, the appropriations have not been made as authorized, so virtually no new projects are proceeding under this program at the present time.

On the basis of our review of the three available grant programs for which Berrien County might qualify, we conclude that the county should proceed under the assumption that no grant support is available. The county should nevertheless monitor grant programs and modify its financing to the extent grants actually become available prior to construction.

This segment comes from a report by a consulting engineering firm to a consortium of fourteen communities and proposes a coordinated modification of their water and sewerage facilities. At the end of the report the writers introduce this short segment whose obvious function is to answer the easily anticipated question, "but what about financing these improvements with federal funds?"

The first paragraph states the situation: the communities themselves need to generate $21,575,000 in local funds to support the modifications.

The second paragraph poses the anticipated question: "We reviewed available grant programs to see if federal funds might materially reduce financial requirements" (that is, are federal funds available?); this paragraph also states the answer, the hypothesis: "the outlook is dim."

The next three paragraphs (6.1., 6.2., 6.3.) each identify a federal program which might supply funds. Each paragraph also shows that obtaining funds from these sources is unlikely.

The final paragraph reasserts the hypothesis. It also introduces a recommendation.

The example presents a pattern that you can easily adopt. An outline of a question/answer segment is as follows:

1. Explanation of the question or issue. Pose a specific hypothesis or identify unknowns.
2. Statement of the hypothesis or answer. You may also introduce method and criteria for establishing the answer.
3. Support for your hypothesis or answer, the supports arranged in a descending sequence of significance.

Comparative Segments

We have saved for last the most complex of the six patterns we describe. It is complex because there are variations of the pattern and because the nature of the information presented by the pattern requires you to deal separately with numerous details. This order is employed when your primary purpose is to create for your reader an understanding of the similarities and differences between two or more objects, processes, or ideas so that he can make an informed choice between alternatives. It involves your making any of three points:

A is like B because it resembles B in crucial respects;

A is unlike B because it differs from B in crucial respects;

A is like B in some respects but unlike B in other respects.

The first point is dominantly comparative; the second dominantly contrastive; the third mixed comparison and contrast.

The comparative segment is more nearly a persuasive pattern than an informative pattern. In each of the three situations cited above, the writer is saying, "I conclude that A stands in this relationship to B." In no case is he saying simply, "Here is A; here is B." That is, in each case the writer makes a judgment or assertion about A's relationship to B and defends that assertion on basis of presented evidence. If you are to handle comparative segments well, you must understand this essential emphasis of the pattern upon interpreting and making judgments about the things being compared. You should never leave your reader to sift through the evidence for himself. Clearly emphasize the judgmental element.

To demonstrate just how important it is for you to tell the reader what you conclude, read through the following example. It is an instance in which the writer did not bother to state his conclusions:

negative example

There are two economic aspects to be considered in choosing between solid state control circuits and relay control circuits. These are material cost and labor cost. They are as follows:

	Solid State	*Relay Control*
Material:	$3,800.00	$ 8,600.00
Labor:	4,100.00	3,200.00
Total:	$7,900.00	$11,800.00

Labor costs include the building of the complete control circuit, but not the writing of the program for the wire-wrap machine. This would be an additional one-time cost of $700.

Here the writer sees his role as simply that of information collector. He makes no judgments; he merely presents information. As a consequence, the segment of comparison—if that is what it is—is rhetorically ineffective. The segment is just a price list which leaves the reader to construct his own comparison. A rewritten version of the segment illustrates how effective it is to present a clear conclusion.

> Solid state control circuits are more desirable for us than relay control circuits. This conclusion is based upon the differences in the total installation costs of the two types of circuits. The two types of circuits are expected to have equivalent maintenance costs and lifetimes.
>
> The results of the comparison show that while the solid state control circuits have a 30% higher labor cost and an added $700.00 cost for programming, their total initial cost is about 36% lower than the relay control circuits because of the significant difference in cost of materials. The figures are as follows:

	Solid State	*Relay Control*
Material:	$3,800.00	$ 8,600.00
Labor:	4,100.00	3,200.00
Programming:	700.00	0.00
Total:	$8,600.00	$11,800.00

> The additional labor and programming costs of the solid state control circuits are more than offset by the reduced cost of material. For this reason I recommend the solid state control circuits for the R-S Palletizer.

The first version tells the reader nothing directly. The writer has gathered some data and he passes it along, nothing more than that. The second version, however, is primarily judgmental and persuasive. The presentation of information is only secondary and supportive. Look at the proportion of material: the first two paragraphs and the last state conclusions, interpret data, and restate conclusions—acts completely neglected in the first version. Only the third paragraph presents the uninterpreted information upon which the conclusions are based. Sometimes in comparative segments, as in the second version, the judgmental element is quite explicit ("I recommend the solid state control circuits"). Other times the judgmental element is implicit; the conclusion has a descriptive tone (for example, the writer might have said, "Solid state control circuits are less expensive than relay control circuits"). Although the directness of the judgmental element will vary in effective comparative segments, you must interpret the comparative data for your reader. Without interpretation, a segment is not fundamentally comparative.

To construct a comparative segment, begin with the judgmental element we have been discussing. Tell the reader precisely what you have concluded about the relationship of the things being compared. From his point of view that is the most important thing in the segment; therefore, it should come first. You can also outline the comparison you are about to make.

Next present a point-by-point analysis of the differences and similarities of the things under consideration. Depending on your purpose, follow one of two patterns.

If the segment is dominantly either comparative or contrastive, the structure of the segment would be:

> Conclusion of the whole comparison (or contrast)
>
> Point one for A and B

Point two for A and B

Point three for A and B

. . . .

The sequence of the points is, of course, determined by their significance; they are arranged in a descending sequence of importance.

If the segment is mixed comparison and contrast, the structure of the segment would be:

Conclusion of the whole comparison and contrast

Similarity of A and B (or difference; the sequence depends upon purpose)
  Point one for A and B
  Point two for A and B

Difference of A and B
  Point one for A and B
  Point two for A and B

The similarities precede differences if your purpose is to demonstrate that the similarities predominate despite the differences. The differences precede similarities if your purpose is to demonstrate that differences predominate despite the similarities.

Both patterns help the reader unravel the complexity in front of him by taking it a point at a time. You group the points of similarity or difference. Furthermore, you help the reader see precisely what the points are by keeping them clearly separated. This point-by-point method of comparison is more effective than the method which groups all the points for A together and all the points for B together. For simple comparison, perhaps, the latter pattern may not be ineffective, but for comparisons of any length it becomes genuinely confusing. When possible, routinely structure comparative segments according to the point-by-point method.

Finally, restate your conclusion, unless the segment is very brief and uncomplicated. Without a restatement, you leave the reader with the last point of the comparison, not with the conclusion of the whole. Consequently, you should reintroduce the judgmental element with which you opened the segment.

Here are examples of each of the two patterns of comparative segments. The first illustrates the dominantly contrastive pattern, the second illustrates the mixed comparison and contrast pattern.

The first example is from a preliminary site evaluation prepared for a public power company by a consultant firm. For the most part, the segment is a straightforward contrast to demonstrate the advantages of one site over another site, both in the same general location. A more complex, detailed study will come later if the utility company should settle upon that general location for a new power plant.

Site Area M and Site Area N

From environmental and site development standpoints, Site Area M has several advantages when contrasted to the previously investigated Site Area N. Our judgment is based primarily on population, ecological, land use, location, and transportation factors.

Population considerations. Site Area M has a lower population density and is more remote from the residential and recreational development along the shore of Lake Erie, and from the increasingly residential pattern along Route 13 east of the Green River. Site Area M has a little under 25 persons per square mile; the only community has 20 residences, a service station, and a general store. Site Area N has about 34 persons per square mile, primarily because of Homer with several hundred residents and numerous commercial establishments.

Ecological considerations. Due to its more remote location, Site Area M would have a less visual impact on travelers, tourists, and residents than Site Area N. A rather frequently traveled hard-surface county road traverses a corner of Site Area N.

Site Area M would create no problems due to icing and fogging from cooling towers or a cooling pond. The downwind paths are not as favorable in Site Area N. Neither site would affect wildlife or aquatic ecology, but the game area along the Green River would have to be protected from detrimental plant discharges on either site.

Land Use considerations. Site Area M would be less disturbing to the agricultural economy of Ontario County. Land use on both sites is primarily agricultural, although soils are unfavorable for intensive farming. Over 50 per cent of the cropland on Site Area M now is idle, and about 40 per cent on Site Area N. More of the better farms in the area are on Site Area N, although relatively distant from potential plant sites.

Location considerations. Site Area M offers better potential for development of a cooling pond. Conditions at Site Area M are favorable for a cooling pond up to 6500 acres of surface area, with 60 per cent of more than 10-foot depth. On Site Area N, spray canals would have to be used to supplement available cooling pond waste-heat disposal facilities. On both sites cooling towers could be used, subject to ecological considerations.

Transportation considerations. Site Area M is more accessible by rail. A Chesapeake and Ohio rail line passes through the township in a NW-SE direction. The line is currently in use and could provide the necessary rail access. Site Area N, of course, also is accessible from the same line.

A minor disadvantage of Site Area M is that it is more remote from the Lake Erie water supply than Site Area N. The foundation and drainage conditions also may be less satisfactory at Site Area M. The intensive site selection study should reveal if these disadvantages need to be considered further. Our preliminary conclusion is that neither of these considerations detract from the basic advantages Site Area M has over Site Area N.

In this example, the core paragraph states the conclusion and identifies the five considerations upon which the conclusion is based. The five are then presented in the order identified—an order of decreasing significance. Notice that in each of these contrastive units of the segment, the opening sentence states the point made in that unit. Subsequent sentences specify some of the generalizations made in the opening sentence. The segment concludes by introducing and dismissing other considerations. This comparative segment is effectively arranged because the material is clearly segmented point by point. The reader can quickly read the segment and determine the support on which the conclusion is based. He is not overwhelmed by the particulars, which are subordinated but easily accessible.

The second example is a long segment, approximately half of a proposal by a naval architecture firm to a small company which has asked for bids to design a ship for the company. The firm proposes a preliminary and contract design rather than a more limited preliminary design with outline specifications. The company soliciting bids is unfamiliar with ship design and shipbuilding procedures. Therefore, the writer of the proposal uses a mixed comparison and contrast pattern to convince his reader that contracting for preliminary and contract design is preferable although three times more costly.

Contracting for Ship Design

As we see your present position, the most important thing is for Wagner Associates to purchase the amount of design work which, over a period of the next 20 or 30 years, will give the most economical use of a thoroughly satisfactory ship. You should consider *all* costs and *all* factors from now through the next 30 years. When you do you will realize that contracting with us for Preliminary and Contract Design (Method A) will be far more advantageous than for just Preliminary Design with Outline Specifications (Method B).

Method A will provide you with naval architects who will represent you in an objective manner, always keeping your best interests in mind since we will not be concerned with the shipbuilder's profit or loss. We will enforce construction of the ship as designed. This method will be the most economical even considering only expenditures incurred up to ship delivery.

Method B is weak because preliminary design cannot be developed in sufficient detail to assure that all of the many important characteristics of the ship are defined fully enough to obtain the ship you desire. The shipyards, when quoting on construction, probably will not be bidding on the same product.

Method A and Method B

The ship design process includes (1) preliminary design, (2) contract and specifications design, and (3) detailed construction design.

(1) Preparation of preliminary design consists of developing the basic characteristics of a practical ship to suit your special requirements. During this phase of the work, the basic dimensions, arrangements, and characteristics are determined. This design usually includes special investigations and studies such as: selection of optimum hull form; maneuvering; main and auxiliary machinery selection; etc.

(2) Preparation of contract plans and specifications defines the ship developed during the preliminary design phase in sufficient detail to permit shipyards to submit fixed price construction bids. In addition, on the basis of these specifications the shipbuilder can be made to build the ship that has been designed for the owner by the naval architects. The architect assists the owner during bid evaluation.

(3) Approval of detailed construction plans and inspection of construction of the ship. The shipyard will do a considerable amount of detail design, engineering, and plan preparation. These plans and calculations should be reviewed and approved for use in construction of the ship to assure that it is being built in conformance with the contract plans and specifications.

By Method A, Preliminary and Contract Design, you may engage us to do the first two and insure that the third is done properly. We will work with you in the planning and conceptual design phase of the work, prepare plans and specifications in sufficient detail to obtain fixed price construction bids from a large number of shipyards based upon a carefully designed and defined ship, and represent your interests during construction.

By Method B, Preliminary Design with Outline Specifications, you primarily engage us to do the first design stage. We will prepare a preliminary design and a set of outline specifications; you will use these documents as contract plans and specifications to obtain bids from shipyards and as a basis for inspection and acceptance of the vessel. As you can see, Method B will provide you with far fewer services than Method A.

Method A on Detailed Construction Design

The primary advantage of Method A is that by developing a comprehensive set of contract plans and specifications, the naval architect goes through a process that practically ensures himself and the owner of a well thought-out and complete design. Using these plans and specifications as his inspector's tool, the naval architect can enforce construction of the ship as designed. The shipyard will prepare very detailed specifications for the building of the ship, covering all aspects of materials, equipment, machinery, outfitting, electrical systems, and services to be furnished by the shipyard. Acting on your behalf we will review and approve all plans and calculations to insure the ship is being built in conformance with the contract plans and specifications.

Contracting for our services by Method A is the only way for you to insure that the ship delivered will be the ship you had designed. It also is the only way to insure that the shipyard does not have significant cost overruns during construction. Method B does not get into detailed construction design and inspection.

Method A on Contract and Specifications Design

By Method A you will contract for us to develop the preliminary design in sufficient detail to permit shipyards to submit fixed price construction bids. These contract plans and specifications not only assure that all bids received from the shipyards cover the same ship, but, more importantly, they define the ship sufficiently so that the owner, the naval architect, and the shipbuilder each have the same full picture of the ship. These plans and specifications will be used to insure that the shipbuilder builds the ship designed. By Method A we also will prepare a fairly detailed cost estimate for ship construction so that you can properly evaluate the bids.

By Method A we will specifically provide the following plans:

(a) Lines and offsets
(b) General arrangement—inboard and outboard profiles
(c) General arrangement—decks and superstructure
(d) Arrangement of pilot house and quarters
(e) Structural plans; shell and typical sections
(f) Arrangement of main and auxiliary machinery, shafting, air intakes, etc.
(g) Diagrams: lube oil and diesel oil
salt water cooling
ballast
fire mains
drains
potable water
(h) Power and lighting systems—elementary wiring diagram
(i) Switchboards
(j) List of motors and controls
(k) Navigation and remote control systems
(l) Anchor and boat handling
(m) Heating, ventilation, and air conditioning diagram

In addition we will do the following studies and calculations: all calculations and studies to support the design and requirements for ABS, Coast Guard, Public Health Service, AIEE, etc., including hydrostatic curves and tank capacities; intact and damaged stability; powering; shafting; heating, ventilation, and air conditioning; pump selection; antiroll tank; engine controls, etc.

By Method B you will contract for us to develop specifications in reasonably sufficient detail to define the major systems of the ship. We also will assist you in evaluation of the bids.

However, contracting for our services by Method A is the only way for you to insure that you do not receive a ship designed by the builder in such a way as to minimize his costs at your expense and that of the performance of the ship.

Methods A and B on Preliminary Design

Under both Methods A and B you will contract for us to provide a preliminary design. This involves weighing tradeoffs between alternate solutions to arrive at the design of the basic dimensions, arrangements, characteristics, and systems.

By Method A we will do the following:

Plans

(a) Hull form
(b) General arrangement and profile

(c) Structure: midship and typical sections
(d) Machinery arrangement

Calculations, Studies, etc.
(a) Weight calculation
(b) Stability; intact, damaged
(c) Powering and endurance
(d) Electric load and power analysis
(e) Seaworthiness, antiroll tank, bilge keels
(f) Outline specifications

By Method B we would do:

Plans:
(a) General arrangement and profile
(b) Midship and type section
(c) Machinery arrangement

Calculations, Studies, etc.
(a) Weight, trim and intact stability
(b) Damaged stability
(c) Powering and endurance

Costs

Our quotation for all the necessary plans and specifications for your proposed ship, as explained under Method A and on the basis of a standard Navy type cost-plus-fixed-fee contract, is as follows:

Direct Labor: 23,000 man-hours @ $10.00/hr.	$230,000
Overhead: 23,000 man-hours @ $7.00/hr.	161,000
Direct Charges: Reproductions	4,000
Telephone & Telegraph	500
Estimated Cost to M. Rosenblatt & Son, Inc.	$395,500
Fixed Fee (8%)	$ 31,640
Estimated Cost to Wagner Associates	$427,140

The following quotation is submitted for the required design work using Method B as explained above. We submit it reluctantly as we believe it to be an inferior and more costly approach.

Direct Labor: 8,000 man-hours @ $10.00/hr.	$80,000
Overhead: 8,000 man-hours @ $7.00/hr.	56,000
Direct Charges: Reproductions	2,000
Telephone & Telegraph	340
Estimated Cost to M. Rosenblatt & Son, Inc.	$138,340
Fixed Fee (8%)	11,067
Estimated Cost to Wagner Associates	$149,407

Let us reemphasize the superiority of Method A over Method B. Despite additional initial design costs, Method A will be by far the most economical procedure for Wagner Associates when one considers that overall costs include design, construction and operation over the life of the ship, inclusive of fuel, maintenance, repairs, etc. We believe that this method will even be the most economical if one considers only the expenditures that will be incurred up to the time of acquisition of an acceptable vessel.

In this extended example of a mixed comparison and contrast segment, the writer had to explain in considerable detail but at the same time arrange the comparison to sell his reader on what he labels "Method A" of ship design. In the first unit of the segment, he presents the assertion or conclusion of the comparison. He stresses his positive point. He asserts that cost comparison is not a meaningful criterion by which to choose between the two methods.

The second unit of the segment introduces the two methods, and briefly describes the three points by which they are to be contrasted. Implicit in this unit are the judgments explicitly made in the point-by-point comparison. Notice that this second unit is arranged in terms of the design process, from stage one through stage three. However, if the entire segment had been arranged that way, the writer might not have been able to demonstrate the superiority of Method A as effectively. Consequently, the three units that follow, each presenting one point of the comparison, are arranged so that the most telling points for Method A are presented first.

The last unit of this extended segment introduces the negative comparison between the two methods, the direct cost comparison. The method the writer proposes will cost the reader $427,140 while the other method will cost only $149,407. If the writer had not written an effective comparative segment, the reader might prefer the $149,407 method. However, after the evidence marshalled point-by-point in favor of the $427,140 method, the reader probably will agree with the writer that, all factors considered, it will be the more economical method.

Comparison segments often present information that is quite complex. In a comparative segment you must be careful not to let the details of the comparison obscure your rhetorical purpose—the point of the segment. These two examples illustrate the alternative outlines for comparative segments:

1. Statement of conclusion or interpretation of the material being compared, the judgmental element.
2. A point-by-point analysis of the differences and similarities of the things under consideration, arranged in a descending order of significance. If the segment is mixed comparison and contrast, this is a two-step procedure, each with its own overview. The sequence of the two elements is determined by purpose; the predominant element comes first.
3. A restatement of the conclusion.

In the preceding pages we have discussed and illustrated six fundamental patterns that may be used to organize segments, units, and even paragraphs of your report. We have also indicated why and how these units may be used. Just one more short comment needs to be made about arranging report segments and units.

The six patterns we have discussed are not the only possibilities. They are the most basic, the most rhetorically fundamental of the patterns a report writer needs. However, combinations of these patterns, and even other patterns, are possible. Definition is often a useful pattern, for example. However, most of your segments, units, and even paragraphs can be arranged according to the six patterns we have explained. If you need to vary patterns, we suggest only that you keep in mind the principle that underlies our six: the particular design of any segment is determined by the specific set of rhetorical objectives you seek to accomplish for particular audiences. With that principle in mind, go ahead and construct whatever variations you need.

CHAPTER 8

Editing sentences

Poorly edited sentences are like noise in a communication system. They can interfere with the clear reception of the signal and at times can even block out the signal entirely. Although few writers enjoy the routine work of filtering out this noise, the work must be done. Doing it efficiently means that you must edit sentences both in context and individually.

In the previous chapter we observed that the task of arranging segments, units, and paragraphs occurs at the point between planning and implementation. After planning the basic two-component structure and what goes into the discussion component, you arrange the segments and begin to write. You continue until you produce a rough draft of most of the report except perhaps some of the preliminaries. Although you may revise this draft significantly, at this point your concern is to get your ideas down on paper in some form.

In this rough draft stage you should not worry too much about surface details. In fact, if you worry about making your sentences scrupulously correct in the rough draft stage of writing, you can easily lose track of your ideas and lose control of your arrangement. You should try to transpose your thoughts to paper as best you can. Your sentences will reflect unconscious habits and intuitive abilities, and sometimes they may be unclear because your thoughts are still foggy in some places. You cannot edit and polish sentences until you clarify your thinking as you revise the report. Only during revision, therefore, do you pause to consider individual sentence structures. Effective sentence design is a matter of editing.

In this chapter we discuss this editing stage of report writing. We assume that on the basis of principles we have developed you have planned your report well and have written a draft in which effective structure is carried to the paragraph level. (If those assumptions are not correct, if the report is still in the planning stages or is still badly organized within segments, you are not ready to move on to editing—and no amount of editing will salvage the report.) Our objective is to provide you with procedures you can follow to determine if your rough draft sentences are effective, and if not, to suggest how to improve them. We divide the chapter into two segments: the first deals with contextual editing, that is, editing sentences in terms of their interrelatedness to other sentences; the second introduces single sentence editing, that is, editing one sentence at a time.

CONTEXTUAL EDITING

The individual sentence does not exist independently; its structure follows from the pattern of sentence structures in the entire paragraph, which in turn has a pattern derived from its purpose and even subject matter. Of the many grammatically and stylistically "correct" forms a sentence may take, the form it often takes is determined by the context within which it exists. For this reason, contextual editing is the most important sort of editing you will do.

What do we mean when we say, the form a sentence often takes is determined by the context within which it exists? The following example, from a paragraph of comparison, should make this clear:

> Modern destroyers are "volume-controlled," so the significant design features are those relating to utilization of volume. Formerly, destroyers were "weight-controlled," and the naval architectural characteristics such as buoyancy and stability were those chosen to carry the weight of hull, machinery, ship systems, people, and payload.

Here the form of the second sentence is determined by the form of the first. The main clauses have the same grammatical structure. More significantly, the parallel grammatical slots have almost precisely the same words in them, as the following analysis indicates:

Modern	destroyers	are	"volume-controlled"	so . . . .
Formerly	destroyers	were	"weight-controlled"	and . . . .

The subordinate clauses, despite superficial differences, also have the same parallel structure:

the significant design features	are those	relating to utilization of volume
the naval architecture characteristics (such as buoyancy and stability)	were those	chosen to carry the weight (of hull, . . . payload)

The entire form of the second sentence is determined by the form of the first. Change the form of the first, and you must necessarily change the form of the second. Or, change the form of the second and you have lost the undercurrent of pattern that ties the two sentences together.

The underlying pattern which relates these two sentences derives from the comparative purpose of the paragraph. The comparative purpose requires that the sentences parallel one another in order to establish the relationships of the two things being compared. Comparison which is unsystematic and random will not be clear. If, for example, the second sentence had begun, " 'Weight-controlled' destroyers were those which were formerly designed to carry . . . . ," the pattern would have been broken, the close connection of the sentences lost, and therefore the purpose of the paragraph obscured.

The example illustrates the fact that from an editorial point of view the context is more important than individual sentence forms. In this example, the form of the first sentence clearly determined the form of the second. But what determined the form of the first sentence? Precisely because the sentences are conceptually interrelated, you should try to establish a contextual pattern for the forms of both of the sentences. The contextual pattern should be clear and appropriate to the rhetorical purpose of the paragraph or, perhaps, unit.

When you do not interrelate your sentence structures you confuse the reader by making the sequence of ideas difficult to follow. Your first editing task, therefore, is to determine for each of your paragraphs or clusters of paragraphs (unit) if the sentences are effectively interrelated. To do this you can follow a three-step contextual editing procedure. You should:

> First, determine if there is a core sentence in the paragraph or paragraph cluster to establish a pattern.
>
> Second, determine if this core sentence establishes a pattern appropriate to your purpose and subject matter.
>
> Third, determine if the sentences clearly follow the pattern the core sentence establishes.

Apply this procedure to each unit in the report.

First step: Determine if there is a core sentence in the paragraph or paragraph cluster to establish a pattern.

As we said in the previous chapter, each paragraph or cluster of paragraphs should begin with a core sentence. (Sometimes, of course, the core sentence will be preceded by a transitional sentence.) This sentence controls the structure of the paragraph and sums up the details developed in the subsequent sentences. It is the largest level of generalization presented.

If your rough draft is fairly coherent, you should be able to scan it and quickly identify the core sentence in each of its units. Wherever you cannot, your first editing task is to reconsider the function of the whole unit and then to establish a core sentence. This may require you to rethink and recast the entire unit; editing often will not salvage it. At times, of course, this may be a simple matter of rearrangement because you will have inadvertently written a unit inductively, that is, from particulars to generalization. Most of the time, however, you should be able to identify core sentences.

The following examples illustrate what you can encounter as you scan a report to identify core sentences. Hopefully, you will encounter units like the first two examples. In the first example the core sentence is readily apparent.

> Three sources of nutrient load in West Bay must be identified before we consider how these nutrient sources may be reduced. The natural source of nutrients is the decay of organic materials in marshes, lakes, and topsoil which releases the constituent elements to the ground and surface waters of the watershed. The second source of nutrients is the inorganic fertilizers applied to crops and lawns which is either washed away in surface runoff or leached into the subsoil and eventually reaches the bay. This source is very difficult to control because the fertilizer is applied in many places by a great many well-meaning individuals. The third source of nutrients is the effluent from the wastewater treatment plant. Because it is concentrated at one point, this source is most amenable to control by treatment for nutrient removal before release to the receiving stream.

The core sentence, "three sources of nutrient load . . . must be identified," controls this paragraph. It generalizes by stating that there are three sources of the nutrients flowing into West Bay. Notice specifically how this sentence controls the structure of the paragraph by establishing an enumerative pattern. [There are "three sources": (1) "the natural source," (2) "the second source," and (3) "the third source."] This structure suggests a listing in parallel form. The writer, having originally chosen to put the core idea, "the three sources," in the grammatical subject slot, repeats this subject in each of the following sentences. By doing so the writer maintains the rhetorical focus of the paragraph.

In this second example, the first sentence establishes a pattern of particularization followed in the five subsequent paragraphs.

> 8.2. Second, the published literature on the Salina in Michigan and Ontario indicates that no salt is present at the Fermi site. Statements from four of the most recent such reports should verify this finding.
>
> 8.2.1. K. K. Landes, in the text accompanying the U.S. Geological Survey Oil and Gas Investigations Preliminary Map 40 (1945, The Salina and the Bass Islands Rocks in the Michigan Basin) says of the Salina salt beds, "The F and lower salts disappear in southern Wayne County . . . The D salt is not present in the Macomb County well, and it disappears with the higher and lower salts in southern Wayne County . . . . The B salt, like the D and F salt, ends abruptly in southern Wayne County."
>
> 8.2.2. C. S. Evans, in a paper entitled, "Underground Hunting in the Silurian of Southwestern Ontario" (1950, Proc. Geol. Assoc. of Canada, Vol. 3, pp.55–85) includes a map showing the southern limit of salt a mile or so south of Amherstburg, a town on the east bank of the Detroit River seven miles north of the Fermi site.
>
> 8.2.3. B. V. Sanford (1965, Salina salt beds, Southwestern Ontario, Geol. Surv. Can.: Paper 65–9, 7 pp.) states that the B salt has the widest distribution of any of the saline units. Both his figure showing the distribution of the B salt and the figure showing the total salt thickness within the Salina and its distribution show that there is no salt within the Salina Formation in Ontario south of a point one mile south of Amherstburg.
>
> 8.2.4. In an American Chemical Society Monograph (1960, Monograph 145, edited by D. W. Kaufman) K. K. Landes discusses the salt deposits of the United States (Chap. 5). In the section on Michigan (pages 71–74) he states that the bedded salt deposits of the state are in either the Salina (Silurian) or Lucas (Detroit River Group-Devonian) formations. The Devonian salt occurs only in the northern part of the Southern Peninsula. All salt mining is done from the Salina, and in the Detroit-Windsor area. His Figure 1, on page 72, showing the aggregate thickness of Silurian salt deposits in northeastern United States and Ontario shows the southern limit of Silurian salt to be north of the Wayne-Monroe County line; in fact, according to the figure there is no salt present anywhere in the section beneath any of Monroe County. The natural brines (distinct from the bedded salt deposits discussed above that are either mined or artificially brined) utilized within Michigan are from the Parma, Marshall, Berea, and Dundee formations, all of which occur only to the north and northwest of the Fermi site, and are higher stratigraphically than the rocks that underlie the Fermi site.

The core sentence, "the published literature indicates no salt is present," establishes a pattern with the dominant element being a chain of parallel subjects and verbs. Each of the sentences after the core sentence particularizes the generalization. But notice that each of them exactly repeats the form.

"the published literature . . . indicates . . . ."

"K. K. Landes . . . says . . . ."

"C. S. Evans . . . shows . . . ."

"B. V. Sanford . . . states . . . ."

"Both his figure . . . and the figure . . . show . . . ."

"K. K. Landes . . . discusses . . . ."
"he . . . states . . . ."
"His Figure . . . shows . . . ."

The subordinate element in the core sentence, "no salt is present at the Fermi site," establishes a second pattern, a chain of parallel subjects and verbs:

"that no salt is present . . . ."
"The F and lower salts disappear . . . ."
"D salt is not present . . . ."
"and it disappears . . . ."
"The B salt . . . ends abruptly . . . ."
"Southern limit of salt to be . . . [is] . . . ."
"No salt [is] within . . . ."
"The Devonian salt occurs only . . . ."
"The southern limit of Silurian salt is . . . ."
"no salt present anywhere . . . ."
"the natural brines . . . are . . . ."

This pattern established by the core sentence accounts for the structures of almost every sentence and clause in the unit. It provides a signal that editing may be needed in any sentence whose form clearly departs from the pattern.

In this third example, the rhetorical purpose is to identify the assumptions lying behind a conclusion. However, because the writer has put the core sentence toward the end of the paragraph, neither the purpose of the paragraph nor the pattern controlling the forms of the sentences can be readily determined by a reader.

VII. ANALYSIS OF POTENTIAL SOURCES OF GUEST TRAFFIC

In analyzing the drawing power of a motor hotel at this location, we have assumed that the site will be improved with a first-quality establishment that is in harmony with the present level of development in the Georgetown area. We have also assumed that the recommendations contained in this report will have substantial compliance and that guests of all types will be provided with a modern, well-designed motor hotel offering accommodations currently unexcelled in the Lexington Metropolitan Area. In addition, we have assumed that amenities of the character and type recommended in this report will be incorporated in the development; that the motor hotel will enjoy competent management; and that it will be affiliated with a recognized chain organization in the public accommodations field. Based on these assumptions we identify three potential sources of guest patronage for a motor hotel at the proposed site: business travelers, tourists, and group meetings and small conventions.

This is an instance in which contextual editing is a simple matter of rearrangement. You can move the last sentence to the beginning of the paragraph and slightly recast it to establish the pattern: "We had to make certain assumptions about the design of the motel in order to identify the potential sources of patronage." This core sentence establishes the parallel forms of the following sentences, "we have assumed that . . . ." (The last sentence in its present form might remain to forecast the following units.) No more contextual editing is needed.

In this final example, neither a core sentence nor a pattern can be identified because there is no controlling rhetorical purpose. What is the purpose of this unit? Is it cause? Effect? Process? Or analysis? Because the writer of the example did not understand his purpose, he did not formulate a core sentence to establish a pattern for the unit.

negative example

Clad Collapse

After a gap has formed in the fuel column, inward creep of the Zircaloy clad continues. With no fuel pellets to arrest the inward creep of the clad, complete flattening of the fuel rod can result. Collapsed sections of the rod have varied from 0.5–4.0 inches.

Observations of the Centerville and Doeville reactors have shown that the clad collapse has occurred in the upper ⅓ of the core. The occurrence of gap formation in one location rather than evenly distributed along the fuel rod is attributed to clad creepdown and collapse under conditions of pressure, temperature, and neutron flux. Figure 1 illustrates the collapse time of Zircaloy – 2 at 3500°C in a fast neutron flux of 3.85×10^{14} n/cm^2 sec. Parameters are pressure differentials and thickness/radius ratio (t/R). For t/R = 0.12 and P = 2250 psi, collapse occurs in about 800–1000 hours.

C. R. Leach believes neutron flux is more significant in affecting collapse time than pressure differentials. Internal prepressurizing of fuel rods should not increase the collapse time by more than 1500–3000 hours at an approximate flux of 4×10^{14} n/cm^2 sec. Since irradiation creep is essentially directly proportional to fast flux, a decrease in flux will increase the collapse time. Changes in flux due to interpellet gaps may decrease collapse times due to a local increase in the fast flux.

With a very large gap, 1.5–3.0 inches, another mechanism may control. Complete clad flattening forces may be sufficient to further separate the pellet stack.

Since the central region of the reactor core has the highest neutron flux, the largest gaps will occur near the top of the fuel stack in the PWR. As a consequence, local neutron flux peaking occurs which further increases clad temperatures. The increase in clad temperature is due to pellet gap formation as a function of distance from the fuel pellets. The increased gap temperature along a fuel gap will further enhance the inward creep of Zircaloy.

As you can see, the reader seeking to understand this unit is confused immediately. In the first three sentences alone he receives conflicting signals of rhetorical purpose and, therefore possible patterns. The phrase, "after a gap has formed," suggests a process pattern, a sequence of events occurring over time. The phrase, "complete flattening of the fuel rod can result," suggests a different sort of process pattern, a causal chain. The phrase, "Collapsed sections of the rod have varied," suggests an analysis pattern to particularize the effect (collapsed sections).

This example is an instance in which no amount of editing can make the unit coherent and comprehensible. The unit requires total rewriting. The writer would have to rethink the whole, and decide what the rhetorical purpose of the unit was. Only after having done that could he decide what form the core sentence should take to best present his generalization and establish his pattern.

The first step in contextual editing is to insure that each paragraph or unit has a core sentence to establish a pattern. If there is no core sentence the reason may well be that there is no rhetorical purpose to yield a core sentence and establish a pattern. (If there is a rhetorical purpose, you ought to be able to summarize it in one sentence. The contents of the "Clad Collapse" example cannot be summarized.) This step of the editing procedure identifies any units which need rethinking and extensive rewriting. If your first draft was well planned, however, this step should be a simple matter of skimming the report to verify the presence of core sentences in your units.

Second Step: Determine if this core sentence establishes a pattern appropriate to your purpose and subject matter.

After you have scanned your report to determine if each unit has a core sentence and after you have supplied core sentences for those that do not, you come to the next step of the contextual editing process. In this step you evaluate the core sentences to make certain that each establishes a pattern appropriate to your rhetorical purpose. Determining appropriateness is determining if the focus of the core sentence in the rough draft is consistent with the purpose of the unit. Does the focus of the core sentence establish a pattern that reinforces your purpose? Or, does it establish a pattern that undermines your purpose?

Suppose, for example, you are editing a soil study report you have written and you find this sentence as the core sentence of a paragraph unit:

> It has been concluded that the basic cause of frost heave is the ice layers (and the increase in the water content).

What is the focus of this sentence—which, by the way, comes from an actual report? There are several possibilities, none entirely clear.

First, look at the subject. (Nouns in subject positions receive particular emphasis, so they should be the first things you inspect.) Here there is no immediately clear signal of focus because the grammatical subject is the pronoun, "it." Clearly "it" is not the core idea of the unit, so perhaps the focus might be upon what "it" refers to, the subject of the subordinate clause, "the basic cause of the frost heave." If the focus is on "basic cause," the pattern established should be a cause/effect pattern to separate out the one cause ("frost heave") from among several other possibilities.

Second, look at the verbs. Verbs and verbal constructions focus your attention on actions rather than on things. Therefore, the core verbal concept might indicate the focus of the unit. In this example the first clear signal of focus is the phrase, "has been concluded." If that is the focus, the pattern established should be a persuasive pattern, assertion plus support. Another signal of verbal focus is contained in the expression, "formation of ice layers." If that is the focus, the pattern established should be a process pattern to present a causal chain culminating in the effect, frost heave.

Given these possibilities, what is the focus of this core sentence? Without the rest of the paragraph, you cannot be certain. In this instance, you must read the entire paragraph to discover that the actual focus is on process, that the paragraph is essentially a process unit. In other words, the rhetorical purpose of the paragraph is to explain an action, a process of causation. Thus the focus of this core sentence clearly should be on an action—on the act and consequences of "formation."

Here is a revised version of the core sentence and the paragraph to clarify this concept of focus:

> The formation of ice layers underneath the pavement causes frost heave. Although the temperature deep in the ground remains constant throughout the year, the temperature in the ground near the surface under the pavement fluctuates with air temperature. When the pavement surface freezes, the temperature in the ground under the pavement falls below freezing. The freezing and low temperatures induce capillary tension which sucks up water from the water table below. This capillary tension greatly increases the amount of water in the first zone under the pavement. When this water freezes, the soil expands far more than it otherwise would, and causes the pavement to heave.

The core sentence of this paragraph now clearly signals that the focus of the paragraph is on the causal process. "Frost heave" results from a chain of causal events: "temperature fluctuates" → "below freezing" → "sucks up water" → "increases amount of water" → "expands far more" → "pavement to heave." (The core sentence could have had the passive construction, "frost heave is caused by the formation of ice layers underneath the pavement," and still the focus would have been on the causal process.)

Because the core sentence establishes the focus of a paragraph, your rhetorical purpose should determine the focus and therefore the form of the core sentence. The following example of a core sentence illustrates how rhetorical purpose determines focus:

> Corona discharge is caused by the electric energy in the line.

This core sentence introduces a paragraph in which the focus is on an explanation of an effect, "corona discharge." Because the concept, "corona discharge," is in the grammatical subject slot, the focus of the sentence is upon the effect. If the sentence were recast to put the concept, "electric energy," in the subject slot ("electric energy in the line causes corona discharge"), the focus would be upon the cause. In either form, the core sentence will establish a process pattern. However, the rhetorical purpose, explanation of an effect, meant that the passive construction, "corona discharge is caused by . . . ," was the more appropriate in this case. The rest of the paragraph methodically presents a causal sequence: "electric energy accelerates free electrons"; "the accelerated electrons remove electrons from other air molecules"; "the air molecules become ionized"; and so on.

Here are two more examples where rhetorical purpose determines the focus and form of the core sentence:

> Watch standers were trained to respond to a boiler room fire with this five-step sequence of actions designed to initiate fire-fighting action on the ship. (1) Close the quick-closing fuel supply valve. . . . (2) Using the remote control . . . , turn off the fuel oil pump. . . . (3) Using the switches . . . , turn off the boiler room ventilation system. . . . (4) Using the telephone . . . , inform the Officer of the Day. . . . (5) Using the portable fire extinguishers . . . , attempt to contain the fire. . . .

This core sentence introduces a paragraph in which the focus is on the stages of a process and their sequence. Notice that the process, in this instance, is not a causal sequence. Notice also that the enumeration serves to define the sequence, not just to itemize. Because the core sentence establishes such a clear pattern, the subsequent sentences efficiently omit the subject. The pattern established is appropriate to the rhetorical purpose.

> The first program for which Berrien County might qualify is a program administered by the Department of the Interior.

This core sentence introduces a paragraph which focuses on explaining the details of a program. The form of the core sentence (which puts the concept, "program," in the subject slot and then modifies it, ". . . is a program . . . ,") signals an analytical pattern which will present the details of the program. If the core sentence had said, "the program developed" or "the program will enable," the pattern established would have been a process pattern. The rhetorical purpose determined the focus of the core sentence and of the paragraph pattern appropriate to that focus. (Notice that the introductory phrase, "The first," indicates that this paragraph is the first in a parallel series of paragraphs following an enumerative pattern. The pattern of this paragraph, therefore, is actually a pattern within a pattern.)

In contrast to the examples above, the following example presents a group of sentences which form an inappropriate pattern:

negative example

> Particle mobilities depend, therefore, on the amount of retarding force offered by the strip. Cellulose acetate strips are uniformly porous, allowing a minimum of variation in the retarding force throughout the strip. For this reason, clear, distinct electrophoregrams are possible.

In this instance the subject matter and form do not match well at all. Each of these three sentences has a different subject concept ("particle mobilities," "cellulose acetate strips," and "electrophoregrams"), and there is no direct object. Yet, the writer describes a causal sequence which produces a distinct electrophoregram. His failure to establish a consistent subject-object series demonstrates his failure to grasp the form of the process involved. Furthermore, his failure to use active verbs demonstrates his failure to pattern his material according to his rhetorical purpose.

Core sentences, then, should establish patterns appropriate to rhetorical purposes. As we said, the second step in contextual editing is to determine if core sentences establish appropriate patterns. How do you determine appropriateness? You decide if your purpose is to describe, to explain a process, to identify causes or effects, to compare, to enumerate, to analyze, or to particularize. Then you look at the combinations of sentences to determine if the pattern of sentence structures complements the rhetorical purpose. In other words, you must match what you are trying to say with the way you are trying to say it.

Remember, the core sentence of a paragraph creates expectations in a reader. "Site Area M would be less disturbing than . . .", "Our conclusion is that . . .", "The three sources of nutrient load . . .", "The published literature indicates . . .", "Elimination of these tubes would result . . ."—each of these phrases suggests probable development in subsequent sentences. These sentences will be expected to derive their content and form from that initial sentence. Whether the core sentence is a "good" sentence or not, whether it says what you intended, it will establish a pattern and create expectation. For that reason, you should verify that the form of a core sentence focuses on your core idea.

Third Step: Determine if the sentences clearly follow the pattern the core sentence establishes.

We have said that the core sentence sets up a pattern. This final step of the contextual editing procedure is simply a matter of trying to work out the pattern with a reasonable degree of clarity and consistency.

As you were writing your rough draft, you were groping along one sentence at a time. After putting one idea on paper, you moved on to another idea, hoping it would follow from the previous and lead to the next. You wrote "horizontally."

Now you should think "vertically." That is, you should mentally arrange the sentences one below another so that the subjects, verbs, and objects fall into vertical columns. Then check to see if you have maintained focus by being consistent.

When you think vertically, your primary concern is to examine the fundamental architecture of the cluster of sentences. Here, for example, is a comparative paragraph as it appeared in rough draft.

The steel system can be constructed more rapidly than the masonry system. Steel columns can be erected much faster than concrete columns, which must be allowed to cure in their forms for two or three weeks. Steel wall panels also can be placed much faster than block walls can be erected, so the prospect of delays caused by the masonry union can be eliminated. Insulation can be sprayed more rapidly than styrofoam panels can be glued in place. Thus, a steel system can be constructed in one-third to one-half the time a masonry system can be erected.

Although this paragraph was written "horizontally," one sentence at a time, it certainly is not ineffective. Vertical thinking makes it evident that the writer maintained the relatively consistent focus as he worked his way along from sentence to sentence:

core sentence:	steel system	can be constructed	masonry systems
sentence two:	steel column	can be erected	concrete columns
sentence three:	steel wall panels	can be placed	block walls
sentence four:	insulation	can be sprayed	styrofoam panels
sentence five:	steel system	can be constructed	masonry system

The writer has kept his subjects, verbs, and objects parallel to maintain a generally clear focus on the comparison being developed. However, vertical thinking suggests that the paragraph could be improved with modest editing of sentence four, which blurs the focus slightly. Its subject, "insulation," does not repeat the pattern of subjects; consequently, the reader must hesitate to decide if insulation is sprayed on steel walls or on masonry walls. This minor interruption of the pattern can be eliminated by editing the sentence to fit the context: "Steel walls can be insulated more rapidly by spraying than masonry walls can be insulated by gluing styrofoam panels in place."

Here is another example paragraph as it appeared in rough draft:

Railroads in Michigan are being abandoned at an increasing rate. Since 1965, 500 miles of track have been abandoned, most of it in the Upper Peninsula. Railroads have applied to the ICC for permission to abandon an additional 545 miles. According to newspaper accounts, the MPSC is considering abandoning another 915 miles. This totals 1960 miles, or 29% of Michigan's 6,750 miles of railroad.

This is also a fairly effective paragraph. The core sentence focuses on an action, abandonment of railroads. And vertical thinking makes evident the pattern of particularization followed by the subsequent sentences. The subsequent sentences, except the last, maintain the relatively consistent focus on the pattern of verbs.

core sentence:	Railroads	are being abandoned
sentence one:	500 miles of track	have been abandoned
sentence two:	Railroads	have applied . . . to abandon
sentence three:	MPSC	is considering abandoning
sentence four:	This	totals

Every sentence except the last focuses attention on the action; the subjects change, but the sentences are interrelated effectively by the pattern of verbs. The last sentence, however, clearly breaks the pattern. Yet this sentence can be easily edited to fit the pattern by changing it to: "Thus, 1,960 miles, or 29% of Michigan's 6,750 miles of railroad, will soon have been abandoned." This minor editing to fit the last sentence into its context considerably strengthens the paragraph because it reasserts the focus of the core sentence.

Of course, the above examples are much more easy to edit than most paragraphs in any report. We introduce these two examples to suggest how vertical thinking can help you identify sentences which need editing. Most paragraphs you write cannot be edited so easily. They may have several patterns operating simultaneously; they may subordinate elements which momentarily but necessarily interrupt a pattern; and they may have sentences whose forms derive from the patterns of larger contexts. As an example of the sort of paragraph you may encounter more often, the following paragraph is more complex than the previous two:

> (1) The first program for which Berrien County might qualify is a program administered by the Department of the Interior. (2) It provides for grants to certain communities for construction of waste treatment plants and certain intercepting sewers. (3) This program grants funds to qualifying communities for 30% of costs of constructing waste treatment facilities. (4) Under this program, the Village of Stevensville has been offered a grant to cover 30% of treatment costs currently proposed for construction. (5) The program authorized $18 million for Michigan in the current fiscal year but the federal government has curtailed this program so that only $1.23 million is available. (6) Although you might qualify for this source of funds, the small amount of money available does not suggest much likelihood of your receiving a grant.

As we have indicated, this is an effective paragraph which maintains a clear focus on explaining the details of a grant program—an analytical pattern. The paragraph, however, is not uncomplicated. The fourth sentence in the paragraph does not advance the analysis of the program; it momentarily interrupts the pattern by changing subjects to illustrate a point made in the previous sentence. The sixth sentence in the paragraph derives from the larger persuasive pattern of the segment in which this paragraph exists. Despite these complications, contextual editing by vertical scanning should reveal the coherence of the paragraph.

Determining if the sentences clearly follow the pattern the core sentence establishes is largely a matter of vertical scanning to verify consistent focus. This is also a matter of verifying that the words and phrases you have used to connect the sentences in the pattern are appropriate. The phrases, "on the one hand" and "on the other hand," clearly connect sentences in a contrastive pattern. "As an example" particularizes. "First," "second," and "third" enumerate. "On the heat extractor end" and "on the platform surrounding the basin" describe. In working out any pattern you will use connectors to tie the sentences together. Verify that the connectors you have used are sufficient and appropriate.

Just how effective connectors are is illustrated in Figure 8-1. In this paragraph, which you have seen before (p. 105), we have deleted most of the content words to highlight the connectors the writer used. Even without the substantive words, you can still see that the paragraph has a process pattern. Further, you can see signals of subpatterns ("For example" signals particularization; "therefore" signals an effect.) Finally, you can see that the writer has been consistent.

The example in Figure 8-1 illustrates why contextual editing is the most important editing for you to do. The effectiveness of your text does not depend upon the correctness of the sentences individually. Effectiveness depends upon the sentences being interrelated appropriately in terms of rhetorical purpose. Our three-step procedure for contextual editing should assist you in identifying and improving the patterns of your sentences.

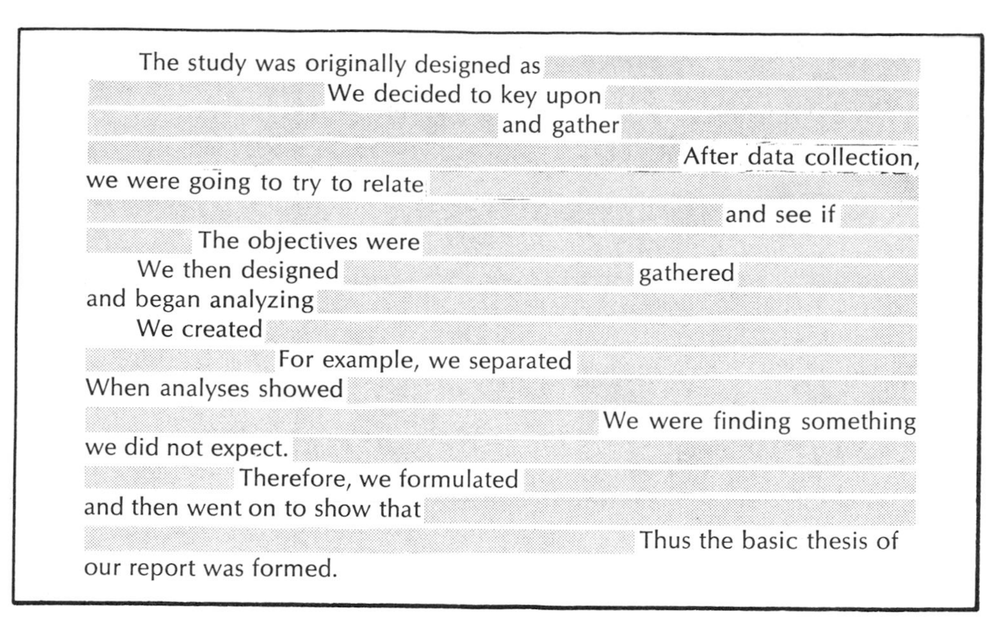

The study was originally designed as
We decided to key upon
and gather
After data collection,
we were going to try to relate
and see if
The objectives were
We then designed gathered
and began analyzing
We created
For example, we separated
When analyses showed
We were finding something
we did not expect.
Therefore, we formulated
and then went on to show that
Thus the basic thesis of
our report was formed.

Figure 8–1. Paragraph with substantive words deleted to highlight connectors.

SINGLE SENTENCE EDITING

We have suggested that contextual editing is fundamentally more important than single sentence editing because sentences do not exist independently. However, in almost any rough draft many single sentences can stand improvement. They can be edited for directness, efficiency, and clarity independently of the context. Single sentence editing, then, is a matter of tradeoffs between the demands for contextual coherence and the demands for directness, efficiency, and clarity.[1]

In the remaining pages of this chapter we will suggest some editing tactics you can use to improve individual sentences. However, we will not exhaustively discuss single sentence editing because, unlike contextual editing, it is amply and well discussed in a number of handbooks. If you find our editing suggestions are not sufficient to make your report presentable, we encourage you to consult these sources. The suggestions we make are for manipulating sentences to achieve the qualities of directness, efficiency, and clarity.

Directness

You should make your sentence forms directly reflect the subjects and actions of the idea you are trying to convey. The grammatical subject of a rough draft sentence can be different from the core concept. The grammatical action of a sentence can be different from the verbal concept. If you cannot immediately put your finger on the core subjects and actions of rough draft sentences, they need to be edited. When you edit, identify the subjects and verbs of your rough draft sentences, and change them if they are not the core subjects and actions.

[1] Although we discuss single sentence editing after contextual editing, in practice single sentence editing is done simultaneously with contextual editing. Single sentence editing is not—as the arrangement of this chapter could suggest—the last editing you do.

In the following example, you cannot immediately identify the core subject. Neither the first word, "it," nor the fourth word, "responsibility," which fill the grammatical subject slot, is the core subject of the sentence. The grammatical subject should therefore be changed.

> It is the responsibility of the Department of Highways and Traffic to investigate all fatal automobile accident sites in the city.

The core subject is, "Department of Highways and Traffic," so the sentence should be edited to read:

> The Department of Highways and Traffic is responsible for investigating all fatal automobile accident sites in the city.

In the following example you cannot immediately identify the core action. The word, "involves," fills the verb slot, but that word suggests a focus on the stages of the filtering process. Clearly the core verb is "separates," which focuses on what filtering does. The grammatical verb should therefore be changed.

> The filtering technique involves separation of signals of different frequencies.

The sentence should be edited to read:

> The filtering technique separates signals of different frequencies.

Or,

> Filtering separates signals of different frequencies.

Or,

> Signals of different frequencies are separated by filtering techniques. (The passive voice, which some handbooks suggest you should always avoid, is actually necessary at times to maintain contextual focus.)

Any of the revisions above improves the sentence. Which is most effective depends on context.

In this group of additional examples (all of which come from actual reports) we illustrate how to achieve directness. We have edited the first four sentences to focus on the core subjects:

> *Draft version:* After a brief meeting with Jim Blackman it was decided that four tests would be performed on the samples.
> *Edited version:* Jim Blackman and I decided to perform four tests on the samples.
> *Or:* After meeting with Jim Blackman, I decided to perform four tests on the samples.

From the draft version, the reader cannot determine who made the decision, i.e., who the subject is. This version of the sentence may have resulted from the writer's unwillingness to say, "I decided."

> *Draft version:* In this report an efficiency test applied to a small steam turbine is described.
> *Edited version:* This report describes an efficiency test applied to a small steam turbine.

Draft version: If the antenna is to be used in an urban area, where many different UHF stations exist, it is imperative that the field pattern be very narrow.
Edited version: The field pattern must be very narrow if the antenna is to be used in an area with many UHF stations.

The draft version is weak because the core subject occurs after four nouns and pronouns. Three of these occupy subject slots.

Draft version: An important consideration which is almost always overlooked is that engines in a laboratory receive far more expert attention than the typical engine in a private owner's vehicle.
Edited version: Engines in a laboratory receive far more expert attention than engines in private cars. This fact is usually overlooked.

The core subject, engines, is obscured because the writer has placed a subordinate idea first.

Here is another group of sentences which illustrate how to achieve directness by editing to focus on the core action.

Draft version: Selection of the circulation system will need to be carefully watched to ensure absorption of the proper wave length of incoming radiation to excite the methyl viologen, and chlorophyll if it is used.
Edited version: The circulation system must be carefully selected to ensure absorption . . .
Or: We must carefully select the circulation system to ensure absorption . . .

In the draft version, the grammatical verb, "will need to be carefully watched," has nothing to do with the meaning of the sentence.

Draft version: Data definition languages were finally accepted as a component of information processing systems equal in importance to programming languages with the advent of generalized data base management systems.
Edited version: With the advent of generalized data base management systems, data definition languages became as important as programming languages in information processing systems.

Draft version: The column is still operable when there is a small amount of flooding on the surface.
Edited version: The column still operates when there is a small amount of flooding on the surface.

This draft version illustrates a very frequent indirect structure, in which the core action, "operates," is in noun or adjectival form, "operable."

Draft version: This ion "cloud" is responsible for the corona effects.
Edited version: This ion "cloud" causes the corona effects.

The above examples illustrate how the core action of a sentence can be easily obscured by translating the verb into a noun, and how the core subject can be buried in a subordinate position. The core action and the core subject also can be obscured when they appear in a sentence which should follow a pattern but does not. That is, both verbs and subjects can be compounded; when they are they should be presented in parallel form. If they are not, the sentences become obscure. Here are three examples:

Draft version: After thoroughly investigating this accident site, it was found that the roadway was properly designed, constructed, and was in good condition.
Edited version: The accident site was thoroughly investigated and the roadway found to be properly designed, constructed, and maintained.

The draft version failed to put a series of three parallel actions in parallel form.

Draft version: Clarification was also needed regarding design changes resulting from the transfer and problems associated with steam generator construction.
Edited version: We needed to clarify the design changes resulting from the transfer and from the problems associated with steam generator construction.

The edited version clarifies the parallel relationship of "the design changes" and "the problems."

Draft version: We do not know if the aircraft manufacturers are aware of this problem or what voltage levels are used and the type of transients that are encountered.
Edited version: We do not know if the aircraft manufacturers are aware of this problem, what voltage levels are used, or what type of transients are encountered.

The edited version changes an illogical "a or b" construction into a parallel series.

Directness is primarily a matter of keeping the grammatical form of a sentence consistent with the thought behind the sentence. If the core subject of a sentence is "engines," the grammatical subject should not be, "an important consideration which is almost always overlooked is. . . ." If the core action of a sentence is "select," the grammatical verb should not be, "[a selection] must be made." Whenever possible, then, edit the form of a sentence to directly reflect the idea.

Efficiency

You should make your sentences as efficient as possible. Like directness, efficiency is a characteristic of good instrumental prose. Long and flabby sentences are simply difficult to follow and therefore make for inefficient communication. When you edit, shorten and simplify sentences as much as contextual considerations permit. Some texts suggest that you count words to assure that your sentences do not exceed the theoretically acceptable average of 21 words per sentence. Counting words may be a way to edit, but it has a makeshift aspect because you have no way to determine which sentences to shorten. Short sentences are not necessarily efficient; sentence length is a function of the complexity of the ideas expressed. To shorten sentences when you edit, first make sure that each sentence expresses only one important idea; second, make sure that the idea merits the grammatical weight you have assigned to it.

Rough-draft sentences can be inefficient because clauses containing important ideas are thoughtlessly spliced together when they should stand by themselves. Here are several examples:

Draft version: Most voids are pores which exist on the surface and some voids have no outlet to the surface and are known as dead pores.
Edited version: Most voids are pores which exist on the surface. However, some voids, known as dead pores, have no outlet to the surface.

The writer of this draft version had put two quite distinct ideas into one compound sentence. In doing that, he has lost sense of the meaning.

Draft version: Unless the polymer chemically bonds with the carbon a crack may propagate along the carbon-polymer interface and thus negate most of the strengthening effect of the polymer.
Edited version: Unless the polymer chemically bonds with the carbon, a crack may propagate along the carbon-polymer interface. The crack negates most of the strengthening effect of the polymer.

The draft sentence describes a causal process but puts two effects in one sentence. This is inefficient because the causal sequence is obscured.

Draft version: But since the operating temperature of this process is between 153°C and 100°C, formic acid is going to form and accumulate at the bottom of the column, which presents a highly undesirable situation.
Edited version: Since the operating temperature of this process is between 153°C and 100°C, formic acid will accumulate at the bottom of the column. This is highly undesirable.

The draft sentence has a double focus; it is both cause/effect and evaluative, and in fact subordinates the evaluation. The edited version presents each focus in a separate sentence.

Draft version: Information concerning volume and tempo changes is extracted from the graphic display, but identification of individual notes and instruments requires the isolation of a large number of fundamental frequencies (some of them being the same for different instruments) and the even more arduous task of associating the hundreds of harmonics with their corresponding fundamentals.
Edited version: Information concerning volume and tempo changes is extracted from the graphic display. However, identification of individual notes and instruments requires isolating a large number of fundamental frequencies (some the same for different instruments) as well as associating hundreds of harmonics with their corresponding fundamentals.

The draft sentence has a double focus. The edited version uses two sentences to differentiate between the two different topics, one a contrast, the other a parallel series.

A rough draft sentence can also be inefficient because excess verbiage and indirect constructions lend an idea more weight than it deserves. Often modifying clauses should be reduced to modifying phrases; modifying phrases should be reduced to single words. Indirect constructions usually should be eliminated.

Here are two examples in which editing has reduced modifying clauses to phrases:

Draft version: While this situation does not constitute a problem at this time, it should be explored further. . . .
Edited version: While not a problem at this time, this situation should be explored further. . . .

Draft version: We have determined that we need a simple tension bar in our structure which will support a maximum expected tensile load (p) of 40,000 lbs. (40 kips).
Edited version: We need a simple tension bar to support a maximum expected tensile load (p) of 40,000 lbs. (40 kips)

In this example, the clause, "which will support," easily reduces to, "to support," and the meaning of the clause, "we have determined," is made implicit in the verb, "need."

Here are two examples in which editing has reduced modifying phrases to words:

> *Draft version:* Since this voyage is longer than 24 hours, overnight accommodations in the form of staterooms are required for all of the estimated 800 passengers who will be aboard the craft.
> *Edited version:* Since this voyage is longer than 24 hours, staterooms are required for the estimated 800 passengers.

In this example, the phrase, "in the form of staterooms," becomes a word, and the phrase, "for all of the estimated," becomes "for the estimated." Notice also that a redundancy, "overnight accommodations . . . staterooms," is eliminated, and the redundant clause, "who will be aboard the craft," eliminated.

> *Draft version:* A strength of 8500 pounds per square inch (psi) is required of the concrete.
> *Edited version:* 8500 psi concrete is required.

Here are examples in which editing has eliminated either of the two types of indirect constructions prevalent in rough drafts:

> *Draft version:* It is the opinion of this department that a heat problem will not occur on production vehicles.
> *Edited version:* This department concludes that no heat problem will occur on production vehicles.

The draft version contains an unnecessary "it–that" construction.

> *Draft version:* It has been our contention that the line workers can make a significant contribution to the decision making process. . . .
> *Edited version:* We contend that line workers can make a significant contribution to the decision making process. . . .

Another "it–that" construction.

> *Draft version:* Consequently, it is one that deserves the attention of all those who are in a position to improve the crew's ability to tackle this problem.
> *Edited version:* Consequently, the problem deserves the attention of those in a position to improve the crew's ability to tackle it.

This draft sentence, as the two above, is weakened considerably by an unnecessary "it–that" construction. Notice that the "it–that" construction forces the core subject and action into subordinate slots.

> *Draft version:* The fasteners all showed signs of hydrogen embrittlement which causes high local stress concentrations which resulted in the failure of the fasteners.
> *Edited version:* The fasteners failed because of high local stress caused by hydrogen embrittlement.

The draft sentence contains unnecessary "which" clauses. Eliminating them led the writer to focus on his core action, "failed."

> *Draft version:* As a result of tests which have been completed, . . .
> *Edited version:* As a result of completed tests, . . .
> *Or:* As a result of tests, . . .

This draft version also contains an unnecessary "which" clause. The clause can be reduced to one word or completely eliminated.

Indirect constructions such as these can be easily identified. Just look for "it–that" and "which" constructions; invariably, they should be eliminated.

These examples suggest that editing procedures overlap. Editing to achieve directness brings about efficiency. Editing to achieve efficiency makes sentences considerably more direct. Because they overlap, editing procedures are not performed in sequence. They are done simultaneously.

Clarity

You should make sentences as clear as possible. Eliminate ambiguities. To do so, clarify pronoun references and eliminate dangling modifiers.

Ambiguous pronoun references are prevalent in rough drafts. Rough drafts usually contain excessive numbers of "its," "this's," and "thats," many of them with unclear antecedents.

> *Draft version:* The technical problem concerned itself with designing a recovery facility that would recover DMF from an aqueous stream, so that (it) might be recycled to the new synthetic fibers plant.

In this sentence the circled "it" is totally unclear: "it" can refer to either the DMF or the aqueous stream. Also, "it" can be confused with the "itself." Finally, the sentence has two "thats" (underlined), which confusingly have two different referents.

> *Draft version:* Birefringence is characterized by the diversion of a light vector within the medium into two orthogonal components; moreover, (it) transmits each of these components at a different velocity.

Question: What does "it" refer to?

> *Draft version:* First, it was difficult to insulate the sample enough to prevent arcing to the wall of the oven. This was corrected by placing glass sheets on the surfaces inside the oven. Second, the insulation resistance was low enough to allow enough current to activate the "fault" detection circuit at the higher voltage. This required that the test be restarted.

Question: What do the two "this's" refer to?

> *Draft version:* An AC field (60Hz) was applied to the sample because (it) generates corona on every half cycle and (it) can readily be seen on the oscilloscope.

Question: What do "it" and "it" refer to?

> *Draft version:* The whole habitat will be situated on the barge so that it can be towed and dropped wherever wanted. This entails difficulties in ballasting the barge and the ascent and descent control.

Questions: Is the barge being towed or is the habitat being towed? What does "this" refer to? (Also, "ballasting" the ascent and descent control?)

Draft version: There are many sources of enzymes, but the reason that Vibrio succinogene was selected is that (it) provided us with the most information, not that there is not a better source; this must be determined through further study and experimentation.

This sentence has five pronouns. The third "that" has a different referent from the first two "thats." The "this" has no referent at all as far as we can determine.

The writers of these sentences, of course, had no difficulty determining what the pronoun references were—the writer reading his own rough draft can understand what he meant. However, he must edit so that readers can also understand what he means. Editing pronoun references usually is quite simple. You can even, as we have done, circle pronouns and draw arrows to the referents. If these arrows cannot be drawn, editing is needed. Furthermore, if one pronoun has two different referents in a paragraph, editing is needed. And, usually, if a paragraph is overloaded with different pronouns, editing is needed.

Dangling modifiers also are common in rough drafts, although they are more difficult to identify than vague pronoun references. A participial modifier should modify the first noun after or the noun immediately before the phrase. Again, the question technique may help you edit.

Draft version: In observing the actions of operator 2, there was a definite idle time in each cycle.

Question: Who is idle, the observer or the operator?

Draft version: By preventing the liquid down-flow and by supplying fresh liquid at the same time, a plug of liquid is formed.

Question: Is the plug cause or effect?

Draft version: By using enough electronic circuitry, machines can be built capable of doing arithmetic operations on binary numbers, such as addition and multiplication.

Questions: Who uses the circuitry? What are "addition and multiplication"?

Notice in this example that a dangling modifier is not necessarily a participial phrase. The second dangling modifier illogically attaches to the noun immediately preceding, "numbers."

Draft version: Upon filling out the proper form and sending it to Data Processing this event will appear on the next Schedule Matrix and detail listings.

Questions: Who fills out forms? Who sends them? Who details listings?

Draft version: **Cost and Time Estimate**
Having completed a trial run on 4% of the job, a cost and time estimate can be made to an accuracy of plus or minus 10%. These estimates are all inclusive, assuming the only yield to be the proposed system (including tags) in a fully operational state. Using the same 4 warehousemen as in the trial run, seven weeks is the best estimate of the time requirement. Based on current prices, $12,000 is the estimated cost. Through inauguration of the new floor plan in conjunction with its complementary record-keeping system, worker efficiency should attain or exceed 100%.

Questions: Who completed a trial run? Who assumes? Who uses? Who inaugurates?

This last example illustrates how easy it is for a writer to fall into a pattern. One dangling modifier invites another. Of course, most of the time these dangling modifiers will be more awkward than unclear. The same is true of ambiguous pronouns. In any cluster of sentences, however, small increments of uncertainty accumulate, and force a reader to stumble, slow down, and perhaps stop.

Just as we do not think it necessary to duplicate handbooks on grammatical style, we do not feel it necessary to discuss single sentence editing at length. Single sentence editing is often best done by someone other than the writer. Other people may not be able to improve the basic architecture of a report, and they may not be able to edit contextually, but usually persons other than the writer can identify surface defects the writer himself might overlook. The writer should concentrate on contextual editing. If he is successful in contextual editing, many single sentence editing problems will disappear.

Designing the basic architecture of a report and designing individual segments can be challenging. Meeting the challenges successfully can be very rewarding. Editing—often an uninteresting, annoying task which comes near the end of the entire writing process—comes when the writer is heaving a sigh of relief at having "finished" and is anxious to get the report off his hands. But it must be done. Editing filters out the "noise" in the system. As with any real communication signal, noise cannot be entirely eliminated. Nevertheless, it must be reduced so that it does not interfere with the reception of the signal. You edit for your audiences.

CHAPTER 9

Additional design features: Layout and visual aids

The design of a report is not the design of an idea; *it is the design of a* thing. *The report writer must therefore understand how physical design features can reinforce and clarify his ideas.*

If you genuinely master the concepts we have discussed in the first eight chapters, you should be able to design conceptually clear reports. You should be able to gather and select the material appropriate to your audience and purpose, to arrange those materials in an effective order, and to write a draft which carries clear design right down to the sentences. You should be, in short, nearly finished with the rhetorical task. But not quite.

You must still consider the physical design of the report. For the report's design is not, after all, the design of an *idea*. It is the design of a *thing*. Once written, the report will be typed or printed. It will have shape and appearance. It will array itself on sheets of paper in particular ways. Therefore, just as at every other point in the design process, after the report has been written you must make important decisions: you must decide how you want the report to appear. (You make these decisions even if the report will be turned over to a technical writing and illustrating department for production and distribution.) Whether your secretary or an editorial department prepares the finished report, the draft you hand over usually will be produced without significant changes being made in the text, layout, or visual aids. If you hand in a sketch of a visual aid with inadequate labeling, the finished product will be inadequately labeled no matter how professionally it has been redrawn. It is not the role of the secretary, editor, or illustrator to make your design decisions.

To help you resolve necessary decisions before you can hand over your finished draft for final production, we have a number of suggestions which will make the physical design of the report reinforce and clarify its contents. We suggest ways to design the layout and use visual material.

SIGNAL THE INTELLECTUAL DESIGN OF YOUR REPORT

You should make the internal, intellectual design of your report apparent to even casual readers by incorporating clear design signals into the physical layout.

A technical report should be easy to read. Most readers will page quickly through it, reading a bit here and a bit there as they hunt for the specific segments that genuinely concern them. If you have not given your readers clear design signals, they will not be able to do this efficiently. They may have to grope along on their own or have to read the whole report straight through. Especially when skimming, they may even conclude that the report is badly planned. Therefore you must give your report not only the fact of order but the appearance of order as well.

You have four devices to use: headings, transitional elements, numbering, and white space. Well used, these can make it almost impossible for a reader to miss seeing the intellectual design of a report at a glance. They can guide him quickly to the segments he needs to read; they make evident the strengths that are really there.

Headings

Headings are your best aid to the reader in a hurry. Like a table of contents superimposed on the text, they help him to find quickly what he is looking for in your report. At the same time, they give him a sense of overview and general awareness of what is contained in segments of the report he may not read. But if your headings are to function well they must be well designed. They must be used frequently and must be made as informative as possible.

Too often, writers do not use enough headings to signal divisions in their reports. They use headings to designate only the major segments. However, they should also use headings to signal divisions within segments down to the unit level and sometimes even to the paragraph level. In the following example many writers would have included only the first heading. Yet clearly the headings on the units (here paragraphs) within the segment are useful to the reader. In an intellectual sense they add little to what is there; however, they make the structure of the unit apparent at a glance:

> **Site Area M and Site Area N**
>
> From environmental and site development standpoints, Site Area M has several advantages when contrasted to the previously investigated Site Area N. Our judgment is based primarily on population, ecological, land use, location, and transportation factors.
>
> *Population considerations.* Site Area M has a lower population density and is more remote from the residential and recreational development along the shore of Lake Erie, and from the increasingly residential pattern along Route 13 east of the Green River. Site Area M has a little under 25 persons per square mile; the only community has 20 residences, a service station, and a general store. Site Area N has about 34 persons per square mile, primarily because of Homer with several hundred residents and numerous commercial establishments.
>
> *Ecological considerations.* Due to its more remote location, Site Area M would have a less visual impact on travelers, tourists, and residents than Site Area N. A rather frequently traveled hard-surface county road traverses a corner of Site Area N.
>
> Site Area M would create no problems due to icing and fogging from cooling towers or a cooling pond. The . . . .

As you can see, inserting headings on these second and third levels of your report is essentially a simple editorial task; however, the headings will be much appreciated by a reader in a hurry.

In addition to the fact that writers do not use enough headings, they also use weak headings. Too often they rely entirely upon a stock vocabulary of tired, descriptive terms: *Introduction, Background, Materials, Methods, Results, Conclusions, Recommendations.* These main headings of course have the virtue of familiarity. However, they provide no specific clue to content. Like a scarecrow in the middle of a field, such headings signal that something is planted there, but just what, the reader will have to find out for himself. How much more helpful it would be to use headings such as any of the following to replace or supplement the main descriptive headings:

The Role of Nitric Oxide in Smog Formation
Why Did the Prototype Fail?
Basis I: Same Cubic Volume of Cargo in Fewer Containers
Failure to Rate Vendors Effectively
Why Study the Exercise ECG?
Basic Form of the S/3 Modem—Interface
Benefits and Costs to Merchants
Procedure for Collecting Phytoplankton
Formulation of Speed-Distance Curves
Eliminating the Shock Waves

Headings such as these are particular; they give information. That is, they have to do with what is contained in a particular segment of a particular report. Unlike such terms as *Background* or *Costs,* they cannot be transferred unchanged to another report. It is this very specificity that makes them useful to a reader. Although descriptive headings will sometimes be necessary to sketch out the basic architecture of a report, as much as possible use informative headings and carry them down to third and even fourth levels of division. (Examples of effective use of headings appear in the tables of contents illustrated earlier, p. 69, pp. 101–102, p. 106.)

Transitional Elements

Headings are especially helpful for the report reader who skims. Transitional elements are helpful for the reader who reads the entire report.

Immersed in the details of the discussion, the reader who reads all of the report can easily lose sight of its architecture. He is not sufficiently helped by the headings because for him they are separated by pages of text which he reads slowly and carefully. He needs to be guided along as he moves from segment to segment. Of course, he may be able to make sense of the report without transitional elements, but their absence will force him to flip back to previous overview and forecast segments. Delaying the reader in this way is inefficient. He needs to be guided along by transitional elements.

A transitional element performs two functions. First, it "rounds out" the segment that has gone before, and signals that the segment is finished by restating the core idea. In brief units this can be done with a single sentence. For example, "We can see, then, that there is no chance for getting meaningful results from this test series." Here the word "then" has a conclusive, final quality that tells you this sentence, a restatement of the core idea, is the last in the unit. In longer units restating the core idea can be done with a brief paragraph. Here is one example you have already seen:

> On the basis of our review of the three available grant programs for which Berrien County might qualify, we conclude that the county should proceed under the assumption that no grant support is available. The county should nevertheless monitor grant programs and modify its financing to the extent that grants actually become available prior to construction.

And here is another:

> 8.4. In sum, the evidence indicates no salt underlies the Fermi site. The evidence consists of close visual inspection of the cores taken from the upper part of the Salina for Detroit Edison, four recent reports based on a detailed study of carefully selected well-logs and cuttings, and all the older published reports each based on only a few rough drillers' logs.

As you can see from these examples, this transitional function is essentially the same restatement function we discussed at length in chapter seven.

The second function of the transitional element is a forecast function. The forecast function reestablishes the larger pattern in which the segments exist. (The restatement function is internal within a segment, and completes the structure of the segment.) The forecast function reestablishes the pattern by making explicit the reasoning behind the arrangement. This transitional function explains why the segment to follow comes next. Ordinarily the transitional function can be performed very simply at the end of a segment. (Remember that the next segment will open with a forecast of its own.) Something like this will suffice: "We have seen what causes the interference; now we will see how the interference can be eliminated." Or this, "Before examining the effectiveness of the vendor quality index, I must explain when it will be used." Or this:

> We have begun the series of tests outlined above. The testing is incomplete but has given some promising results. The results of the initial tests are the topic of the next section.

As you can see in these examples, these transitional touches are not absolutely necessary. Although the reader could decipher the arrangement for himself, by being explicit you relieve him of that necessity.

Numbering

Like headings, this device signals segmentation. Like transitional elements it makes explicit the pattern of arrangement in a report. Yet like both devices, numbering is a device used by report writers either in a limited way or not at all. We believe that because it is an easy device to use and an extremely effective one, you should routinely use it in most reports.

Numbering is an easy way to signal segmentation and parallel or hierarchical relationships in a report. For example, even without headings, the segmentation and parallel relationships between these two paragraphs (from a larger unit of the report) are clear:

> 8.2.1. K. K. Landes, in the text accompanying the U.S. Geological Survey Oil and Gas Investigations Preliminary Map 40 (1945, The Salina and Bass Islands Rocks in the

Michigan Basin) says of the Salina salt beds, "The F and lower salts disappear in southern Wayne County . . . The D salt is not present in the Macomb County well, and it disappears with the higher and lower salts in southern Wayne County . . . . The B salt, like the D and F salt, ends abruptly in southern Wayne County."

8.2.2. C. S. Evans, in a paper entitled, "Underground Hunting in the Silurian of Southwestern . . ."

Numbering is also an effective way to signal hierarchical relationships among segments of a report. At a glance a reader can find the controlling and important ideas. In the example above, under the number 8 he will find the core idea. Under the numbers 8.1., 8.2., 8.3., and 8.4., he will find the main supporting ideas. Under the numbers 8.2.1., 8.2.2., and so on he will find the details. If he is in a hurry, numbering tells him which units he may safely skip and which he must read.

In addition, numbering precisely indexes the segments. The paragraphs in the example above were not indexed in the table of contents; however, the heading for the segment containing the paragraphs was indexed: "8. *Lack of evidence of rock salt deposits.*" A reader can locate, or relocate, this particular segment and even those particular paragraphs very quickly. In this case the fact that the paragraphs are not in the table of contents is irrelevant.

In the example above, notice that the writer used an arabic numbering system (8., 8.1., 8.1.1.) rather than the traditional roman numeral, capital letter, arabic numeral outline system (VIII, A, 1). If you do not already use the all-arabic system, we suggest you begin. The system is increasingly common and is required in most federal government reports. The information retrieval potential of this system insures that it will eventually replace the traditional system.

White Space

If you have ever seen a speaker quiet a noisy audience simply by stepping up to the podium and not saying a word, you know the effect that saying nothing can sometimes have. In a room full of noise, the speaker's silence stands out as different and therefore attracts your attention. The fact that he is quiet signals something; he is either about to begin or about to pass from one unit to another. So you pay attention. Silence is a communication signal to which all audiences respond.

"Silence" has the same effect in writing. In a page full of print, a block of unprinted lines—white space—stands out immediately. The reader knows that the space signals something. It can tell him that one division, segment, unit, or paragraph is ending and another beginning. Or it can tell him the writer is moving from one level of importance to another.

White space used to signal movement from one division to another is vertical white space; that is, unprinted lines are used according to some consistent spacing pattern. This pattern should reflect the hierarchy of relationships represented by the all-arabic numbering system discussed above. The simplest illustration of this pattern is the double spacing usually found between paragraphs of single-spaced text. Figure 9-1, *Table for vertical spacing in formal reports*, illustrates two patterns for using white space, one for single-spaced text, one for double-spaced text. Understand of course that these are arbitrary patterns which may be modified as situations and company formats demand. You should, however, prepare your own table for vertical spacing if this table of ours is not appropriate. You should also prepare a similar table for informal reports.

White space used to signal movement from one level of importance to another is

	Vertical Spacing For Single-Spaced Text	Vertical Spacing For Double-Spaced Text
To signal divisions between the two report components, the segments of the opening component, and the segments of the appendices:	New Page	New Page
To signal divisions between segments of the discussion (e.g., Background, Materials, Results):	8	10
To signal divisions between units within a segment (e.g., 3.1, Plant Noise Levels, and 3.2, Sources of Noise):	4	6
To signal divisions within a unit (e.g., 4.3.1, Stress Measurements, and 4.3.2, Density Measurements):	3	4
To signal all other divisions (e.g., between paragraphs):	2	2

Figure 9–1. Table for vertical spacing in formal reports.

horizontal white space; that is, block indentation is used according to some consistent pattern. Like numbering, horizontal white space signals the relative importance of material within a segment. Figure 9-2, *Horizontal white space used within a segment to signal subordinate relationships,* illustrates this use. Notice that this writer uses only horizontal white space; that is, he has used only double spacing between paragraphs, units, and segments. The two patterns for using white space can of course be combined.

We have said that white space must be used according to a consistent pattern. More importantly white space must be used meaningfully. A page of paragraph fragments is no easier to read than a page black with type. Use white space but use it carefully and deliberately.

Headings, transitional elements, numbering, and white space are four devices you can use to signal the intellectual design of your report. (Of course, these devices are not a cosmetic that will hide fundamental ugliness.) If your report is well designed, without these devices your report will be no less structured. However, these four devices will make the design apparent to your readers.

Before we move on to discussing visual aids, the next device you can use to help make your report clear to your readers, we need to add one more observation about the four design signals we have just been discussing. We have discussed them separately, but in fact these signals are ordinarily used together, as some of our examples illustrate. Because two of the signals are verbal (headings and transitional elements) and two are "mute" (numbering and white space), the four

8. **Lack of evidence of rock salt deposits**

Our study indicates that there is no salt present in the Silurian Salina Formation which underlies the Enrico Fermi site. This conclusion is supported by visual inspection of cores taken from the upper part of the Salina Formation at the site, by published reports on the Salina in Michigan and Ontario, and by well-logs and drillers' reports from the area.

8.1. First, on the basis of a visual inspection of the core from the several deep borings made for Detroit Edison both at the Fermi site and the Monroe site I can report that no salt was encountered in that portion of the Salina penetrated (approximately 210 feet at Fermi and 150 feet at Monroe). As the Salina . . .

8.2. Second, the published literature on the Salina in Michigan and Ontario indicates that no salt is present at the Fermi site. Statements from four of the most recent such reports should verify this finding.

8.2.1. K. K. Landes, in the text accompanying the U.S. Geological Survey Oil and Gas Investigations Preliminary Map 40 (1945, The Salina and Bass Islands Rocks in the Michigan Basin) says of the Salina salt beds, "The F and lower salts . . .

8.3. Third, well-logs and drillers' reports do not indicate that salt may be present at the Fermi site. A careful reading of over 100 well-logs for the Monroe County area turned up only two in which salt was even mentioned, and in neither case is the evidence persuasive that salt might be present at the Fermi site. The one report mentioned salt at Milan, where one might expect salt to be present. The other mentioned salt near Lambertville; however, the latter report, not backed up by samples reviewed by a geologist, is perhaps questionable.

8.3.1. The first mention of salt in the logs reviewed was at Milan, Michigan, first reported on by A. C. Lane in the Michigan Geological Survey Annual Reports for 1901 and 1903. The log for this well, based on samples showed 5 feet of rock . . .

Figure 9–2. Horizontal white space used within a segment to signal subordinate relationships.

signals effectively complement one another. In other words, they are redundant devices which you should use simultaneously.

COMPLEMENT THE TEXT WITH VISUAL AIDS

You should complement and clarify your text with liberally used, well placed, and well designed visual aids.

The first thing to say about visual aids is that most report writers do not use enough of them. Underestimating the worth of visual aids, and leaving them last, many writers depend entirely too much on prose—often torturously long-winded—

to create word pictures of things and ideas that could be more easily and more effectively represented by a visual aid. Take this passage of market analysis, for example:

> Analysis of the size of selected segments of the markets for gauging and assembly equipment in 1969 indicates a potential sales of one hundred and eight million dollars. Of these projected sales, coarse assembly machines will account for sixty million dollars, precision assembly machines will account for twenty million, automatic gauging machines for ten million, automatic soldering equipment for ten million, and In- and Post-Processing gauging components for eight million.

The prose in the example conveys a vague picture of the potential markets. Certainly we understand the numbers of dollars and their sources, but the explicit comparison is lost in the writer's linear prose presentation of the comparison. A much more effective comparison might have been made in graphic form. For example, with a few words of introduction the following revised version seems much more effective than the prose version:

As the following figure makes evident, coarse assembly machines offer us the largest potential for sales in 1969:

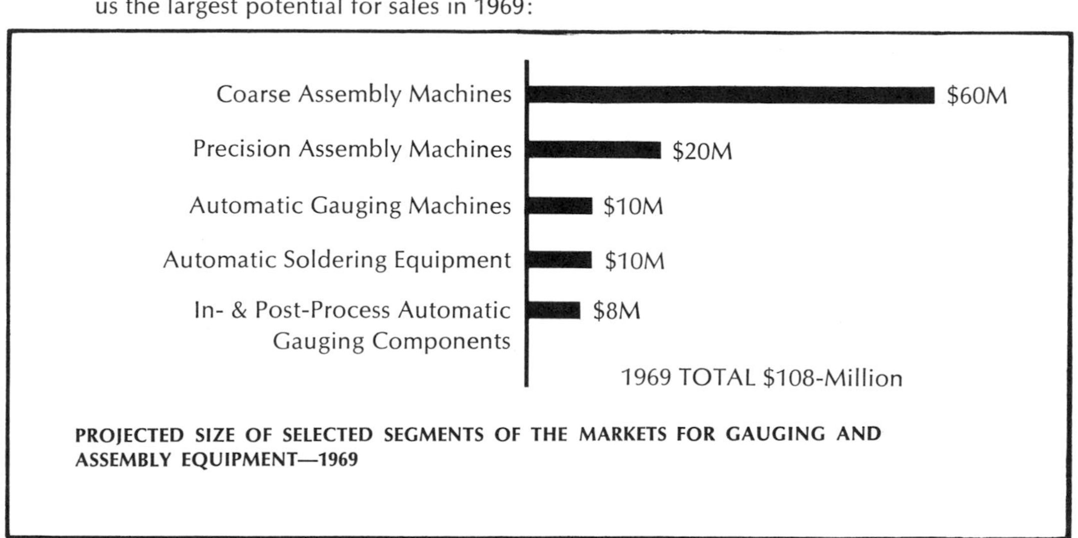

PROJECTED SIZE OF SELECTED SEGMENTS OF THE MARKETS FOR GAUGING AND ASSEMBLY EQUIPMENT—1969

Figure 9–3. Example of visual aid used to achieve economy and clarity.

The prose version is over seventy words long, but it is less explicit and contains less information than the visual representation of the same information. That is a fact novice writers often tend to overlook.

The example above is simple and obvious. In any report, however, there are a number of instances in which a sketch, a cut-away drawing, a photograph, a flow chart, a bar graph, or a table will be far more economical and exact than words. Some instances we have already mentioned. In chapter seven, "Arranging Report Segments and Units," for example, we suggested that both descriptive segments and process segments can often be made more efficient with a visual component. Both require a reader to visualize things, physical objects arrayed in space—something more easily done with pictures than with words. But those are just two of many instances in which a visual aid might help. In the example above, the figure represents a comparison; however, it could just as easily have represented an effect, a bit of persuasive evidence, or an analysis. There will be times when pictures are more efficient than words in any of these patterns. And that is the time to put in the visual aid. Here, for example, is a figure from a cause/effect segment in which a visual aid seems the optimum way to clarify an effect:

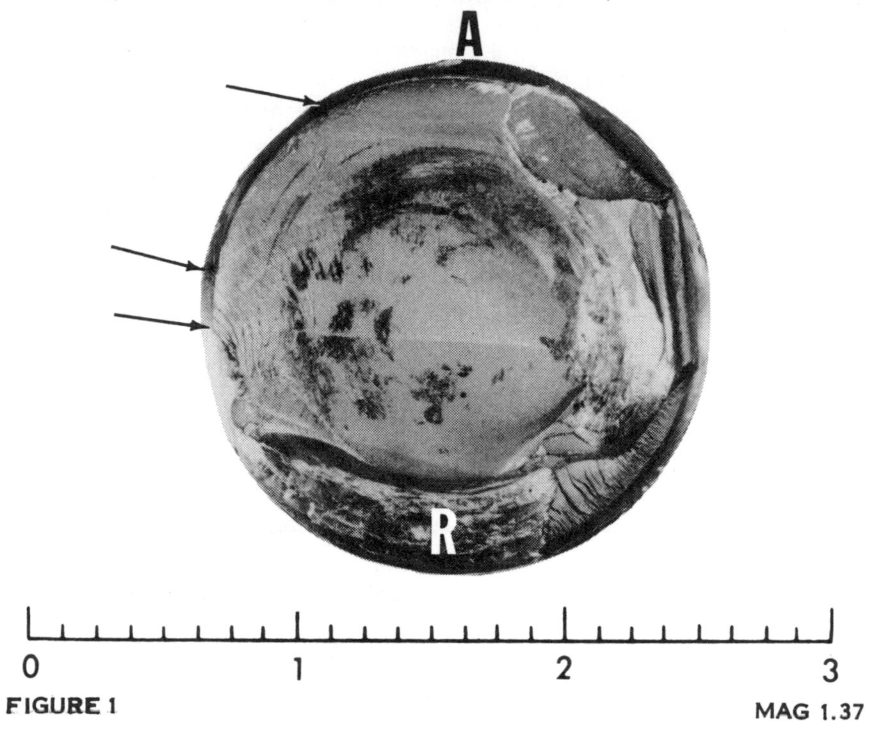

Figure 9–4. Visual aid used to communicate an effect clearly.

In this case the effect could not be made more specific or vivid with words.

Visual aids are often appropriate for use in the text of reports, and you should watch for those instances. But remember, your objective is not to make your task easier by substituting pictures for words. We are not saying, "put in a visual aid when it is easier for you than writing text." Instead we are saying, "use visual aids to make your text clear and economical." If anything, visual aids may require extra work of you, but because they communicate efficiently, they should be used much more often than many report writers tend to use them.

A moment ago we said that the first comment to be made about visual aids is that most report writers do not use enough of them. The second comment to be made is that many report writers do not locate visual aids effectively even when they do use them.

Many reports place visual aids ineffectively. The discussion forms a bloc of prose almost entirely uninterrupted by visual aids. That is, page after page continues without a chart, a sketch, a flow diagram, or anything visual. Scattered throughout this bloc of prose, however, are numerous notes in parentheses—see Figure 2; see Table 4; and so on—referring the reader to the appendices, which serve as a dumping ground for all of the explanatory charts, tables, and figures. Consequently the appendix consists almost entirely of visual aids: page after page with scarcely a line of prose. This arrangement is as absurd as an illustrated magazine with all the pictures at the end.

Watch a reader flipping through a report and you can see the problems with such an arrangement. Trying to mark his place with one hand, he awkwardly flips through the appendix with the other, hunting for Figure 2. Once he finds it, he flips back to the text to reread a line or two and refresh his memory on exactly what he was hunting for. Now back to the appendix again for a closer look at the figure. And back again to the text, trying to put the illustrations and the text together. This is a cumbersome and wasteful procedure which the writer should not force upon the readers.

As a writer, therefore, you must determine which visual materials to put in the appendix and which to put in the text. Those visual aids which are likely to be necessary or even very useful to the readers should be inserted in the text. Do not send readers scurrying off to the appendices with a "see Figure x." Instead, give them those important visual aids at precisely the places they will need or appreciate visual aid. Figures 9-5 and 9-6, *Visual aid necessarily integrated into the text* and *Visual aid supplementing the text to make the reader's task easier,* illustrate these two usages.

(A) A Transistor Curve Tracer Circuit.

To generate a family of transistor curves, one must vary two signals simultaneously while measuring a third. For each curve drawn, the I_b is constant and the Vcc supply from the first D/A converter varied over its entire range while Vce is measured from the A/D converter. Subtracting these two values and dividing by the resistance gives the current I_c. For a family of curves, then, I_b is stepped through a series of values and Vcc varied for each value. This I_b signal is derived from the second D/A converter. The completed test circuit is shown in Figure 3.

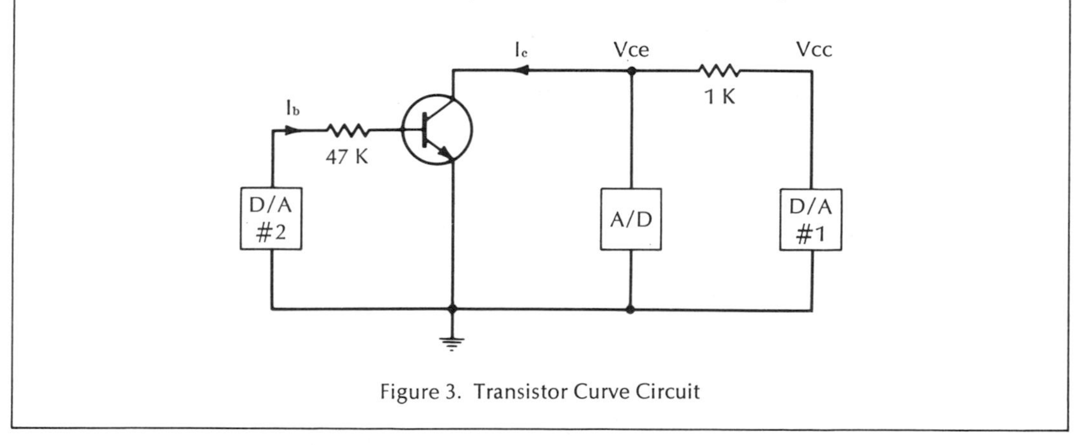

Figure 3. Transistor Curve Circuit

Figure 9–5. Visual aid necessarily integrated into the text.

In the case of Figure 9-5, the visual aid is one that the reader would almost certainly want. If it were in the appendices he would probably stop to look it up; since it is in the text he can continue reading uninterrupted. In the case of Figure 9-6, the sketch of the seventeen state area is not really necessary, but most readers would find it useful. If, on the one hand, the sketch were in the appendix, almost no one would bother to look it up. If, on the other hand, the writer eliminated the sketch he would be forced to list the seventeen states covered by the investigation. Neither solution is as convenient for the writer—or for the reader—as this simple sketch in the text.

Understand, of course, that there is no fool-proof rule for knowing when a visual aid should be put in the text and when it should be in the appendices. You just have

THE STUDY

During the summer of 1966, the Forest Service in cooperation with Doane Agriculture Service conducted a study of farm building construction activities for the years 1963 through 1965. The study included a sample of commercial farms[3] in the Central and Appalachian Regions of the United States (fig. 1). Questionnaires were mailed to all 1,600 members of the Doane Farm Panel, which is carefully selected by stratified sampling techniques to represent the commercial farm market. The panel is periodically checked and rebalanced on the basis of the Census of Agriculture. To insure that the panel is properly maintained, annual characteristic surveys are conducted and checks of reporting accuracy are made regularly.

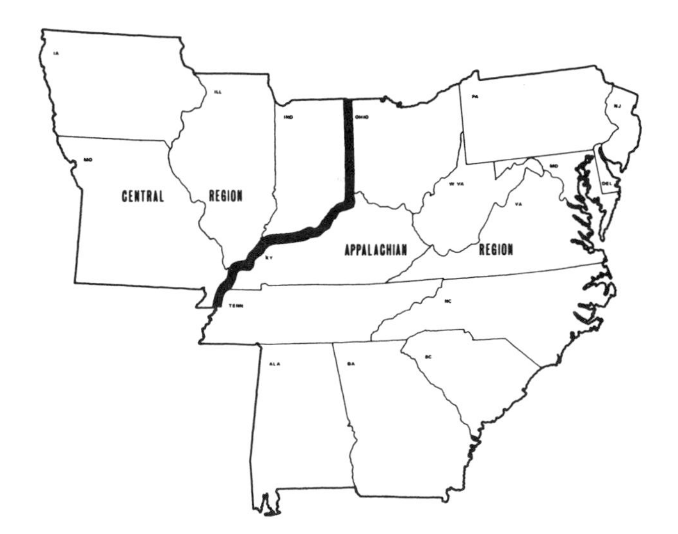

Figure 1.—Seventeen-State study area including the Central and Appalachian Regions.

Results from 1,348 usable questionnaires, obtained from farmers in the 17-State area, are presented in three general divisions: permanent buildings, farm building repair, and portable structures.

[3] *Commercial farms are defined for this study as farms having total farm products sales of $2,500 or more in 1964.*

Figure 9–6. Visual aid supplementing the text to make the reader's task easier.

to be receptive to the notion that some visual aids are necessary or useful in the text. As you design a report, before yielding to the usual "see Figure x" and consigning all visual aids to the appendices, ask yourself if the aid in some form might not be one to integrate into the text. Some of the time the answer will be "yes."

Assuming that you follow our first two suggestions on visual aids, using them liberally and placing them appropriately, our third suggestion is that to be effective your aids must be well designed. We have four guidelines for designing them.

Integrate Every Aid into the Text of the Report by Interpreting it Verbally

A visual aid must be made a part of the text, not an ornament to it. This means that a visual aid must always be verbally interpreted; that is, the reader must be explicitly told what the aid accomplishes, what he is to see in it, why it is there. Without verbal interpretation in the text, a visual aid, because it does not contain many words, is open to misinterpretation. Different viewers will perceive different details in the aid and interpret it differently. Unless you put into words what you want the reader to understand, you cannot be certain the aid will be understood as you intend. Your verbal interpretation is usually a matter of only several sentences. You must introduce and explain the aid: perhaps an introductory sentence before the aid to give the reader a notion of what to look for, then a few explanatory sentences after the aid to state what the reader should conclude. Or perhaps all of the introduction and explanation before the aid. Or perhaps introduction and brief explanation before the aid, with additional explanation after. Figures 9-7, 9-8, and

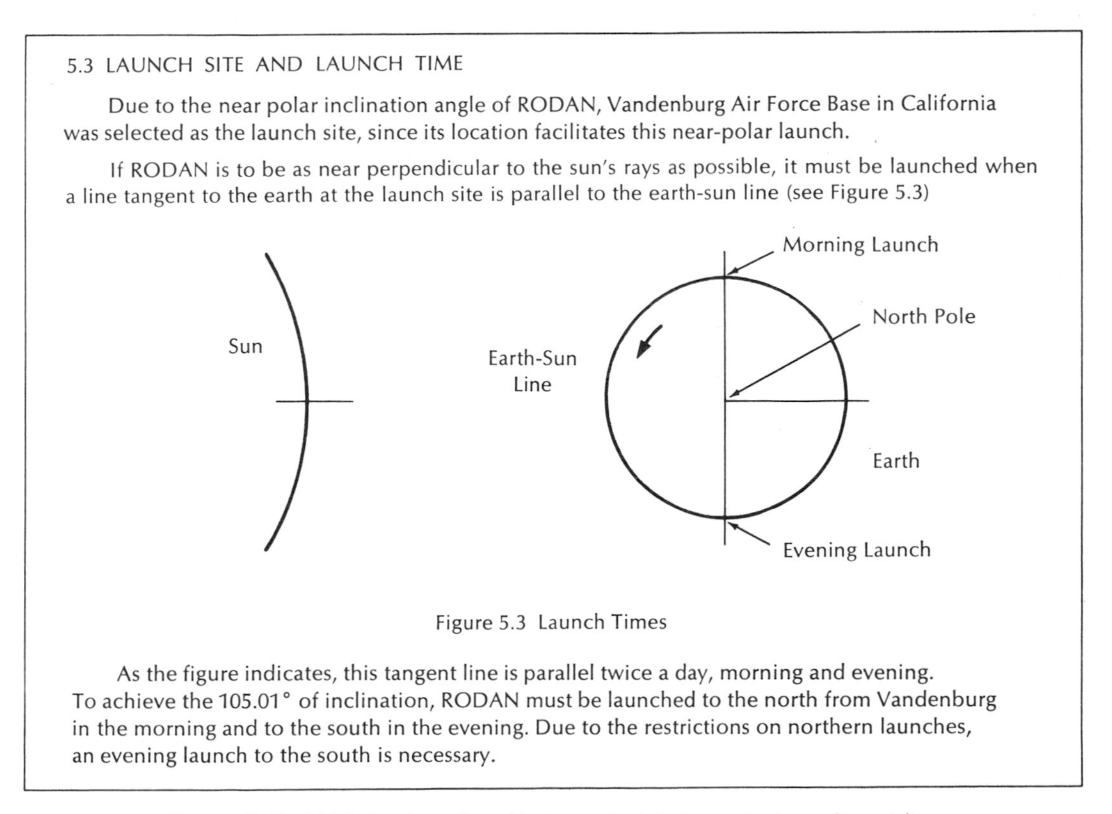
5.3 LAUNCH SITE AND LAUNCH TIME

Due to the near polar inclination angle of RODAN, Vandenburg Air Force Base in California was selected as the launch site, since its location facilitates this near-polar launch.

If RODAN is to be as near perpendicular to the sun's rays as possible, it must be launched when a line tangent to the earth at the launch site is parallel to the earth-sun line (see Figure 5.3)

Figure 5.3 Launch Times

As the figure indicates, this tangent line is parallel twice a day, morning and evening. To achieve the 105.01° of inclination, RODAN must be launched to the north from Vandenburg in the morning and to the south in the evening. Due to the restrictions on northern launches, an evening launch to the south is necessary.

Figure 9–7. Aid introduced and interpreted (interpretation after aid).

Other background variables showed that the Type II operator was notable. There was no significant difference between groups with the numbers of operators who had at one point had their license to operate a motor vehicle under suspension but legal histories showed that the Type II operator was much more likely to operate his vehicle without a valid license and be apprehended ($p < .01$) (see Table 8). There was also a non-significant trend showing that the Type II operators had a greater number of arrests for speeding.

Table 8 – Previously Charged with Driving without a License

	No	Yes	Total
Type I	67 (91%)	7 (9%)	74
Type II	27 (69%)	12 (31%)*	39
Type III	51 (86%)	8 (14%)	59
Total	145 (84%)	27 (16%)	172

*Significantly greater proportion of Type II drivers had previously been charged with driving without a license than Type I or Type III drivers.

($x^2 = 8.29$, 1 df, $p < .01$)

A factor of related interest is that the operator of the Type II vehicle had a considerably greater number of suicide attempts in his history ($p < .05$) (see Table 9). This factor was correlated significantly only with the Type I operators when compared with age ($r = 0.303$, $p < .05$).

Table 9 – Documented Suicide Attempt History

	No	Yes	Total
Type I	62 (87%)	9 (13%)	71
Type II	27 (69%)	12 (31%)*	39
Type III	52 (91%)	5 (9%)	57
Total	141 (85%)	26 (15%)	167

*Type II drivers suicide attempt history significantly greater than Type I or Type III drivers.

($x^2 = 5.34$, 1 df, $p < .05$)

The Type II operators were evaluated to have had the greatest amount of other-than-alcohol drug

Figure 9–8. Aids introduced and interpreted (interpretation before aid).

PERMANENT BUILDINGS

Number and Type of Buildings

During the 3-year study period, 247,000 new permanent farm buildings were constructed within the 17-State study area – 131,000 were built in the Central Region and 116,000 in the Appalachian Region. The survey classified 25 basic types of permanent farm buildings according to 4 basic frame construction types. Under each frame type, several types of exterior wall construction materials were listed.

In the entire 17-State area, grain storage and machine storage buildings were the most frequently constructed types, accounting for about 29 percent of all new buildings (fig. 2).

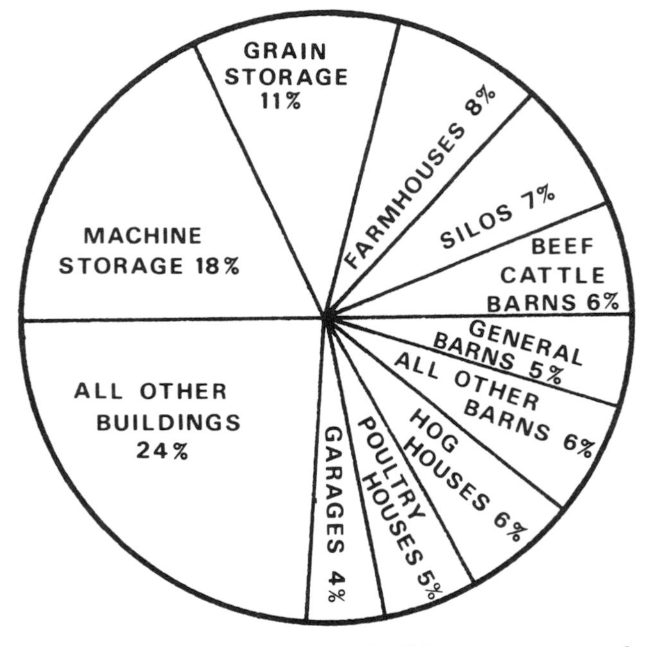

Figure 2.–Percentage of buildings in general categories.

Farmhouses were third, accounting for 8 percent of all new construction, but no other type accounted for more than 7 percent of the total buildings. Machine storage buildings accounted for high percentages in both the Central (20 percent) and Appalachian (16 percent) Regions. Grain storage buildings, however, accounted for 17 percent in the Central Region, as opposed to only 4 percent in the Appalachian Region. This difference in the number of grain storage

Figure 9–9. Aid introduced, interpreted, and further interpreted (interpretation before and after the aid).

9-9 illustrate these three ways of verbally integrating an aid into the text. Figure 9-7 places the introduction before and the interpretation after the aid. Figure 9-8 places the introduction and all the interpretation before the aid. Figure 9-9 places interpretation both before and after the aid. Notice that in each of these aids there is introduction as well as interpretation. Every aid should be introduced and interpreted. Although the methods will vary, you should verbally integrate every aid into the text.

Keep Visual Aids Appropriate and Simple

Use only visual aids which will be helpful to your audiences: an aid which may be clear to you may be entirely unclear to your readers. If you are to help your readers, you must prepare appropriate types of aids and must keep them from becoming too complex or unselective. You might even provide two versions of the same aid: a simplified version for the discussion, accessible to nontechnical audiences, and a more detailed or technically conventional version for the appendix, accessible to technical audiences.

Preparing two versions of one visual aid is a reasonable approach often used. Figures 9-10 and 9-11 show two versions of one aid from a report by a consulting firm of civil engineers. Figure 9-10 is nontechnical enough to be accessible to the Planning Commission being addressed by the report. A fully technical version of the same aid, Figure 9-11, appears in the appendix to that report. Obviously the more complex aid would baffle nontechnical readers, and would not clarify the text; however, it poses no problem for civil engineers on the Planning Commission staff, and in fact supplies the details they may well want. In providing two versions of this one aid, the report writer considered his different readers' needs.

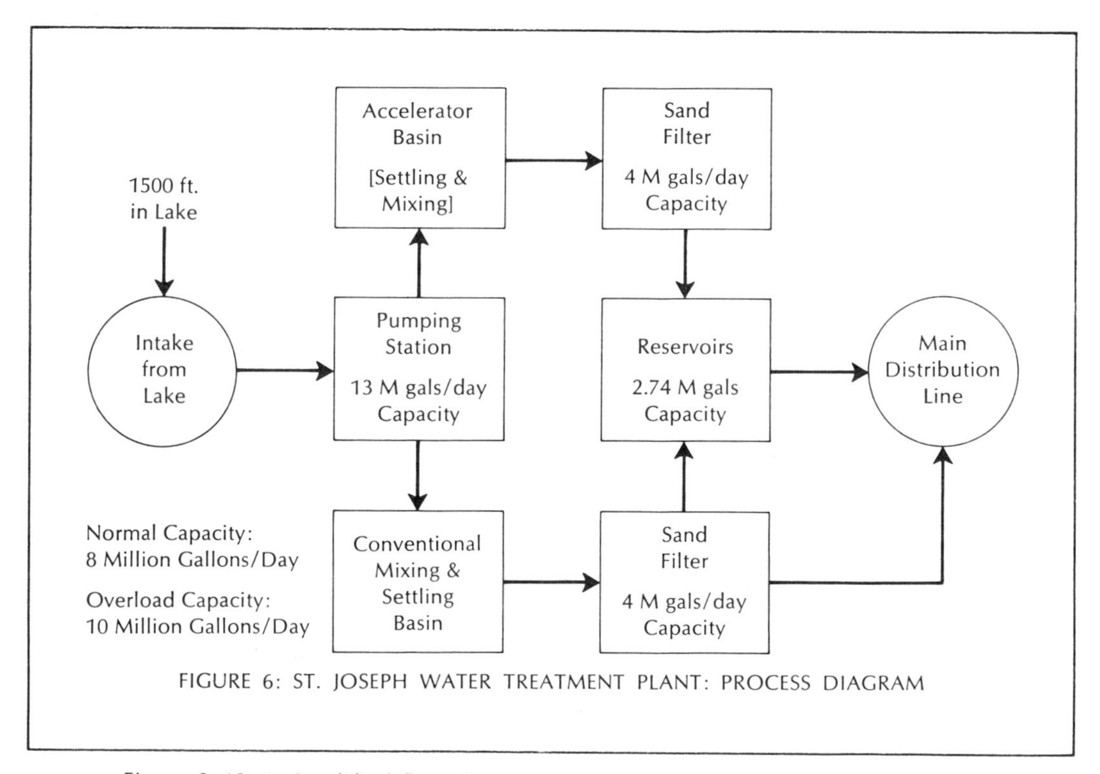

Figure 9–10. A simplified flow diagram appropriate for nontechnical audiences (as it appears in the discussion).

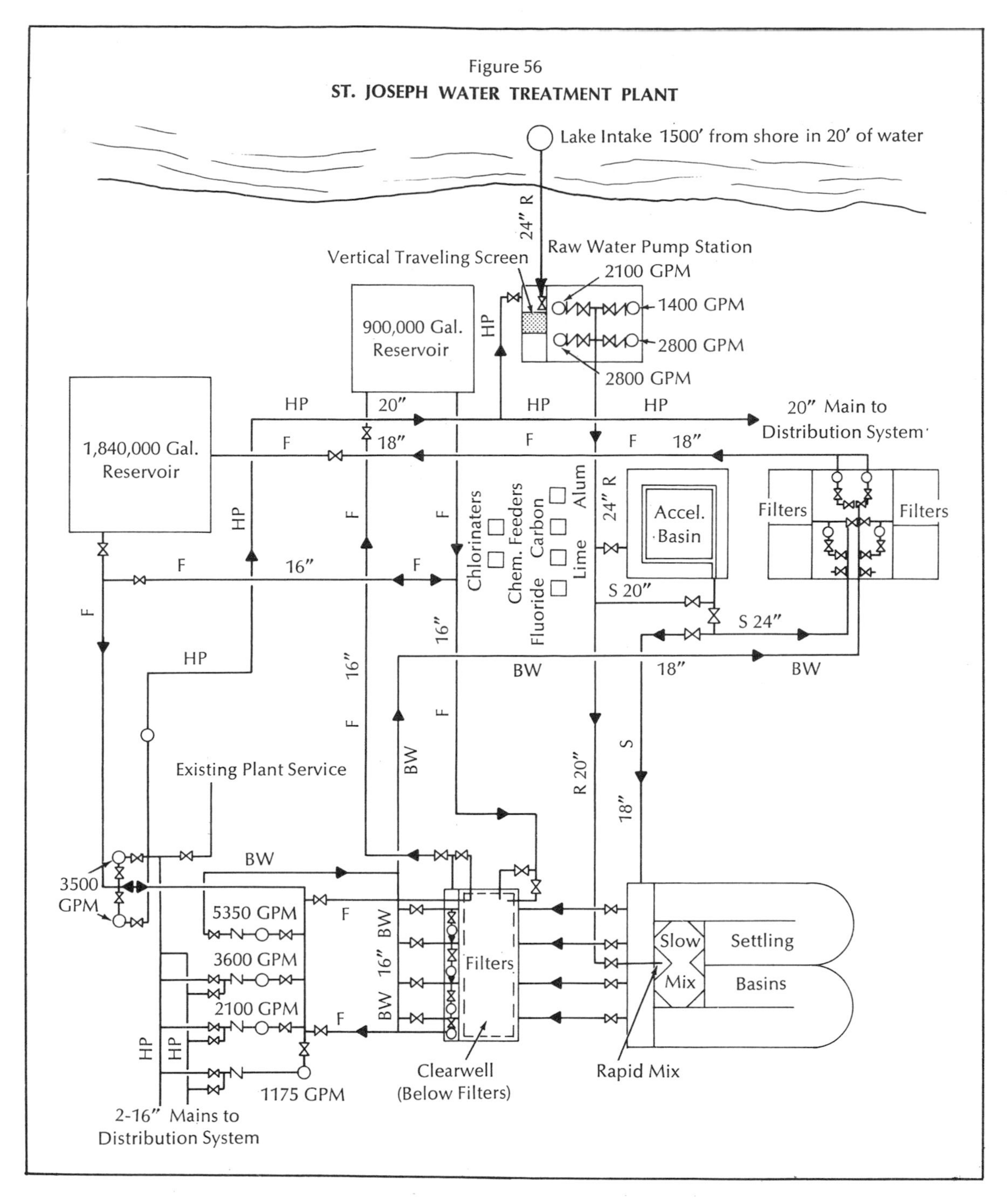

Figure 9–11. A flow diagram appropriate for technical audiences (as it appears in the appendix).

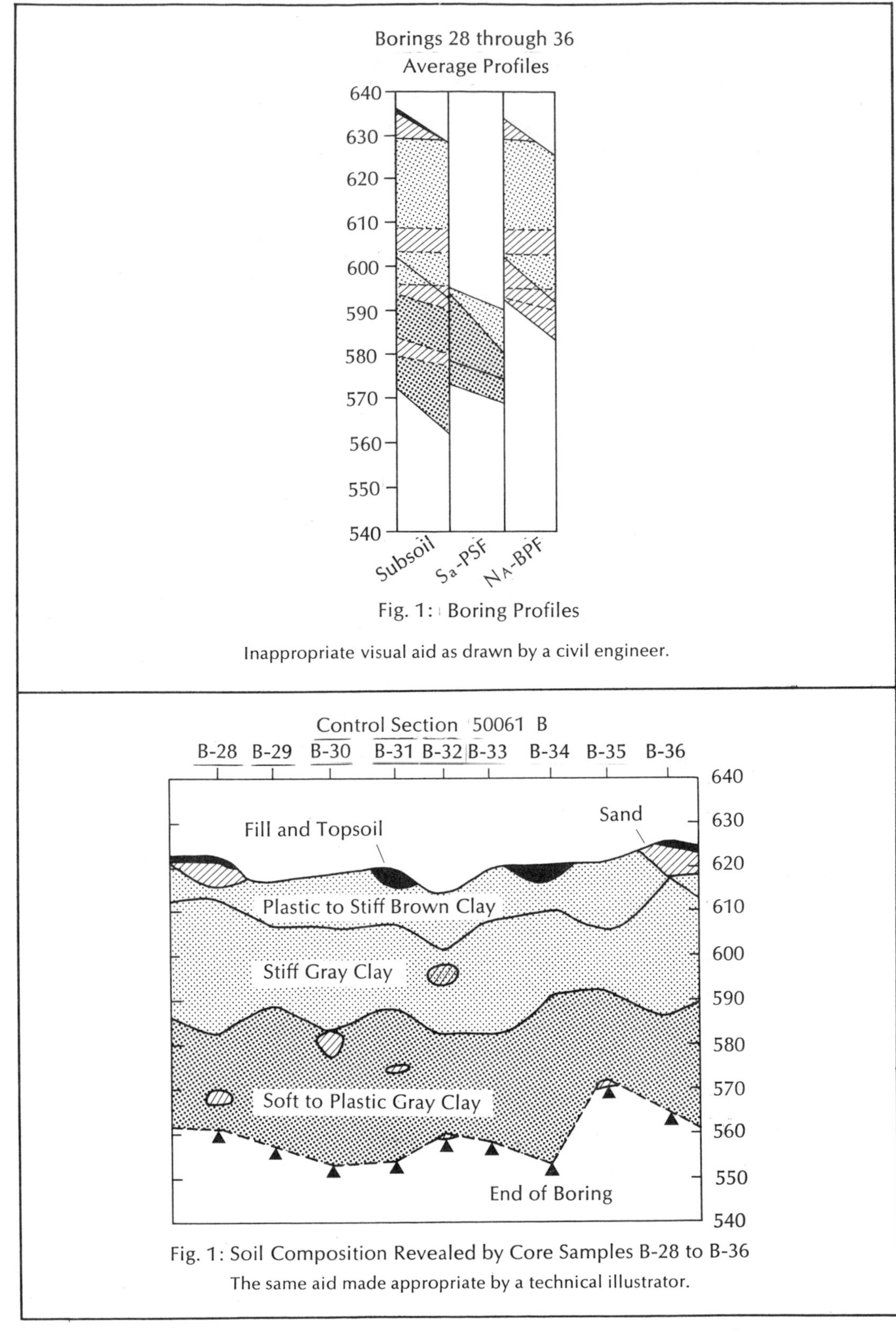

Fig. 1: Boring Profiles

Inappropriate visual aid as drawn by a civil engineer.

Fig. 1: Soil Composition Revealed by Core Samples B-28 to B-36

The same aid made appropriate by a technical illustrator.

Figure 9–12. Inappropriate and appropriate versions of an aid illustrating soil composition.

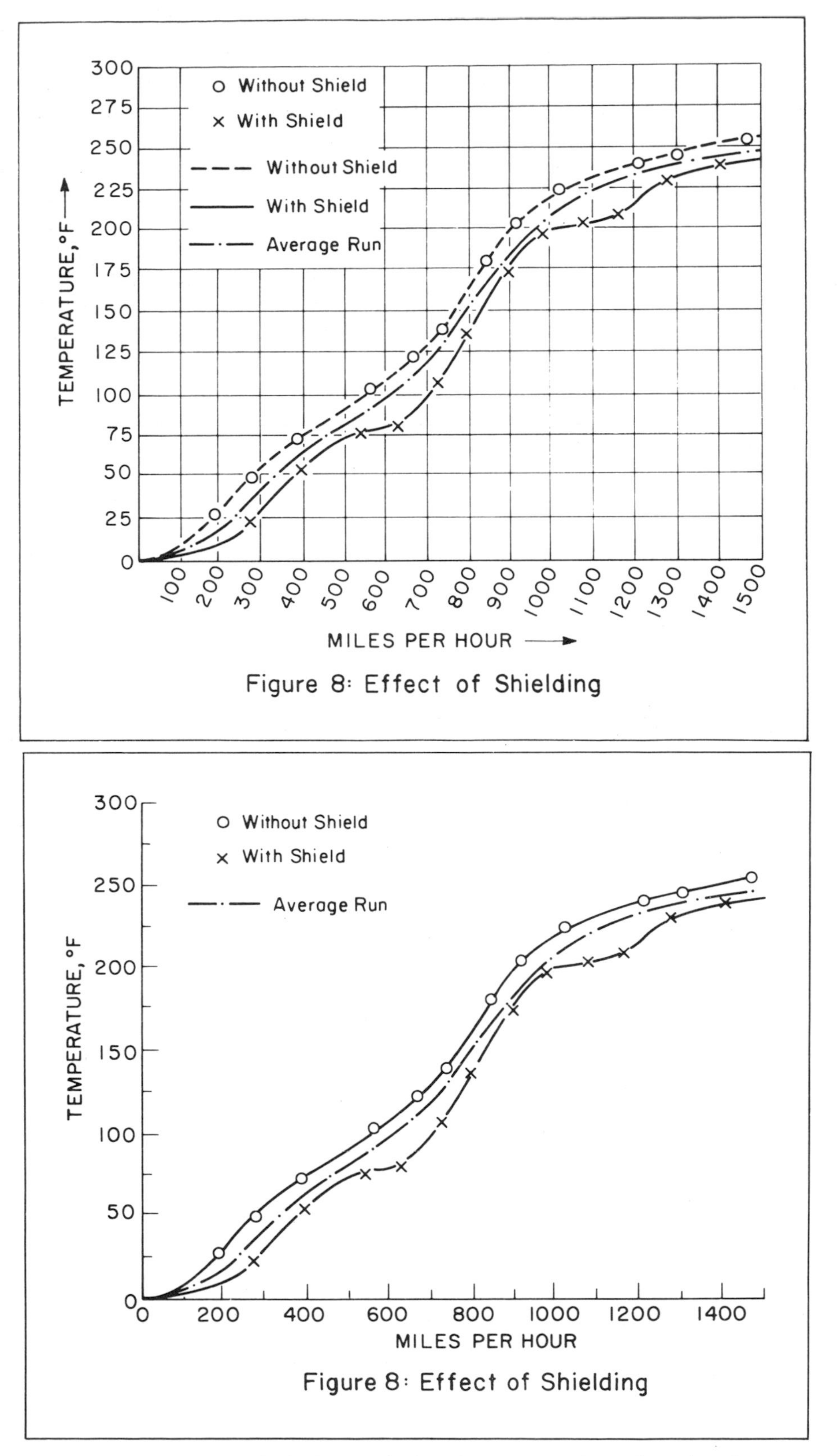

Figure 9–13. Graphs before and after elimination of needless detail.

In preparing visual aids, you should realize that methods of visual presentation which are conventional within a technical discipline may be unclear to someone outside that discipline (even other engineers). For reports addressed to diverse audiences, you may have to use unconventional and even technically imprecise methods of visual presentation to clarify the discussion. In Figure 9-12, for example, the figure at the top of the page shows a visual aid prepared by a soils engineer to illustrate the findings of core samples taken at a construction site. The figure might have been clear to him, and perhaps even to another soils engineer who might take the time to interpret the graph and compare the samples to each other. However, to most readers the figure at the bottom of the page, prepared by a technical illustrator for the report, which addressed technically diverse audiences, gives a much clearer notion of the soil composition. The revised figure, though technically imprecise and unconventional, actually is more clear for these audiences and thus more appropriate than the original version.

Often, preparing an appropriate and simple aid requires only elimination of needless detail. Figure 9-13, for example, shows two versions of the same aid. The version at the top of the page is cluttered with unnecessary detail and consequently is difficult to interpret. The relationships of the curves are obscured by the cross-hatching and labels. The simplified version at the bottom of the page eliminates the unnecessary detail. As you look at the simplified version, notice how

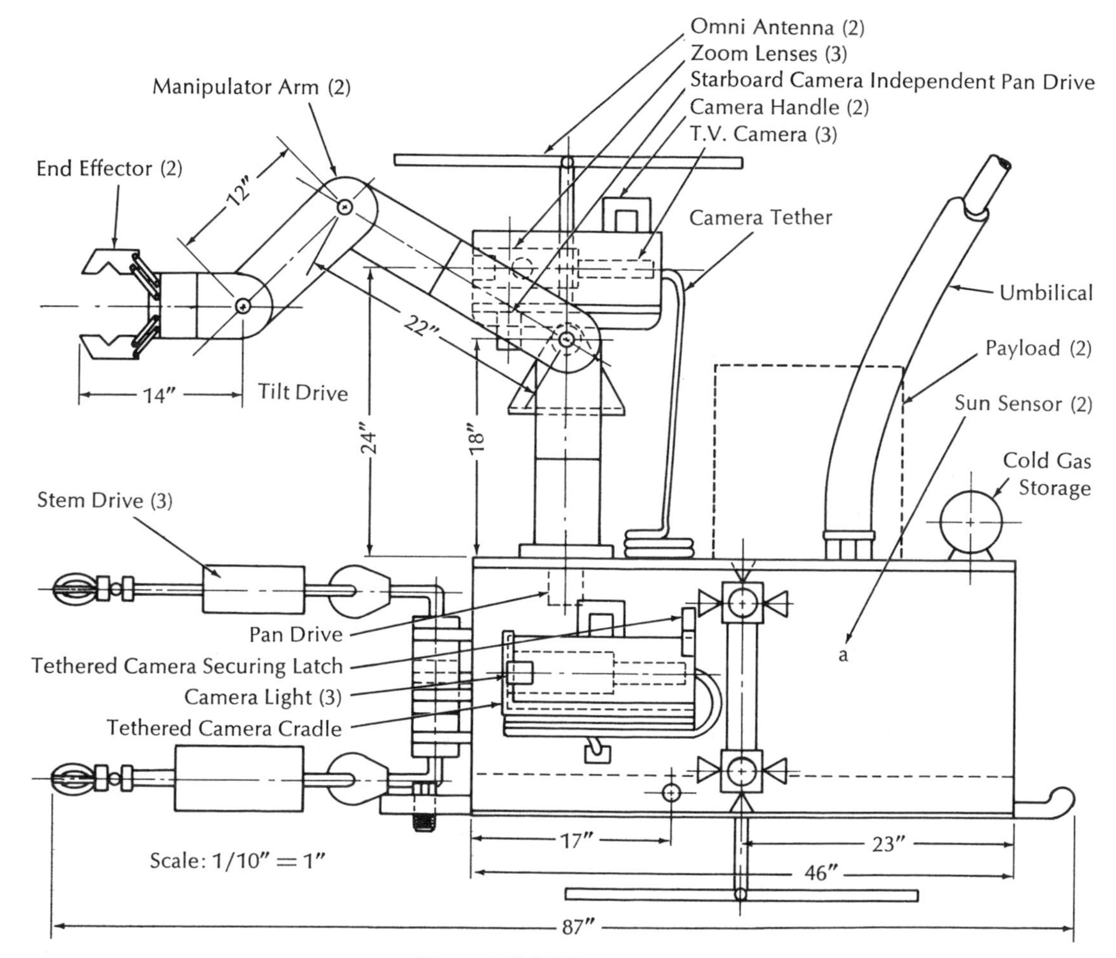

Figure 8: Side View Exterior

Figure 9–14. Visual aid inappropriate for explaining operation of a system.

your attention is focused on the curves themselves—as it should be. The version at the top is the sort an engineer prepares on graph paper to plot the curves as he performs the investigation. The version at the bottom is the sort a writer prepares by tracing as he writes the report. Had he incorporated the investigation version into the text, he would have been inconsiderate of his readers and blunted the effect of the comparison he was making.

Preparing a visual aid that is both simple and appropriate requires you to remove your engineer's hat. For example, Figure 9-14 was used by an engineer in an attempt to explain to an audience of nonspecialists how a manipulator subsystem aboard a Skylab might eliminate "EVA" tasks for astronauts. He wanted only to explain the essential features of the system. His drawing would doubtlessly satisfy him and others like him because he could translate this visual description into a functional explanation. Just by looking at the system *he* could understand essentially what it will do. But what about his audiences? Could they be expected to understand the operation of the manipulator system from this drawing? Could they be expected to ignore all the superfluous detail and see beyond the clutter of his scales, labels, and arrows? In addition to the fact that the drawing is too detailed and technically complex for his purpose, it is not even the appropriate type of aid. It would have been much more effective for him to have used a simplified isometric view of the device such as in Figure 9-15.

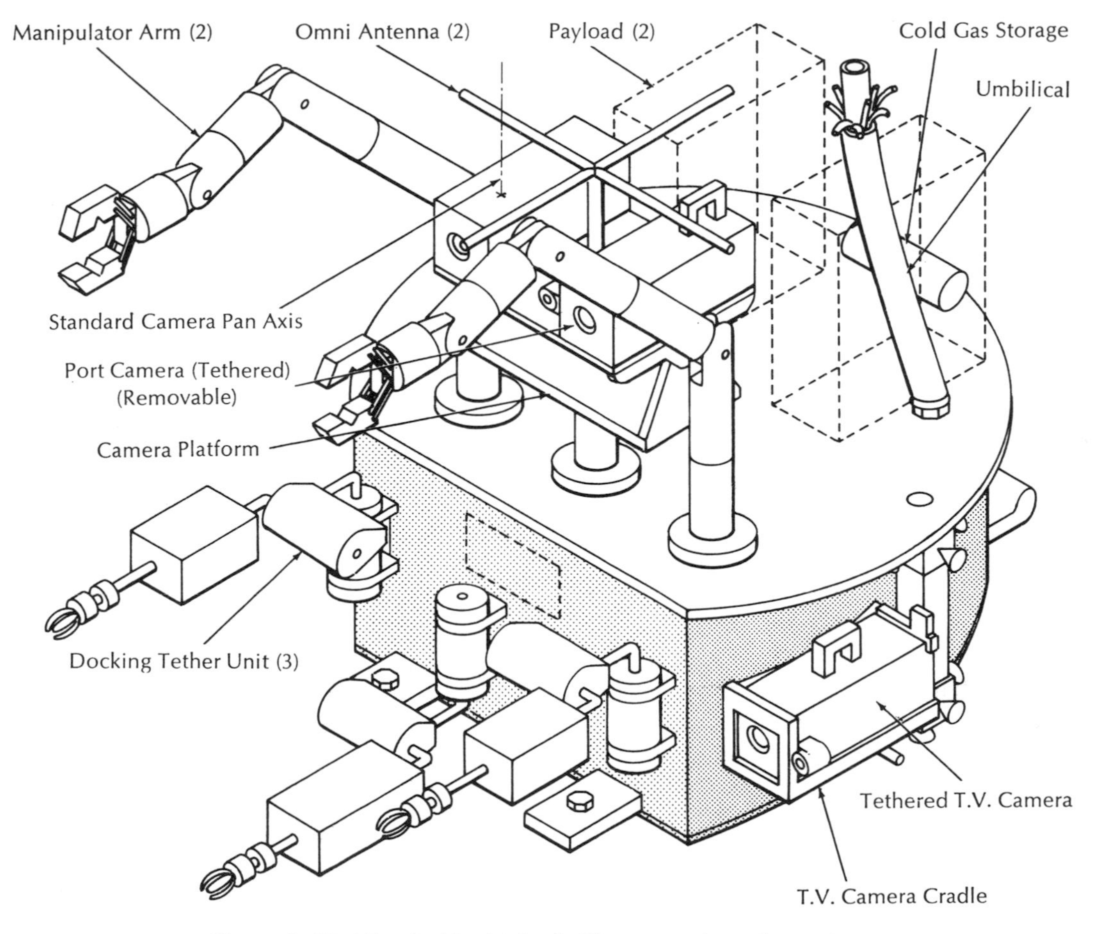

Figure 9–15. Visual aid which clarifies operation of a system.

Remember, then, to design aids effective for your particular audiences you have to think like your audiences. Sometimes that means you will have to remove your engineer's hat.

Clearly identify both the aid and its component parts with labels and a heading.

The purpose of visual aids is to increase comprehension. Without clear labels and headings, however, they become too dependent upon the text and therefore do not clarify the text. In an extreme case, an aid without labels and headings even becomes counterproductive. The following aid (Figure 9-16), for example, would require the reader to stop to reread text in order to understand the aid. Thus instead of clarifying the text, the aid must be clarified *by* the text. The aid no longer functions as a tool.

A less extreme but nonetheless very common example is Figure 9-17, which attempts to illustrate the principal steps in a squeeze casting process. Notice that on the aid neither the stages of the process nor the parts of the casting equipment are identified, although there is a heading. Thus to understand the squeeze casting process being illustrated, the reader must either turn back to the text for help or take the time to carefully examine the aid and construct his own interpretation. A reader might, of course, be able to do it, but he should not be forced to. (For this example the reader would have been forced to construct his own interpretation because the text did not even explain what the numbers "1, 2, 3, and 4" in the aid identify. In this case what "labeling" there was was less effective than no labeling at all.)

Just as it is important to identify the parts of a visual aid, it is imperative to title it. Without a heading the aid cannot stand on its own and thus is open to

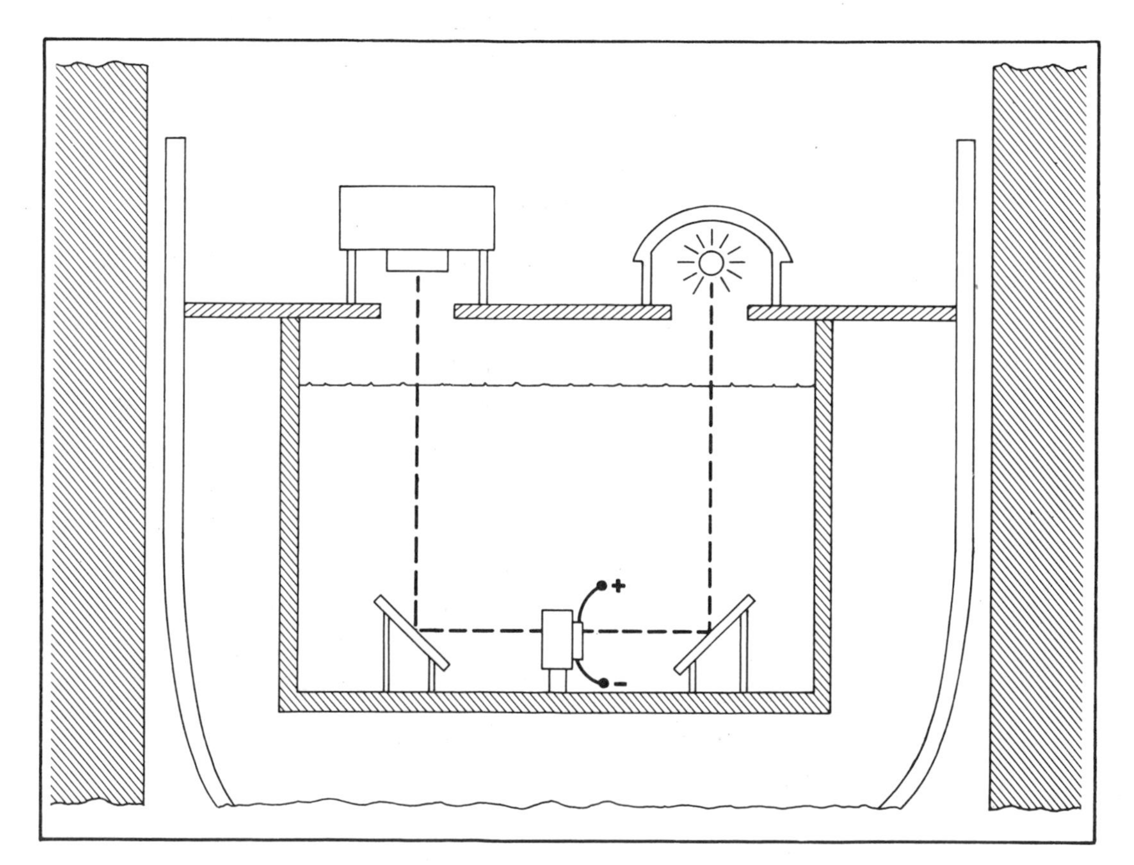

Figure 9–16. Aid without labels or a heading.

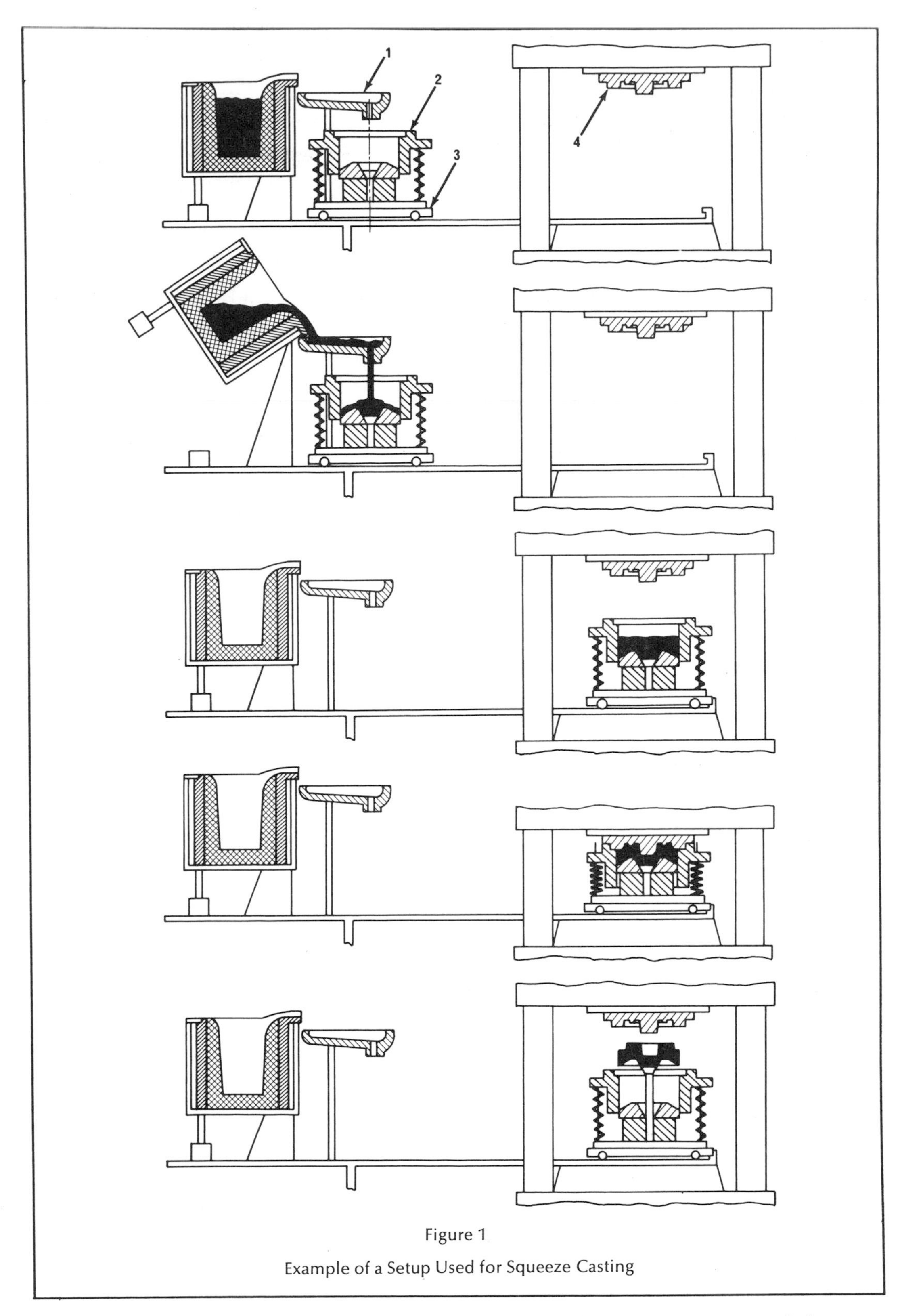

Figure 1

Example of a Setup Used for Squeeze Casting

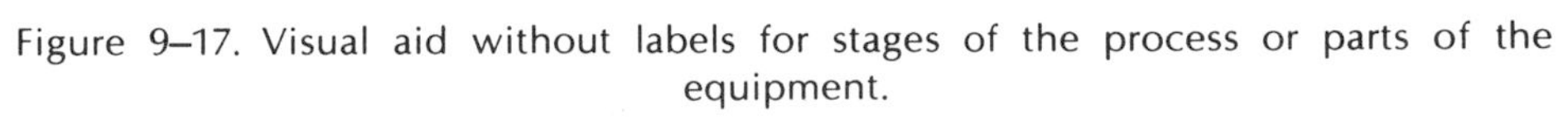

Figure 9–17. Visual aid without labels for stages of the process or parts of the equipment.

misinterpretation. Every visual aid should have an explanatory heading placed, along with the figure number, directly beneath the aid. The heading should explain the aid sufficiently for a reader to be able to understand it even without the text. Further, the heading should enable a reader to relocate a particular aid efficiently, without having to hunt back through the text of the report.

An example of an aid which, because it has no heading, is meaningless without the text, is shown in Figure 9-18. The figure shows that in twelve days something grew to approximately seven millimeters. Question: What? Butterflies? Similarly, Figure 9-19 shows that 47 per cent of those in the 70–60 age group compare to 12 per cent of those in the 29–20 age group. Questions: Per cent of what? What do the age groups represent? In the context of our discussion these examples of figures without headings and labels may seem absurd. Unfortunately, however, they are not at all uncommon. A writer, having discussed the aid in the text, tends to omit information from the aid itself. Be aware of this tendency and make a point of supplying appropriate headings and clear labels for every aid in your report. As much as possible, make the aid capable of being generally understandable without any reference to the text. Try looking at the visual aid without looking at the associated text. Would it be generally clear? If not, supply the necessary information in the aid.

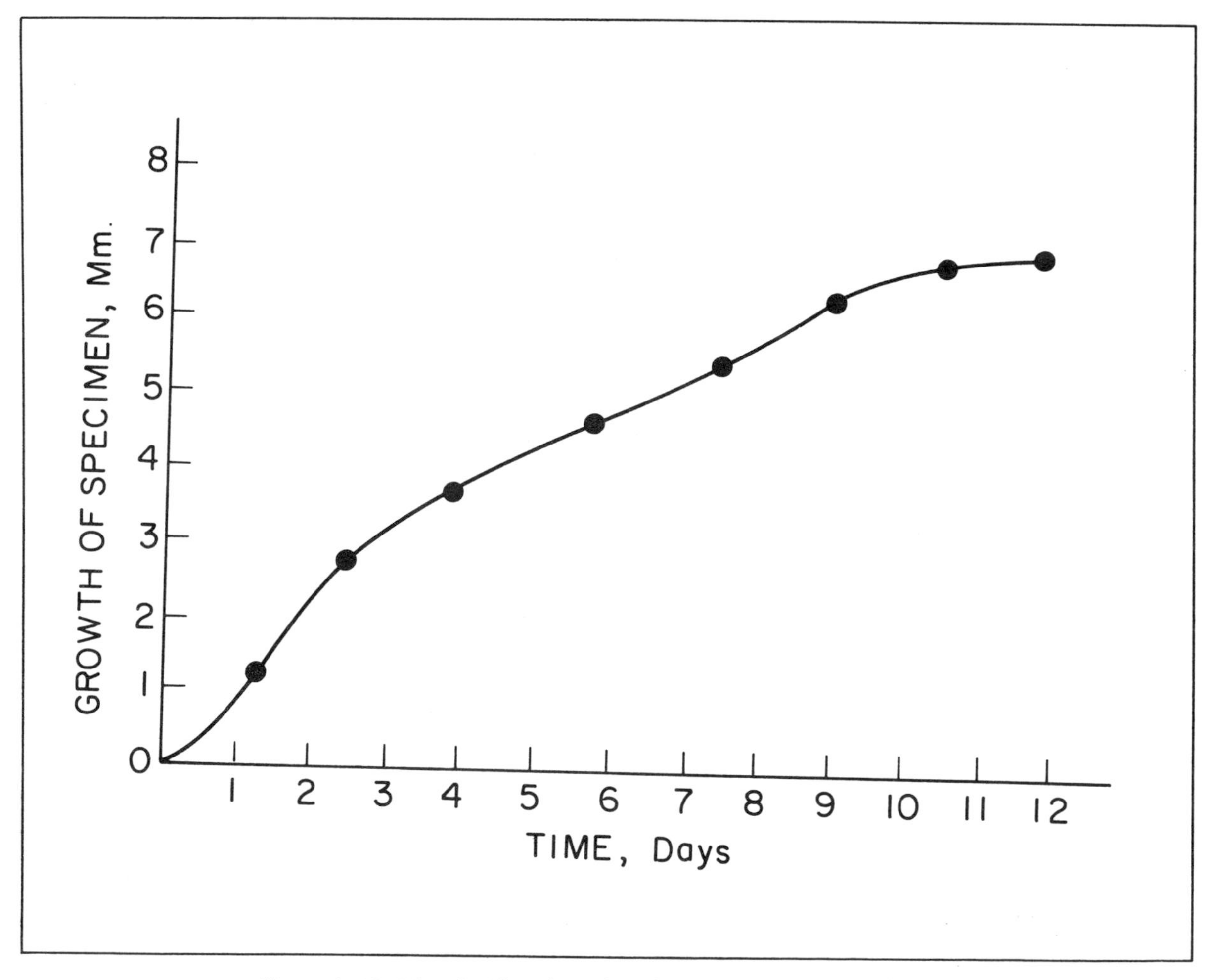

Figure 9–18. Visual aid without heading to identify the subject.

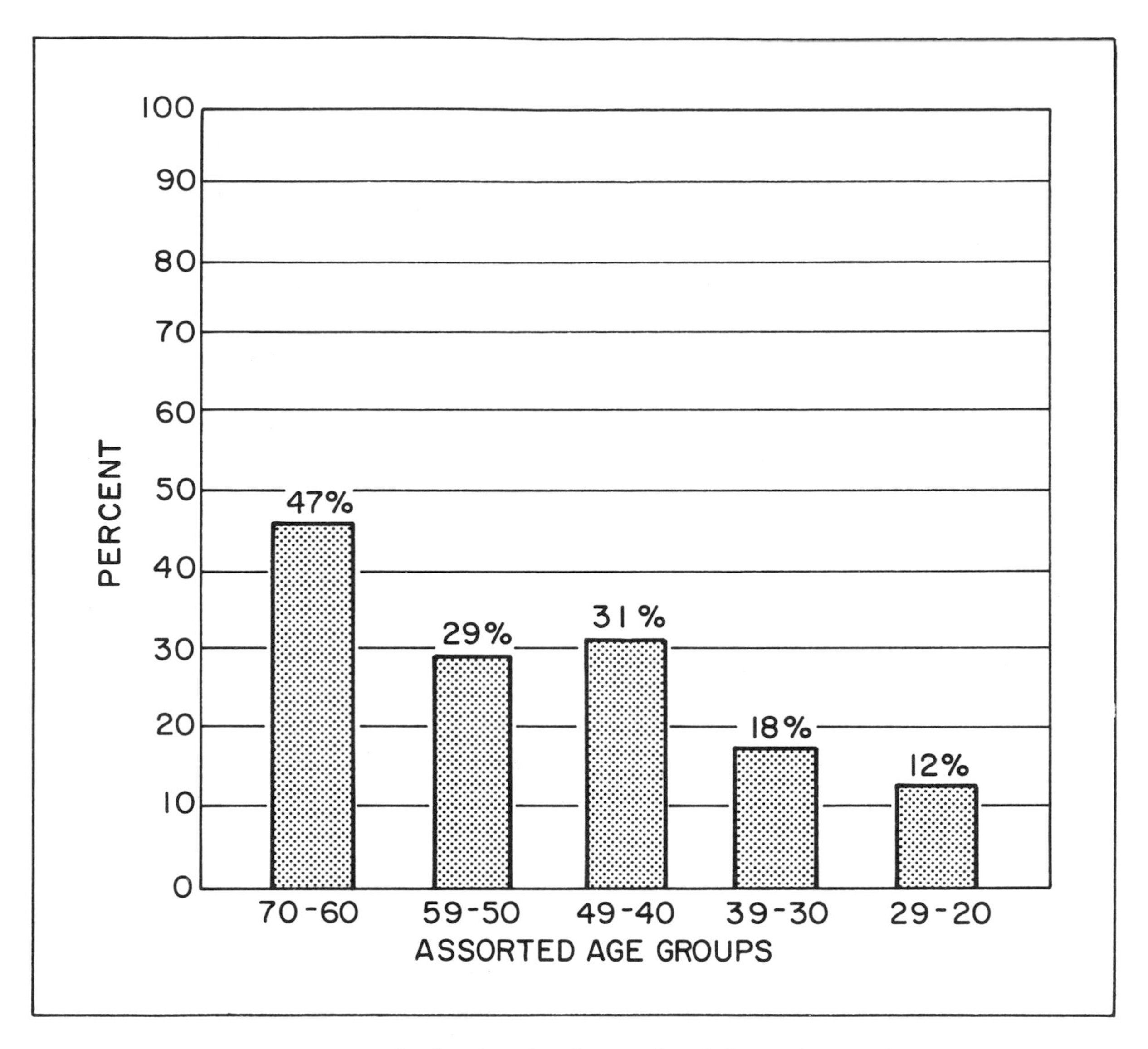

Figure 9–19. Visual aid without heading to identify basis of comparison.

Use Accurate and Appropriate Scales

When you prepare visual aids, particularly graphs, take care to present information accurately. It is surprising how easy it is to lose sight of the obvious fact that the objective of a visual aid is to clarify an idea, not to create a visual sensation. It is easy, for example, to prepare graphs with deeply plunging lines and towering bars. Graphs such as these might be appropriate for advertisement, but not for reports. While the graphs may look impressive, they can also be completely misleading. More accurate scales might make the same graphs a little less impressive, but the graphs will convey the ideas without distortion.

The most common sort of unintentional distortion comes from the device of the "suppressed zero." On the vertical axis in the following figure (Figure 9-20), for example, the scale logically would start at zero. However, to economize, because he wanted to put several graphs on one page, the writer who prepared the drawing started the scale at 10 per cent and went up only to 60 per cent. Thus visually, the graph exaggerates the thermal efficiency achieved at higher compression ratios (the curve peaks at the top of the scale, suggesting perfect thermal efficiency). The redrawn figure, Figure 9-21, indicates how misleading this visual impression of higher thermal efficiency is. If it is inappropriate to carry one of the axes to zero, indicate that you have deliberately truncated the scale. Use a broken line, as illustrated in Figure 9-22.

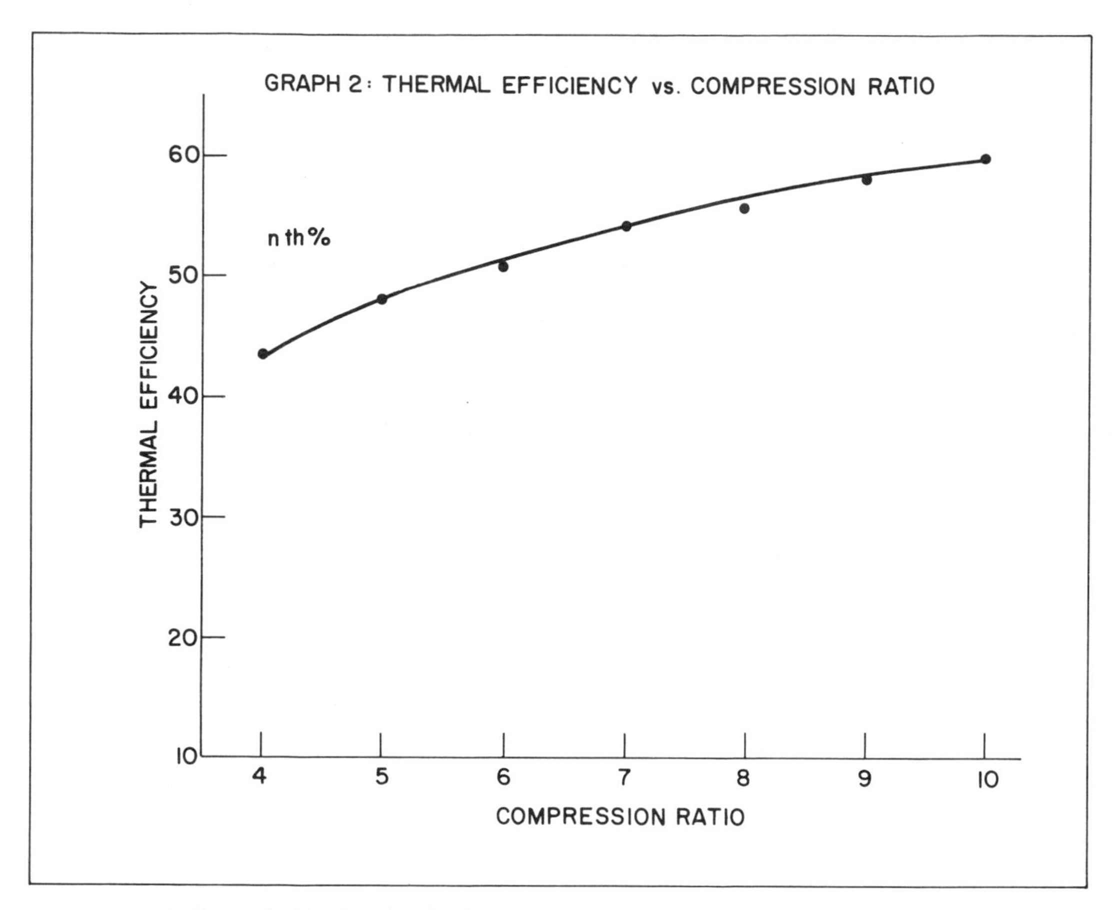

Figure 9–20. Graph which exaggerates because of suppressed zero.

An example which nicely illustrates how unintentional distortion of scales in visual aids can be totally misleading is provided by Figure 9-23. The text accompanying this visual aid compares records of precipitation patterns with records of runoff into a river. The purpose was to predict future runoff. As the text says, this figure "reflects the relationship between precipitation and runoff." But notice how entirely distorted that relationship becomes in this figure. The two graphs use different scales on the vertical axes even though the two axes represent the same thing—inches of water. In addition, the graph at the bottom of the page suppresses the zero while the graph at the top of the page exaggerates by using a misleadingly large scale. A glance at this pair of figures reassuringly suggests that the amount of runoff to be expected will equal the amount of precipitation. Thus, the bars of the top graph are the same size as the bars of the bottom graph. But as the redrawn figure, Figure 9-24, clearly indicates, the relationship is not at all what it might seem. The runoff can be expected to be about 25 per cent of the

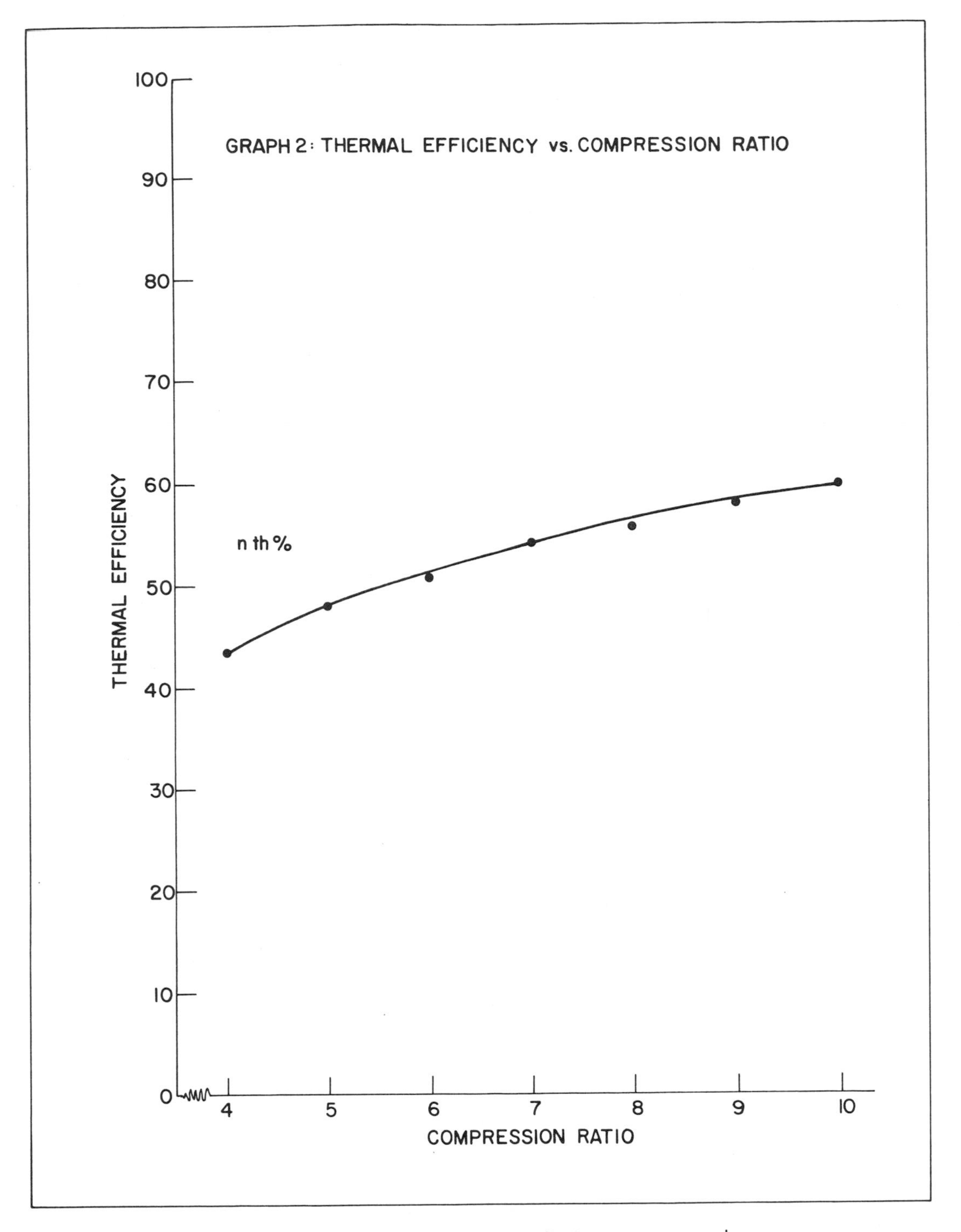

Figure 9–21. Figure 9–20 redrawn to eliminate suppressed zero.

precipitation. (Notice also that the redrawn figure combines the graphs, which makes comparison much easier than does two separate figures.)

The example on page 189 shows how aids can mislead a reader. But it also illustrates another point. The writer who prepared these two graphs might have thought he was being considerate of his readers by liberally using visual aids. He

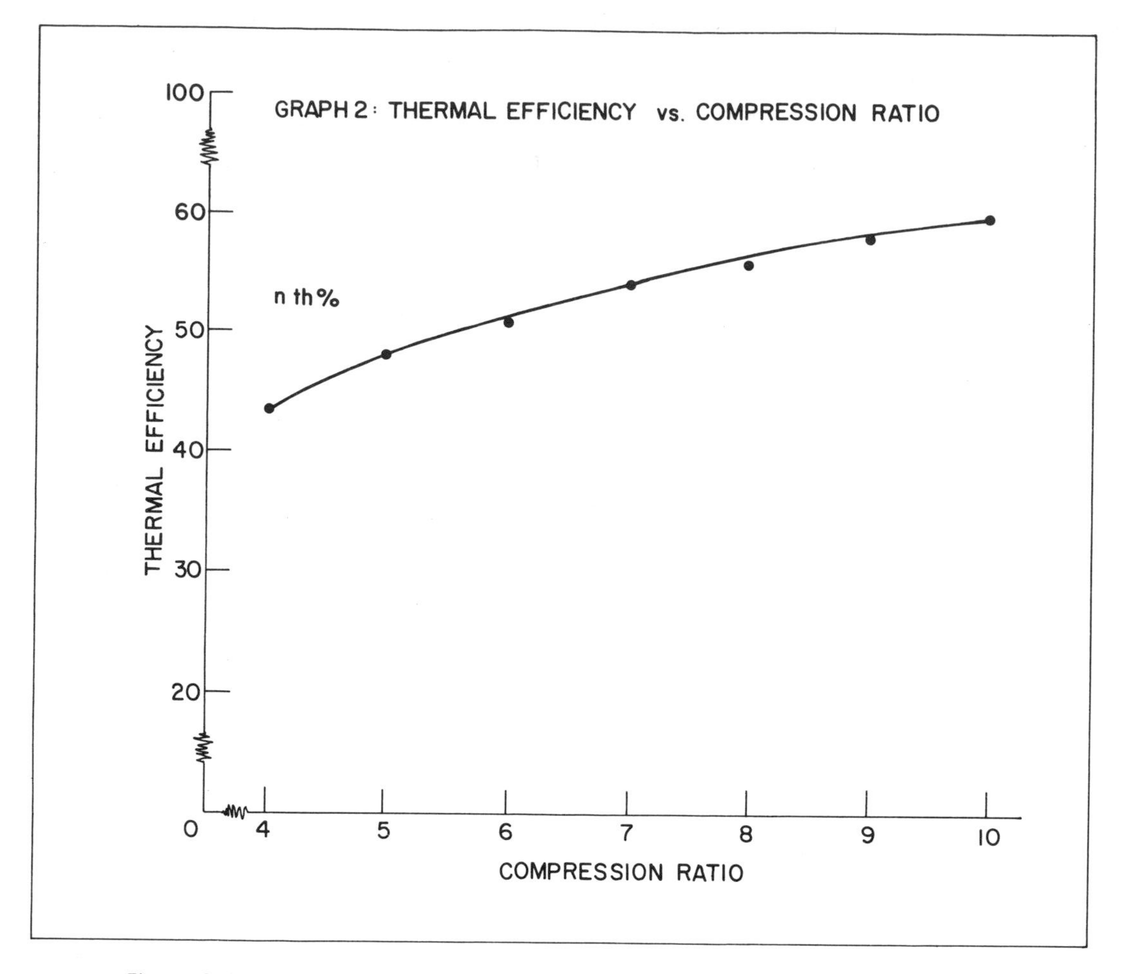

Figure 9–22. Figure 9–20 redrawn, using broken axes to indicate truncated scales.

might even argue that he was following our first suggestion. But clearly he was thinking more of his own effort than of his reader's. Having prepared the graphs separately, he found it easier not to redraw them. Liberal use, then, and appropriate placement are not enough to assure that visual aids will help the reader. Visual aids must also be well designed. All three characteristics are necessary.

Effective layout and appropriate use of visual aids requires hard work and thoughtful consideration by the writer. Since this design effort comes near the end of the report design process, the writer may tend to be slack in this effort, thinking of his own time rather than of his reader's. But he should be thinking of his reader's efforts, not his own. The report design is ultimately translated into a physical thing. Therefore, the writer should plan the form of the report as carefully as he has planned its content.

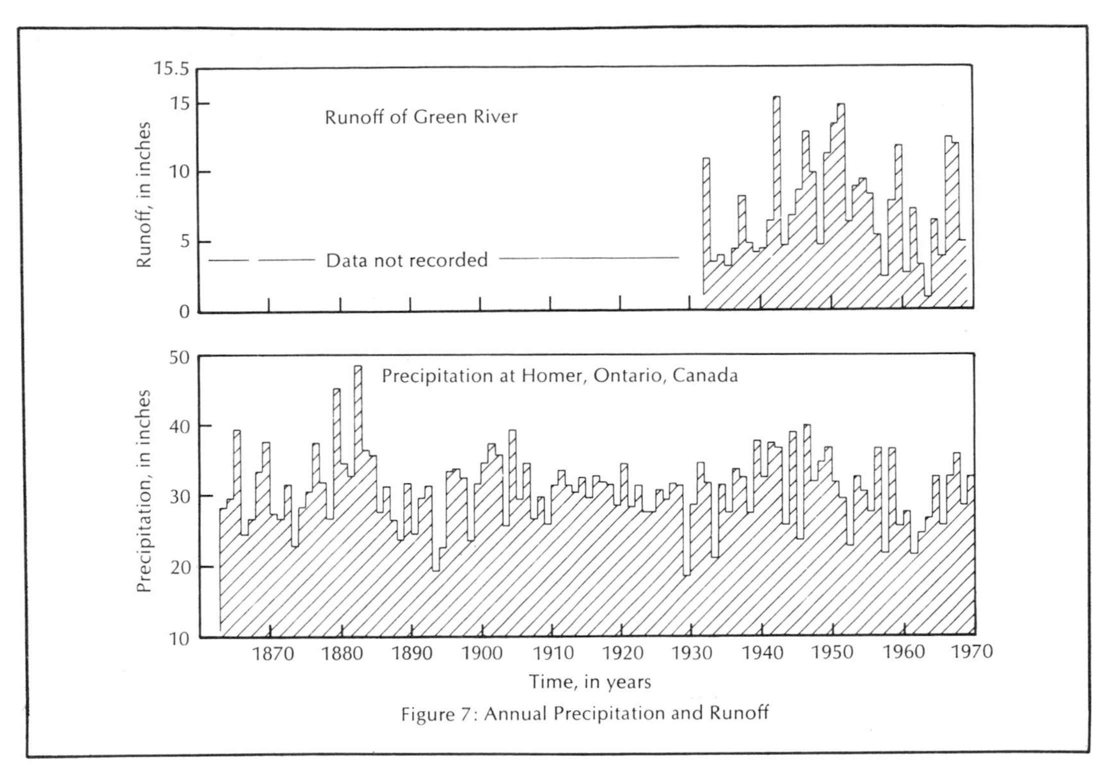

Figure 9–23. Visual aid which distorts because of different scales and suppressed zero.

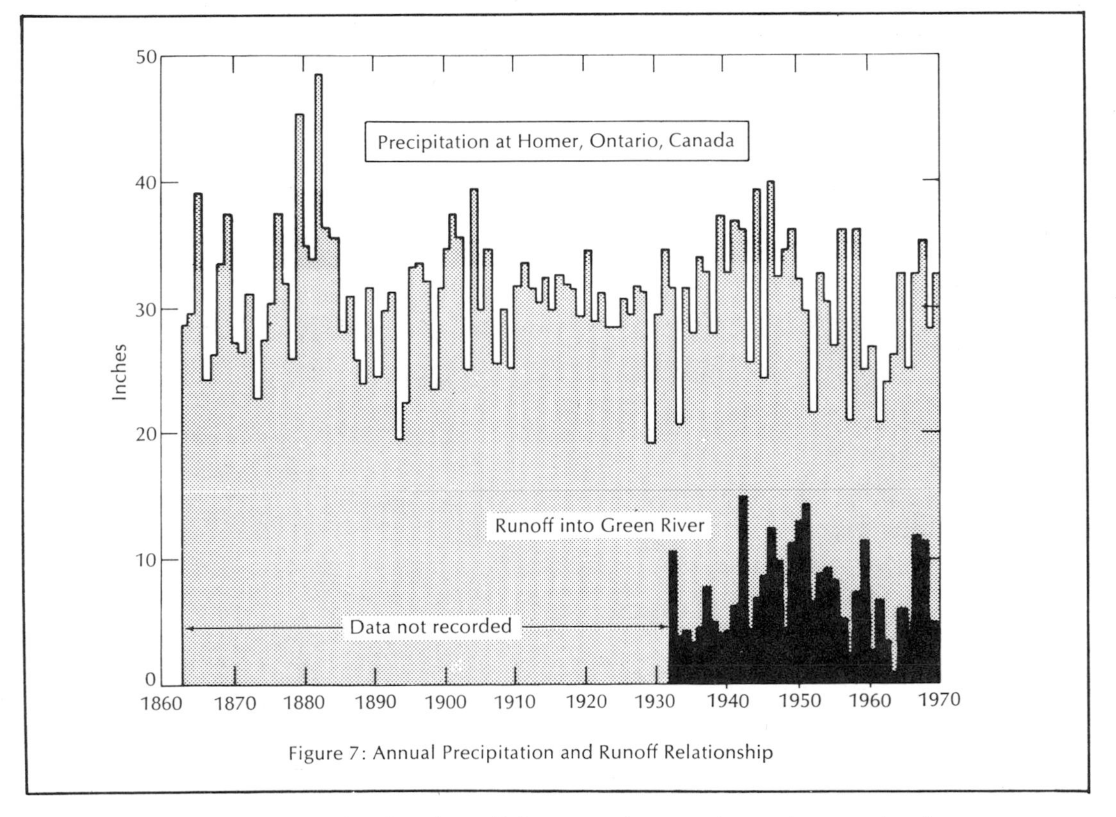

Figure 9–24. Visual aid with explicit comparison made on the same scales.

CHAPTER 10

Report design: Guide and checklist

Report writers never have enough time to write; reports are often team-written; and reports are always written between telephone calls, interruptions, and last minute crises. For those reasons the engineer/report writer needs a systematic procedure to carry him from basic planning through a prototype draft to a report ready for publication.

Throughout this book we have been discussing specific stages of the report writing process at length. For purposes of analysis we have divided that process into three fundamental stages, each with three component steps:

Determining the function of the report in the system

The technical communication process
Audience analysis: the problem and a solution
The problematic context: the purpose of the report

Designing the report

Designing the basic report structure
Designing the opening component
Designing the discussion component

Writing and editing the report

Arranging report segments and units
Editing sentences
Additional design features: layout and visual aids

Our objective has been to explain both what you must do to write a good report and why you must do it. We have stressed the "why" more than the "what" because obviously company report formats differ and report writing situations vary. We have presented a rhetorical theory and applied it to particular types of reports. You can apply the theory to other types of reports as the need arises.

Applying rhetorical theory, however, is never as consciously analytical, neat, or easy as our discussion undoubtedly suggests. Report writers never have enough time to write; reports are often team-written; and reports are always written between telephone calls, interruptions, and last-minute crises. Because report writing always takes place in this blur of hectic activity, you will never have time to review each of the chapters in this book to make sure you are performing the process efficiently and effectively. Accordingly, we finish our discussion with this guide and checklist. We give you practical, manageably condensed advice to help you translate the rhetorical theory of the first nine chapters of this book into effective practice. As the title suggests, the chapter is both a writer's advisor and a self-appraisal system to be used whenever you have a report to write.

When you receive an assignment or voluntarily undertake a technical investigation, first schedule adequate time for both the investigation and the report. If you fail to schedule adequate time for the whole project, you will almost certainly end up stealing your report writing time to complete the investigation. Yet in terms of your function in the organizational system, the report writing time is instrumentally your most important time. One consulting engineer stresses the importance of this—and at the same time indicates how difficult it is to schedule realistically. He says, "I usually allow about three days to write a report at the end of a thirty day consulting assignment. For a ninety day assignment I might allow a week. It is never enough. I always end up coming to the office after dinner and on weekends to get the report out on time. And I end up paying my secretary overtime to get the typing done. You would think that after eight years as a consultant, I would finally learn how long it takes to knock out the report, but I still haven't. I have never sent out a report that I thought was really finished."

A good policy, then, is to prepare and stick to a written schedule which realistically allows time for the report writing stage of an investigation. Figure 10-1, *Typical schedule for a two year research project,* illustrates a schedule

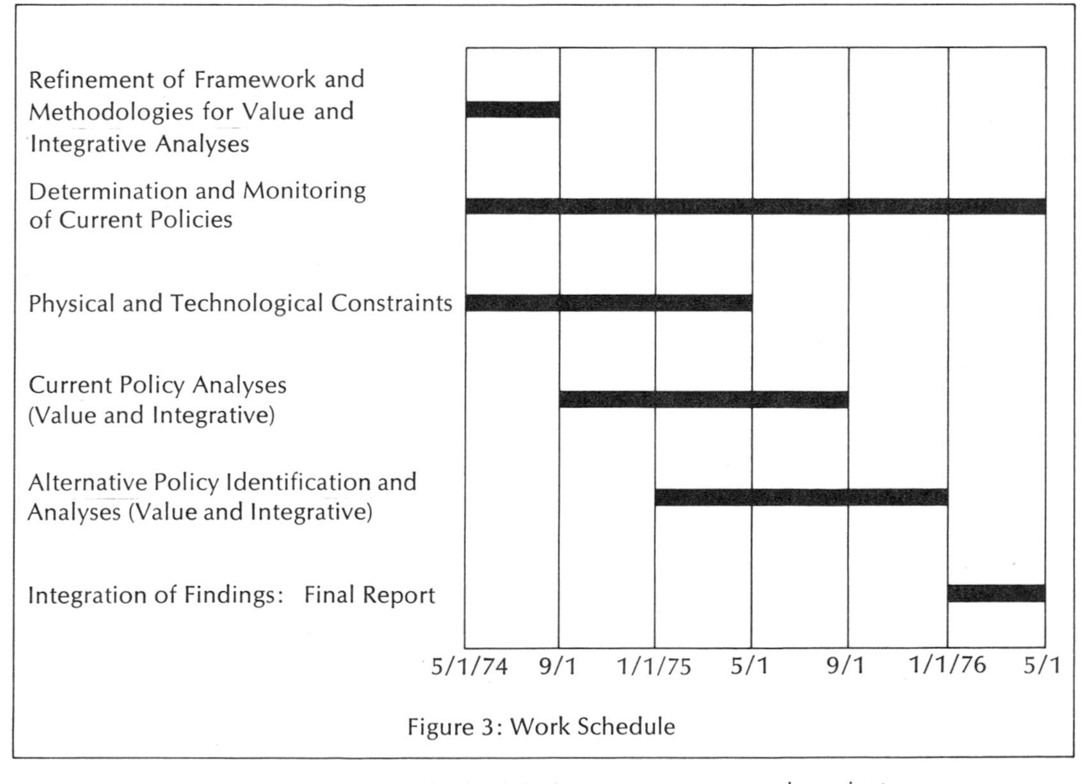

Figure 10–1. Typical schedule for a two year research project.

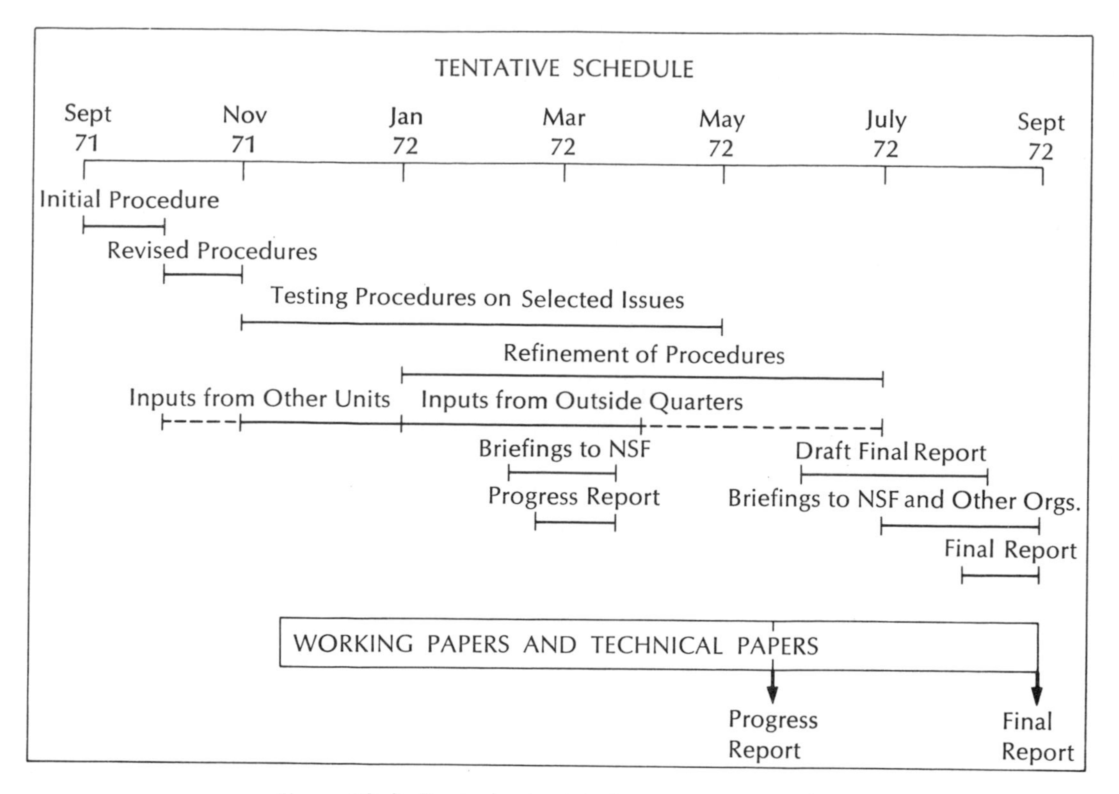

Figure 10–2. Typical schedule for a one year project.

established at the beginning of a two-year team project. Notice that four months were allotted specifically to writing the report. Figure 10-2, *Typical schedule for a one year project,* illustrates a schedule established at the beginning of a one year project. Notice that three months were allotted for drafting and revising the final report. Figures 10-3 and 10-4 show plausible schedules for shorter projects, three months and one month respectively. In both cases, time allowed for the report writing activity should be about 20 per cent of the total project time.

It would be impossible for us to establish a formula for allotting report writing time; but as a rule of thumb, expect that anything less than 15 per cent or 20 per cent or even 25 per cent of the total project time is almost certain to be inadequate. For a team-written report expect from 20 per cent to 30 per cent of the total time to go to pulling the elements of the report together because the act of writing the report is likely to be an important intellectual stage of the investigation.

We have just introduced the fact that many reports are team written. Although throughout this book we have talked as if there were only one engineer/report writer, much report writing is done by engineers functioning as members of project groups. This introduces another complication with which the engineer and the other members of his group must cope: his group must establish a systematic method for integrating the efforts of its members and for producing the project report.

If your company does not have a systematic procedure for integrating group projects, you might propose a system similar to one described to us by the head of the technical writing and publications department of one large company. The system calls for five stages of group interaction:[1]

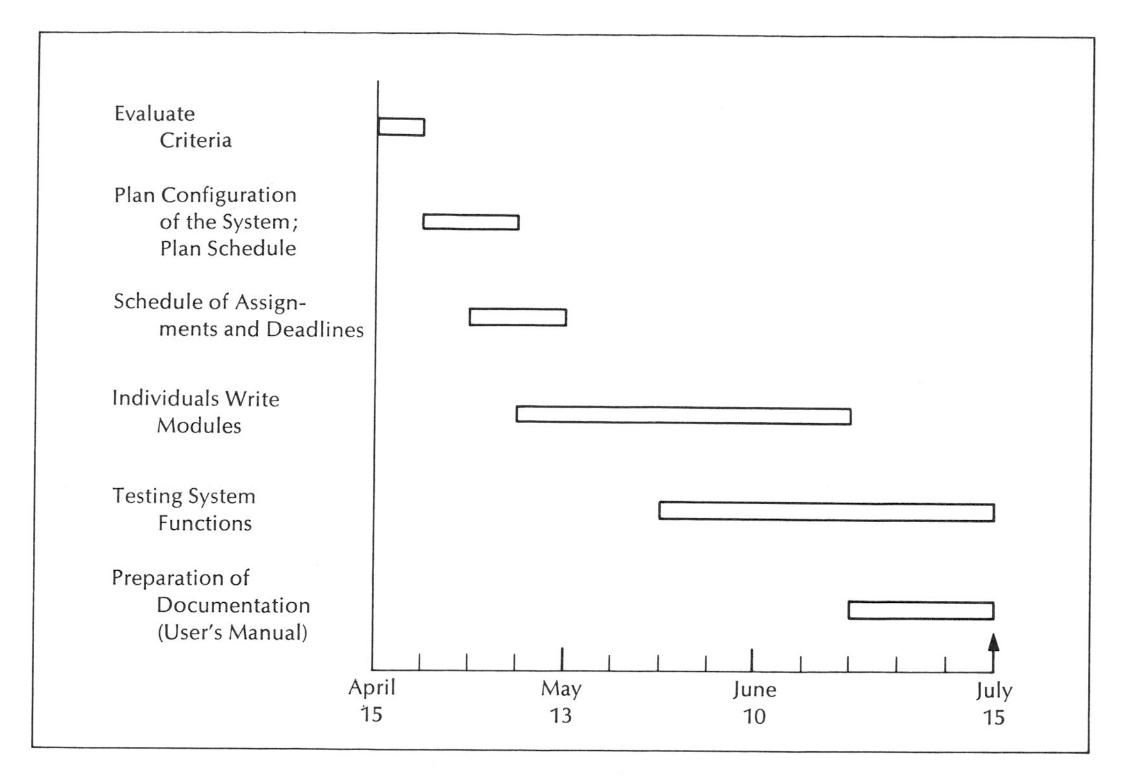

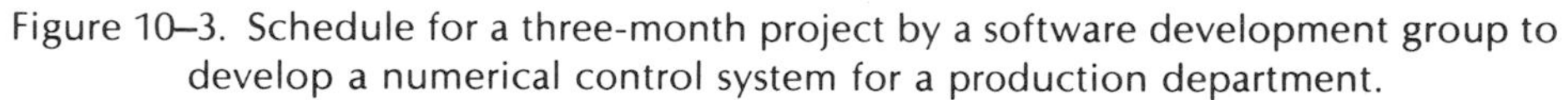
Figure 10–3. Schedule for a three-month project by a software development group to develop a numerical control system for a production department.

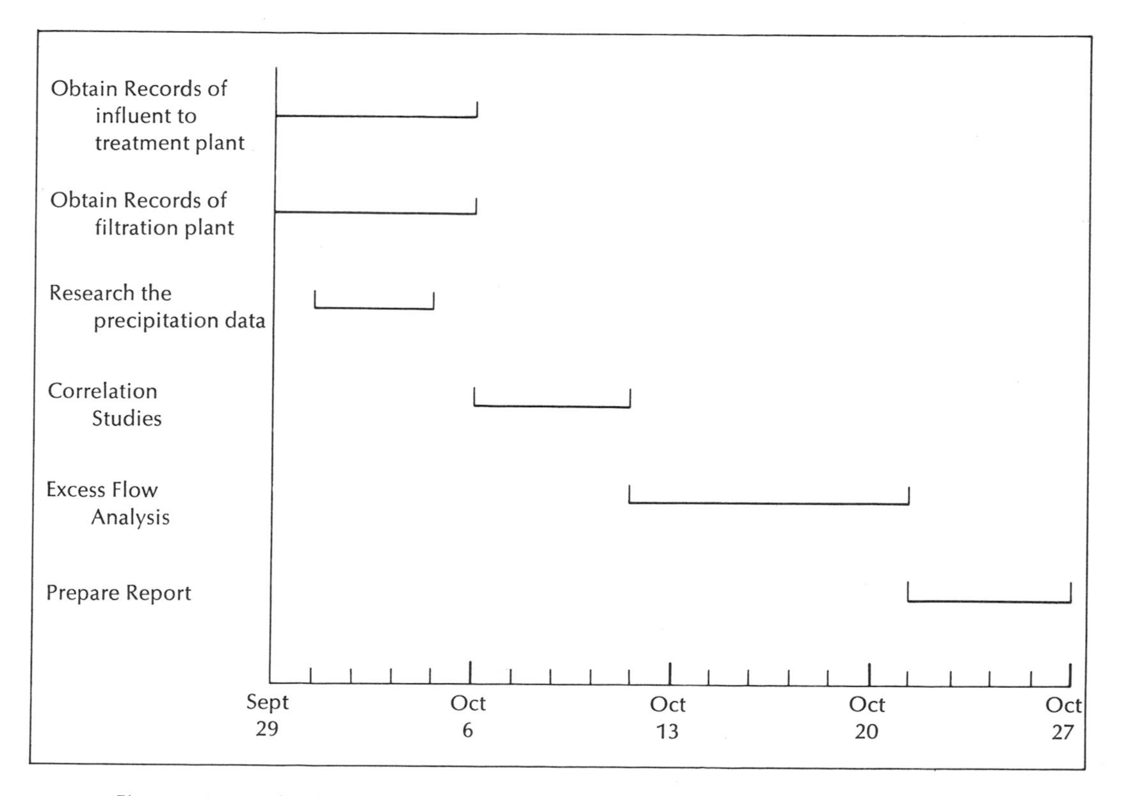

Figure 10–4. Schedule for a one-month project: an inflow/infiltration analysis of a city sewerage system by a consultant.

1. *Planning:* This stage is to review the problem and establish the purpose and theme of the report, prepare a topic outline, and assign topics. This review insures that contributors emphasize the same points regardless of their assigned subject matter. (See Figure 10-5, *Form for preparing topic outline for a team-written report.*)
2. *Preparation of detailed outlines for each topic:* In this stage each writer prepares a detailed sentence outline of the segment or unit he is to write. He states his thesis, itemizes the topic of each paragraph, and identifies visual aids to be incorporated. (See Figure 10-6, *Form for preparing detailed outline for a segment.*)
3. *Review of detailed outlines:* This stage involves a group critique of the detailed outlines submitted by all members of the project group. It also involves reassessing the original plan. (See Figure 10-7, *Completed form with review session suggestions.*)
4. *Writing the segments:* In this stage each writer drafts the segments he has been assigned. Again he uses a standard company form.
5. *Assembling and editing the segments:* This stage involves creating a systematic format, acquiring professional editing, and producing the final report. (Notice that in this company a team of editors, technical writers, and illustrators completes the report writing process.)

PAGE 2

IEFCS FOR BOEING 747/3639

PROGRAM TITLE LOG NO.

BOOK, VOLUME, PART, ETC.	SECTION	SUBSECTION	TOPIC	SUBHEADS	VISUALS	ENGINEER	PAGES
	SYSTEM DESIGN DESCRIPTION						
		GENERAL CAPABILITIES					
			PRINCIPAL FEATURES OF THE PROPOSED SYSTEM				
			CONFIGURATION OF THE BASIC DUAL FAIL-PASSIVE				
				SYSTEM			
			GROWTH TO TRIPLEX FAIL-OPERATIONAL STATUS				
			SUMMARY OF THE APFD OPERATING MODES				
			SYNCHRONIZATION OF THE APFD COMPUTERS				
			ENGAGEMENT IN THE OPERATING MODES				
			SYSTEM INTERLOCKS IN ACCORDANCE WITH				
				SCD REQUIREMENTS			
			APFD OUTPUTS FOR AIRCRAFT ANNUNCIATOR				
				DISPLAYS			
		PITCH AXIS FUNCTIONAL DESCRIPTION					
			IMPLEMENTATION OF THE PITCH AXIS				
				FUNCTIONS			
			PITCH ATTITUDE CONTROL THROUGH DRIVEN				
				ATTITUDE REFERENCE IMPLEMENTATION			
			DYNAMIC PRESSURE SCHEDULING OF ALTITUDE				
				ERROR FOR ALTITUDE HOLD CONTROL			
			ALTITUDE CAPTURE TO WITHIN 30 FEET				

Figure 10–5. Form for preparing topic outline for a team written report.

[1] Keith B. Roe, Manager, Engineering Publications, Sperry Flight Systems. He tells us many companies use this system, which was developed by James R. Tracy, Head, Writing Service, Information Media Dept., Hughes Aircraft Company.

VOL:

SECTION:

SUBSECTION:

STORY BOARD FOR

PUBLICATION NO:

TITLE:

AUTHOR:

EXT:

TOPIC:

THESIS SENTENCE:

STATE POINT OF EACH PARAGRAPH:

1–

2–

3–

4–

5–

FIGURE CAPTION:

Figure 10–6. Form for preparing detailed outline for a segment.

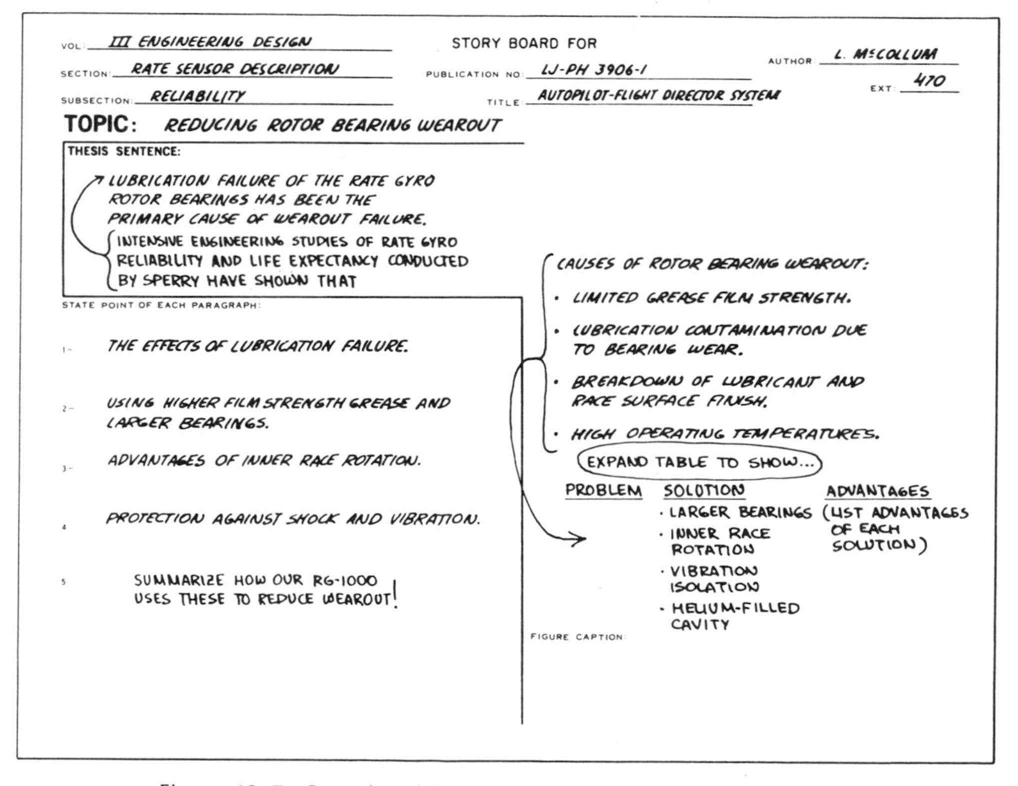

VOL: III ENGINEERING DESIGN

SECTION: RATE SENSOR DESCRIPTION

SUBSECTION: RELIABILITY

STORY BOARD FOR

PUBLICATION NO: LJ-PH 3906-1

TITLE: AUTOPILOT-FLIGHT DIRECTOR SYSTEM

AUTHOR: L. McCOLLUM

EXT: 470

TOPIC: REDUCING ROTOR BEARING WEAROUT

THESIS SENTENCE:

LUBRICATION FAILURE OF THE RATE GYRO ROTOR BEARINGS HAS BEEN THE PRIMARY CAUSE OF WEAROUT FAILURE.

INTENSIVE ENGINEERING STUDIES OF RATE GYRO RELIABILITY AND LIFE EXPECTANCY CONDUCTED BY SPERRY HAVE SHOWN THAT

STATE POINT OF EACH PARAGRAPH:

1– THE EFFECTS OF LUBRICATION FAILURE.

2– USING HIGHER FILM STRENGTH GREASE AND LARGER BEARINGS.

3– ADVANTAGES OF INNER RACE ROTATION.

4 PROTECTION AGAINST SHOCK AND VIBRATION.

5 SUMMARIZE HOW OUR RG-1000 USES THESE TO REDUCE WEAROUT!

CAUSES OF ROTOR BEARING WEAROUT:

- LIMITED GREASE FILM STRENGTH.
- LUBRICATION CONTAMINATION DUE TO BEARING WEAR.
- BREAKDOWN OF LUBRICANT AND RACE SURFACE FINISH.
- HIGH OPERATING TEMPERATURES.

EXPAND TABLE TO SHOW...

PROBLEM	SOLUTION	ADVANTAGES
	· LARGER BEARINGS · INNER RACE ROTATION · VIBRATION ISOLATION · HELIUM-FILLED CAVITY	(LIST ADVANTAGES OF EACH SOLUTION)

FIGURE CAPTION:

Figure 10–7. Completed form with review session suggestions.

No matter what procedure would be most appropriate for your own situation, your group should establish a system which provides for collection of data, formulation of report objectives, appropriate development and integration of conclusions, and efficient preparation of a report with a consistent format. If your project group does not have a system for producing reports, the chances for the group's producing a good report are radically diminished.

A SYSTEMATIC METHOD FOR PREPARING A REPORT

Team-written reports require a systematic method if they are to be done efficiently. Individually written reports also require a systematic method. This is particularly necessary because reports are written in the midst of the daily routine, telephone calls, and interruptions. Based upon the principles we have established in this book we suggest the following method which will carry you from basic planning through a prototype draft to a report ready for publication. We assume you will produce two typewritten drafts of any report you write. The method (outlined in Figure 10-8) consists of the four basic stages explained in the following pages.

1. Preparing the Prototype Draft

The object of this stage is to produce a draft which can be turned over to a typist who will prepare a prototype of the report. In subsequent stages, this prototype will be revised and edited to prepare the finished master copy of the report. Most writers produce first drafts in handwriting. Others use a typewriter, a dictaphone, or a stenographer. Any of these renditions will produce the draft from which the prototype draft is prepared.

Preparing the prototype draft consists of eight steps.

1.1. Defining the problem

This step should be performed both before and after the technical investigation. Performed before the investigation it helps to guide the inquiry. Performed after the investigation it helps to redefine the problem, which in the process of the investigation may have become more clear or more complex. Defining the problem calls for identifying the problem or conflict at issue in the organization and stating the specific technical questions or tasks to be addressed by the technical investigation. (This is discussed in chapter three, pp. 29–38; and Guide 1.1., p. 209.)

1.2. Performing the investigation

From the report writer's perspective, three tasks are necessary during this step. First, you must keep accurate records of what you are doing. Second, you must explain why you are doing it that way. Third, you must organize and compile your data and results in forms that will be useful in the finished report. If they are done carefully at the time of the investigation, these records will provide most of the appendices. If they are not done carefully, later they will have to be reconstructed and redone. (See pages 29ff.; and Guide 1.2., p. 209.)

1. Prepare the Prototype Draft
    1.1. Define the Problem
    1.2. Perform the Investigation
    1.3. Define the Audiences
    1.4. Formulate the Rhetorical Purpose
    1.5. Design the Discussion Component
    1.6. Draft the Segments
    1.7. Draft the Opening Component
    1.8. Send the Rough Draft to the Typist
2. Prepare the Finished Master Copy
    2.1. Verify the Basic Design
    2.2. Edit Sentences in Context
    2.3. Edit Individual Sentences
    2.4. Edit the Format
    2.5. Prepare the Finished Visual Aids
    2.6. Write the Abstract
    2.7. Final Check
    2.8. Send the Edited Prototype to the Typist
    2.9. Proofread the Master Copy
3. Solicit Evaluation by Someone Else
4. Send the Report into the System

Figure 10–8. A systematic method for designing, writing, producing, and distributing a report.

1.3. Defining the audiences

The principal objective of this step is to identify your primary audiences and to determine how they will use your report. In addition you must account for other readers and other uses. (See chapter two, pages 14–23; and Guide 1.3., p. 210.)

1.4. Formulating the rhetorical purpose

This step calls for reexamining the problem and technical questions identified in step 1.1. and for establishing the rhetorical purpose of the report. (Especially important is that you distinguish between the technical purpose and the rhetorical purpose.) Ask yourself what you want to accomplish with the report. What do you want people in the organization to do? (See chapter three, pp. 24–42, particularly 29–31; and Guide 1.4., p. 210.)

1.5. Designing the discussion component (body plus appendices)

The primary objective of this step is to choose the most effective strategy for segmenting, selecting, and arranging the material that will make up the discussion component. You must decide whether the report is to be organized in terms of rhetorical purpose, in terms of the intellectual problem-solving process, or in terms of its subject matter. At the end of this step you should have a clear idea of the report's basic architecture, an approximate idea of how long the report will be, an indication of all the material you will need, and a general idea of how you will apportion the material between the body and the appendices. The end result of this step, then, is a detailed outline which establishes, tentatively at least, the segments, the headings, and the numbering system to be implemented in the prototype of the report.

At this stage you break up the report into manageable segments. If you plan well, you can work on these segments between telephone calls and in almost any convenient order. You might even start your secretary organizing and typing the appendices materials. (See chapter six, pp. 97–110; and Guide 1.5., p. 210.)

1.6. Drafting the segments

The objective of this step is to produce a draft of the complete discussion component. It is probably best to start with the introductory segment because that provides you with the controlling purpose statement (see pp. 93–95). Each of the subsequent segments should be written independently. The structure of each is determined by its rhetorical purpose (see chapter seven, pp. 113–141). This means, of course, that you will have to outline and draft segment by segment rather than outline all of the segments before you start writing. (This will require amplifying and modifying the preliminary outline produced in the previous step.)

As you write these segments try to schedule your time so that you can draft one complete segment without having to put it aside to do something else. If there are other tasks that demand your attention, schedule them between segments. (Simple tasks perhaps might be scheduled between units within the segment.) You schedule your time this way so that once you start drafting a segment you can complete it without losing your sense of purpose or your control of details. (See chapter seven and Guide 1.6., p. 211.)

As you write these segments begin selecting the visual aids that will be adapted from the appendices or prepared independently for the discussion component (see chapter nine, pp. 168–189). Do not bother making anything more than quick, rough sketches. When the prototype draft is prepared, have the typist leave space for the more finished visual aids you will have later. To spend time now doing careful drafting will deflect you from your present task.

Reexamine the records you prepared during the investigation (step 1.2.). See if they are still appropriate for the appendices. If not, modify them and prepare any additional appendices you now see as necessary (see pp. 109–110). Usually you will not submit these appendices for typing in the prototype stage. They will be typed for the first time in the final draft.

1.7. Drafting the opening component

The objective of this step is to produce a mini-report for audiences interested in generalizations about the problem and solution, and for audiences interested in reading only selectively in the discussion. Be careful in this component to address the important audiences organizationally most dissimilar from you.

The secondary objective is to provide all the heading information—title, distribution list, references, dates, et cetera—necessary to introduce the report into the organization. (See chapter five, pp. 59–82; and Guides 1.7.1., p. 214, 1.7.2., p. 215, 1.7.3., p. 216.)

1.8. Sending the rough draft to the typist

The objective here is to assemble all of the parts, organize them as they will appear in the finished report, and give the typist all the necessary format instructions to produce the prototype draft. In this step you make all your preliminary decisions about how the finished master copy should appear: you must prepare a mock-up of the title page and table of contents; you must choose spacing patterns and even type styles. Do not leave these decisions to the typist.

2. Preparing the Finished Master Copy

Now you have a prototype of your report. Probably at this point you get your first real sense of the whole report. In the previous stage you have been so buried in preparing individual segments you could not see how the parts were fitting together. Now, except for some of the attachments and visual materials, it is all there in front of you approximately as it will appear in the finished report. The object of this stage is to produce a polished draft of the report which can be turned over to a typist for preparing the finished master copy. (Depending upon the situation, this master copy may yet be subject to revision and editing by superiors and colleagues; however, presumably these changes will not require retyping much of the report.)

Preparing the finished draft consists of nine steps.

2.1. Verifying the basic design

The objective of this step is to satisfy yourself that you have accomplished what you set out to do. You read straight through the prototype draft. If necessary, take the phone off the hook and isolate yourself; but read the entire report from title page through appendices. You must satisfy yourself that you do not need to turn back to redo any of the steps under stage one. Sometimes after you have finished this step you turn the report over to subordinates. Other times you turn the report over to technical writers, illustrators, and editors. You have relinquished control of the report. Therefore, ask yourself, "Am I willing to let the report as it now stands be a measure of my technical and intellectual accomplishment?" (See Checklist 2.1., p. 219.)

2.2. Editing sentences in context

During this step you examine each of the report's independent segments. These may be segments, units, and paragraphs in the discussion component. Your objective is to verify that each has a controlling idea, and moves from general to particular. If it does not, state the controlling idea or rearrange the segment. You also revise sentences to carry out the pattern established by the core sentence or paragraph. (See chapter eight, pp. 143–153; and Checklist 2.2., p. 219.)

2.3. Editing individual sentences

During this step you eliminate ambiguity and inefficiency. You also correct spelling, punctuation, and typographical errors and mistakes in diction and grammar. Ordinarily it is a good idea to get someone else to help you with this step because sentence problems can be invisible to the person who produced them. When you do this step entirely by yourself, do not try to do it all at one sitting. And try reading segments or pages out of sequence so you do not get caught up in what is being said. (See chapter eight, pp. 153–161; and Checklist 2.3., p. 220.)

2.4. Editing the format

Two things happen during this step. If you were careful to specify format for the typist during the prototype stage, the first objective of this step is simply to verify that no mistakes have crept in. However, expect mistakes in the way lists are indented, headings inserted, figures referenced, et cetera. In reality the prototype never turns out exactly as you envisioned it.

The second objective of this step is to reappraise basic choices of format made earlier. Your idea for a heading might have looked fine in handwriting but not so good in type. Reevaluate all format choices. (See chapter nine, pp. 163–168; and Checklist 2.4., p. 220.)

2.5. Preparing finished visual aids

Reconsider and redo the sketches of visual aids you prepared for the prototype. The objective of this step is to obtain the finished visual aids you will incorporate into the master copy of the report. If possible, get someone else—a professional illustrator—to do the aids. But be certain you give an illustrator good sketches of exactly what you want. A good secretary may clean up your text for you, but illustrators will redraw what they are given and thus reproduce any errors or deficiency in your sketch. (We mean no slight to professional illustrators; their role does not ordinarily permit them to second-guess their sources.) (See Chapter nine, pp. 168–189; and Checklist 2.5., p. 221.)

2.6. Writing the abstract

During this step you condense the discussion component of the report to perhaps 200 words. Be especially careful to make your abstract present information rather than describe what is in the report: briefly explain the problem, your methodology, results, conclusions, and recommendations. Your objective is to provide a retrieval device for potential audiences who have no first-hand contact with your report. Remember, the abstract is separated from the report and exists by itself in an information retrieval system. For that reason the abstract must not serve foreword or summary functions for your report.

2.7. Final checking

During this step you systematically inventory the report's components. Just as a pilot uses a standard preflight checklist to assure himself that all systems are functioning effectively, you use an appropriate final checklist to assure that all the segments perform their functions. During this step you are not really reading for content or even for format. You are reading only to verify the presence of the information and design features you intended each segment to have. (This is particularly important for team-written reports.) In other words, you perform this step more in a descriptive than an evaluative manner: you check items off a list; they are either there or not there. Presumably they should all be there; if not, you must turn back to an earlier step in the process.

To perform this step you need specific checklists for particular types of reports. We provide two—Checklists 2.7.1., pp. 222, and 2.7.2., p. 223—based upon the report formats we have discussed in this book. If these should not fit your company report format, develop your own checklists, perhaps by modifying ours.

2.8. Sending the edited prototype to the typist

The objective of this step is to deliver to your secretary or to the editorial department the revised and edited prototype draft so the master can be produced from it.

2.9. Proofreading the master copy

Once you have the master copy of the report in hand, you must identify and correct any minor errors that may have appeared. You are not reading for content or for format, only for correctness.

3. Soliciting Evaluation by Someone Else

Many readers tend to be defensive about their own material and thus avoid asking for help in evaluating it. This is especially true of students, for whom getting outside help is sometimes even regarded as unethical. However, we believe you should always get someone else to help you evaluate what you have written. The objectives are to verify the soundness of your content and basic rhetorical decisions, and to identify any errors you may have missed while proofreading. Obviously evaluation of content and basic rhetorical decisions must be done during the previous stage of the writing process after the prototype draft is prepared. Proofreading of course should be done after the master copy of the finished report is available. A warning: when you choose an evaluator, your tendency will be to choose someone who will pat you on the back and say "well done." Avoid that tendency. Choose someone who is capable of making sound judgments, candid enough to express them, and receptive to views other than his own.

4. Approval, Publication, Distribution: Sending the Report into the System

The objective of this stage is to get the report into the system. To do this you must obtain any approvals of associates and superiors necessary before the report can be released. It is wise to verify your report with all those people directly or even indirectly involved in the investigation. In the long process of revising and editing the report, you may have slightly misrepresented something. Or you may have overlooked some minor detail. Even if you are the principal investigator, remember that you must continue to work with the others involved in the project. To ignore their final approval is at least to invite hurt feelings and may even result in a significant error in the report's content. In addition to this approval by associates, of course, official authorization by one or several superiors is often needed as well. It is both tactful and wise to seek their approval even when approval by signature is not explicitly required.

Seeing the final master copy through production and distribution is just a matter of checking back to make sure nothing remains undone or is done incorrectly. Do not let yourself get trapped at this late date by some breakdown in the processing and distribution system. Since you know better than anyone else where the report should be sent, do not rely solely upon standard distribution lists. And since you know better than anyone else when the report should arrive, do not rely upon routine distribution procedures. Double-check everything.

AFTERWORD

We started this book by observing that the communication process, like the process of solving technical problems, should be systematic. We said, "If an engineer is able to use rigorous, systematic procedures to resolve technical problems, he or she should be equally able to apply systematic procedures to resolve communication problems." Throughout this book, therefore, we have presented a logic upon which to build a systematic procedure, and from that logic we have derived principles, suggestions, guides, and checklists. The report writer can use these to help him proceed systematically. For us, as we said at the outset, the logic behind the procedure is fundamentally more important than the procedure itself. We have begun with basic design principles and concentrated upon them. Only after he understands the "why" behind a procedure can an engineer/report writer effectively apply the "what" of principles, guides, and checklists.

For the writer faced with producing a report, however, both the logic behind our procedure and the procedure itself may seem cumbersome. To him it may seem inefficient to spend time formulating a problem statement for himself, analyzing his audiences, considering alternative arrangements for his discussion, and so on. After all, these activities take time, and to the report writer with a deadline, time may seem his most pressing need.

Yet we believe that if the report writer looks at the communication process—and at his own role in the system—from the perspective of the organization in which he writes, the logic we have presented and the procedure we have derived from it are genuinely efficient. In fact, we can even defend this assertion in terms any organization can understand: dollars and cents.

A very nice example of that came our way recently from a technical editor in a publications department of a large company doing government contract work. Her department has developed a formula to compute the cost of producing reports. This department receives prototype copies of reports from engineers, which technical writers, editors, and illustrators then prepare for final publication. At subsequent review sessions, the engineer is brought in to review and validate the efforts of the editorial and production staffs. Before editing the prototype, on the basis of the number of pages of text, number and types of illustrations, number of equations, et cetera, the publications department calculates the cost of a report. The department uses this formula to budget its resources and to prepare cost estimates for a writer's office, since his office is billed for the total cost of producing the report. At current rates, this publications department figures the cost of a typical report to be twenty-one dollars per page. And that figure does not include the engineer's work; it includes only editing, illustrating, and production.

That figure no doubt sounds high, but the formula for calculating the cost of a report does not end there. Even though a report writer might assume the production department will take over the burden of turning out a good report—no matter how hasty or ill-designed his prototype—the quality of that prototype is a significant cost consideration.

To each engineer/report writer, based upon detailed logs of past experience, the publications department assigns a "cost constant" which they incorporate into the final cost calculation for a report. These logs document the nature and extent of editing activities characteristically required for each engineer's reports. In two particular cases, the results of incorporating the cost constant into the formula for computing the cost of a report are striking. To one engineer, whose prototype drafts are always well designed, well written, and well edited before he hands them in, the department has assigned a cost constant of .6. To another engineer whose prototype drafts are characteristically crude and hastily done, the department has assigned a cost constant of 2.7. The second engineer, unlike the first, always has inaccurate computations, frequently wants changes in his text, and must often be called in to clarify his explanations; therefore the disproportionate figure.

To illustrate how important the quality of a prototype draft is, the technical editor explained how her department would compute the cost of a thirty-page prototype report handed in by each of these two engineers. Before allowing for their cost constants, the department would calculate the cost of a typical 30-page report at $630.00. ($21.00 per page × 30 pages.) After allowing for the cost constants, however, in the case of these two engineers, the department would calculate the cost of a 30-page report as follows:

First engineer: $21.00/page × 30 pages × .6 = $ 378.00
Second engineer: $21.00/page × 30 pages × 2.7 = $1,701.00

As you can see, then, the quality of the prototype draft is not insignificant. Even in a large company, a difference of $1,323.00 is noticeable. The department can produce four reports by the first engineer for the cost of one report by the second engineer. When these cost figures are reported back to the engineer/report writers' supervisors—as they are—the quality of a writer's prototype drafts is obviously of paramount importance to the engineers themselves.

From this example we draw an axiom with which we will end this book. It makes no difference whether you prepare your reports entirely yourself, have minor assistance, or are relieved of all preparation tasks after the prototype draft stage.

Since your name is on the line, since the quality of your contribution to the organization will be measured by your reports, you must systematically prepare the best reports you are capable of preparing. Once again we will say, as we did in the beginning, an engineer who cannot communicate is in trouble; a manager who cannot communicate is finished.

Guides and Checklists

AN INDEX OF GUIDES AND CHECKLISTS

Guides

Checklists

Guides 1.1.-1.8.

1. PREPARING THE PROTOTYPE DRAFT

Guide 1.1. Defining the Problem

[See pp. 29–38.]

Answer the following questions:

1. Specifically what were you asked to do? (Suggestion: if you do not have written instructions, write out your own instructions and ask your supervisor to verify their accuracy. Get it in writing.)
2. What is the conflict or issue of concern to the organization? (This may require some digging. Your supervisor may not have explained and may not even know the conflict.)
3. What is the technical problem?
4. What are the specific technical tasks or design objectives you must address? (Itemize them; if possible, state them as questions.)

Guide 1.2. Performing the Investigation

1. What are you doing? (Explain the objective of your procedure and itemize its stages, i.e., define your methodology.)
2. Why are you doing it that way? (Justify the methodology. In addition, explain any unorthodox procedures or unfamiliar equipment.)
3. Itemize all the types of information you will need in the report, especially in the appendices. (Are the forms for tabulating the data and recording the results going to yield all the information you will need to include in the report? If not, prepare additional forms and tabulating procedures. You should not have to redo appendix material later.)

Guide 1.3. Defining the Audience

1. Prepare an egocentric organization chart which identifies the specific individuals who will transmit and receive this report. *[See pp. 15–18.]*

2. Characterize the individual report readers. Pay particular attention to operational characteristics. *[See pp. 18–19.]*

3. Classify audiences.
   Who are the primary audiences?
   Who are the secondary audiences?
   Who are the immediate audiences?
   [See pp. 19–23.]

Guide 1.4. Formulating the Rhetorical Purpose

[See pp. 29–36.]

Write out the following:

1. A statement of the problem, i.e., the conflict at issue in the organization.

2. The specific technical questions or tasks arising out of that problem, and addressed by the technical investigation.

3. A statement of the rhetorical purpose, i.e., a statement of what the report is designed to do in relation to the organizational problem and the consequent technical questions or tasks.

Guide 1.5. Designing the Discussion Component

1. Structure your report according to its rhetorical purpose. (Write out a scratch outline. Do this even if you are doubtful about its appropriateness.) *[See pp. 98–102.]*

2. Structure your report according to the intellectual problem-solving process. (Again write out a scratch outline.) *[See pp. 102–107.]*

3. Choose between the two basic strategies you have established in steps 1 and 2. If you cannot choose, if neither basic strategy seems appropriate, structure your report according to its subject matter. (Again write out an outline.) *[See pp. 107–109.]*

4. Having chosen the basic design of the discussion component (body plus appendices), write out a fully developed outline for both discussion and appendices. For short reports a two-level outline will do; for longer reports carry the outline to at least three levels. *[See pp. 113–115.]* Using the outline form 1.5.1., prepare your outline.

1.5.1. Outline form.

Segments

Units

Subordinate units and paragraphs

Foreword [To be written after discussion component]

Summary [To be written after discussion component]

The problem statement *(Introduction)* the beginning,
Amplification segment *[if necessary]* pp. 93–95

Adjust and indent these lines as necessary

The Ending segment *(restatement, summary, etc., pp. 95–97)*

Appendices segments

Guide 1.6. Drafting Segments

For each of the segments of the "middle" do the following: *[Guide 1.6.8. addresses beginning segments, Guide 1.6.9. addresses ending segments]*

1. Determine the purpose of the segment. Which is it?

   Persuasive segment
   Descriptive segment
   Process segment
   Cause/effect segment
   Question/answer segment
   Comparative segment
   Other

2. Write the core paragraph or sentence *[see pp. 115–116]* and mentally outline the rest of the segment or unit.
3. Write the units and paragraphs of the segment according to the outline appropriate for the particular purpose of the segment. See the appropriate guide:

   Persuasive segments, Guide 1.6.1.

   Descriptive segments, Guide 1.6.2.

   Process segments, Guide 1.6.3.

   Cause/effect segments, Guide 1.6.4.

   Question/answer segments, Guide 1.6.5.

   Comparative segment, Guide 1.6.6.

   Other, Guide 1.6.7.

Guide 1.6.1. Persuasive Segments

[See pp. 117–121.]

1. State the conclusion or hypothesis of the segment.
2. Support the conclusion, arranging your evidence in a descending sequence of significance.
3. Anticipate and refute objections.
4. Summarize the support and restate the evidence if the segment is long or complex.

Guide 1.6.2. Descriptive Segments

[See pp. 121–124.]

1. Present an overview of the object's function.
2. Verbally describe the appearance of the object.
3. When appropriate, provide a selective visual aid to complement the verbal description.
4. Explain how the object functions.

Guide 1.6.3. Process Segments

[See pp. 124–129.]

1. Present an overview of the objectives of the process.
2. Forecast and identify the number of stages in the process.
3. Explain any physical objects necessary for the readers to understand the explanation of the process which follows.
4. Present a stage-by-stage explanation of the particulars of the process or event.

Guide 1.6.4. Cause/Effect Segments

[See pp. 129–132.]

1. State the fact in question and the nature of the question (cause or effect).
2. Forecast the structure of the explanation and perhaps summarize the causes or effects to be explained.
3. Present support. For causes, arrange the supports from most significant to least. For effects, arrange them from most probable to least.
4. Anticipate and refute alternatives.

Guide 1.6.5. Question/Answer Segments

[See pp. 132–134.]

1. Explain the question or issue, posing specific hypothesis questions or identifying unknowns.
2. State the hypothesis or answer.
3. If appropriate, introduce the method or criteria for establishing the answer.
4. Support your hypothesis or answer, arranging your supports in a descending sequence of significance.
5. If appropriate, restate the conclusion or answer.

Guide 1.6.6. Comparative Segments

[See pp. 134–141.]

1. State the conclusion or interpretation of the comparison or contrast.
2. If the segment is dominantly comparative or contrastive, present a point-by-point analysis of the similarities or differences. Arrange these points in a descending sequence of significance.

   If the segment is mixed comparison and contrast, present the dominant element first. (That is, if the similarities outweigh the differences, start with the similarities.) Under each of these two elements, similarities and differences, arrange the material point-by-point in a descending sequence of significance.
3. Restate the conclusion.

Guide 1.6.7. Other Segments

[See pp. 115–116; 141.]

1. If the segment is a blend of two or more of the basic types of segments described in the previous six guides, put the guides together and work out an outline which draws upon the outlines for the appropriate segments.

2. If the segment does not seem to be a blend of two or more of the basic types of segments described in the previous six guides, start by presenting a core paragraph or sentence which states the purpose of the segment.
3. Present a forecast or overview of the rest of the segment.
4. Present the particulars, discretely separated and clearly arranged. (Descending significance may not be appropriate, but if you look carefully, there should be a logical order to follow.)

Guide 1.6.8. Beginning Segments

[See pp. 56–58, 93–95.]

1. Present a fully developed purpose statement which identifies the conflict at issue in the organization, the specific technical questions or tasks, and the rhetorical purpose of the report. Here be careful to particularize the technical questions or tasks.
2. Forecast the structure of the discussion to follow. *[See pp. 93–94.]*
3. If necessary, amplify any element of the problem statement by supplying the relevant background material. *[See p. 94.]*
4. If necessary, explain any unfamiliar concepts necessary to understand the discussion. *[See pp. 94–95.]*

Guide 1.6.9. Ending Segments

[See pp. 95–97.]

1. Either, (a) reintroduce the technical problem and summarize the problem-solution process; or, (b) reintroduce the rhetorical purpose and address the organization.
2. If you do (a), do any or all of the following: restate the conclusions; restate the problem; summarize the main points.
3. If you do (b), do either or both of the following: recommend specific subsequent actions on the part of the reader; pose specific questions for subsequent investigation.

Guide 1.7. Drafting the Opening Component

Guide 1.7.1. Heading Segments

[See pp. 61–67.]

1. Formulate a subject line or title and subtitle for the report. Keep it brief, informative, and substantive. Make sure the first noun counts. *[See pp. 61–62.]*

2. Identify yourself by name, role, and department. *[See p. 62.]*
3. Identify your primary audiences by names, roles, and departments. If possible, also identify important secondary audiences by some device such as a distribution list. *[See p. 62.]*
4. Supply reference information necessary to locate the report in time and place. This information should allow for easy retrieval. *[See p. 67.]*
5. For a formal report compile a complete table of contents *[pp. 67–69]*. Make the headings substantive *[see pp. 163–164]*. Use a clear numbering system *[see pp. 165–166]*.

Guide 1.7.2. Foreword Segments

[See pp. 55–58, 67–75.]

1. State the problem, being sure to clarify the organizational context. *[See pp. 29, 32, 70.]*
2. State the assignment; in a manner accessible to general audiences, briefly explain the technical questions or tasks.
3. State the rhetorical purpose of the report clearly and directly. Be sure you state the *instrumental* purpose of the report, not what the report is about. *[See pp. 31, 70.]*

Guide 1.7.3. Summary Segments

[See pp. 75–80.]

1. Briefly, reintroduce the objective of the investigation.
2. If necessary to clarify the conclusions, briefly explain the methodology and important results.
3. Explain the conclusions. Remember, these must be presented in a manner accessible to all of your important audiences.
4. Present your recommendations. (Sometimes steps 3 and 4 are done simultaneously.)
5. Explain the implications for the organization, stressing costs and benefits.

Guide 1.8. Sending the Draft to the Typist

[See chapter nine, pp. 162–189.]

1. Establish a consistent pattern for spacing, indentation, and pagination of the typed draft. Determine:

   margins? (left, right, top, bottom)

indentation of units, paragraphs, quotations, and lists?
single-spaced or double-spaced text?
which segments begin on a new page?
spacing between segments?
spacing between units?
spacing between paragraphs?
spacing around headings, subheadings, tables, and figures?

2. Specify placement and type style for headings and subheadings. (Which are centered, which left margin, which capitalized, which italicized, et cetera.)
3. Specify space to be allowed for subsequent insertion of each visual aid. (Full page, part of a page?)
4. For formal reports, prepare a mock-up of the title page, table of contents, list of figures, tables, etc. (These will not be completed until the prototype draft is typed.)

Checklists
2.1.-2.7.2.

2. PREPARING THE FINISHED MASTER COPY

Checklist 2.1. Verifying the Basic Design

	Yes	No
1. Is the basic design of the report valid now that you see it in front of you?	[ ]	[ ]
2. Are you willing to let the report as it now stands be a measure of your technical and intellectual accomplishment?	[ ]	[ ]

Checklist 2.2. Editing Sentences in Context

	Yes	No
1. Does each segment or unit have a controlling idea—a core paragraph or sentence which establishes a pattern for the segment? (Can you understand it?)	[ ]	[ ]
2. Is the pattern established appropriate to the purpose of the segment? Can you identify the pattern and explain why you chose it?	[ ]	[ ]
3. Do the sentences carry out the pattern by following consistent noun and verb chains? Is the subject slot or verb slot in every sentence filled with an appropriate word?	[ ]	[ ]
4. If not, is the variation justified in terms of a larger context or in terms of amplifying the preceding sentence?	[ ]	[ ]
5. Do connective words and phrases signal the pattern?	[ ]	[ ]

Checklist 2.3. Editing Individual Sentences

At random, pick a page or two of text from the discussion component and use the following questions as a way of isolating problem areas. Once you have isolated problem areas reexamine the entire report in terms of the difficulties you have identified.

	Yes	No
1. Looking at every sentence in the passage you selected, are the primary subjects substantive, important nouns?	[ ]	[ ]
2. Looking at every sentence in the passage, are the verbs strong, active verbs? If not, are all the passive constructions contextually necessary?	[ ]	[ ]
3. Check every "and" or "but" joining clauses. Are all of these compound sentences necessary?	[ ]	[ ]
4. Look at each modifying clause or phrase. Try to eliminate any words you can and still retain the meaning. Try to reduce a clause to a phrase or a phrase to a word. Are all the words necessary?	[ ]	[ ]
5. Identify all "it . . . that" constructions and all "which" clauses. Are they necessary?	[ ]	[ ]
6. Circle every "it," "that," and "this." Can you draw an arrow to every antecedent? (If you find two "its" in the same paragraph, they should have the same antecedent.)	[ ]	[ ]
7. Identify every introductory participial phrase ("By using . . . ," "Upon filling . . . ," "Having completed . . . ," etc.) and then check the first noun in the next clause. Is that noun the source of the action?	[ ]	[ ]

Checklist 2.4. Editing the Format

	Yes	No
1. Are headings consistently used to signal divisions down to the unit level?	[ ]	[ ]
2. Do these headings form a consistent and complete pattern? (The wording and placement of equivalent headings should be parallel and consistent.)	[ ]	[ ]
3. Are the headings, especially second- and third-level headings, substantive?	[ ]	[ ]
4. Are there transitional elements at the ends of most of the discussion segments?	[ ]	[ ]
5. Are the segments and units (as well as perhaps some of the paragraphs) numbered according to a consistent pattern?	[ ]	[ ]
6. Is vertical white space used sufficiently and according to a consistent pattern?	[ ]	[ ]
7. Is horizontal white space used consistently?	[ ]	[ ]

Checklist 2.5. Preparing Finished Visual Aids

	Yes	No
1. Are all visual aids introduced in the text?	[ ]	[ ]
2. Are all aids interpreted in the text?	[ ]	[ ]
3. Are all aids appropriate for the text? (Another detailed version or another kind of the same aid might appear in the appendix.)	[ ]	[ ]
4. Are all the aids uncluttered and simple?	[ ]	[ ]
5. Are all the aids clearly identified by numbers and by substantive headings? (Could the purpose of each aid be understood independently?)	[ ]	[ ]
6. Do the parts of all the aids (and axes, bars, lines, et cetera) have clear labels to identify them?	[ ]	[ ]
7. Are all the aids accurate? (Scales appropriate, no suppressed zero axes, et cetera.)	[ ]	[ ]

Checklist 2.6. Writing the Abstract

	Yes	No
1. Does the abstract present information rather than merely describe what is in the report?	[ ]	[ ]
2. Does the abstract provide an overview of investigation (that is, does it explain the problem, method, results, conclusions, and recommendations)?	[ ]	[ ]
3. Is the abstract sufficiently brief to meet specified length requirements? (Usually 200 words or less.)	[ ]	[ ]

Checklist 2.7. Final Check

Checklist 2.7.1. Final Checklist for Informal Reports, Letter Reports, and Memoranda

1. The heading segment should provide sufficient information for any reader to be able to understand the organizational roles and relationships of the source and audiences. It should also define the subject of the report and fix the report in time by date and by references. Does it? — Yes [ ] No [ ]

2. The foreword segment of the report should identify the specific issue of organizational concern which the report addresses or responds to. It should also make explicitly clear the writer's reason for writing (i.e., the rhetorical purpose). Does it? — Yes [ ] No [ ]

3. The summary segment should provide a condensed statement of the conclusions and recommendations developed in the report; that is, it should summarize how the report responds to the problem which prompted the investigation and the report. Does it? — Yes [ ] No [ ]

4. If the report is relatively long—perhaps over three single-spaced pages—the beginning segment should introduce the discussion by amplifying the technical problem and specifying the technical questions or tasks addressed in the discussion component. Does it? (In short reports the introductory functions are performed by the foreword and summary segments.) — Yes [ ] No [ ]

5. The discussion should be clearly arranged according to either rhetorical purpose, or the intellectual problem-solving process, or subject matter. Is it? — Yes [ ] No [ ]

6. The discussion should contain all necessary—but only necessary—material. That is, it should be selective. Is it? — Yes [ ] No [ ]

7. The discussion should be explicitly segmented by format devices, such as headings, numbering, and white space (some short letter reports omit some of these devices). Is it? — Yes [ ] No [ ]

8. If the report is technically difficult, the discussion should often incorporate simplified visual aids to clarify the text. Does it? — Yes [ ] No [ ]

9. The discussion should be ended by generalizing in some appropriate manner. (For example, restatement of conclusions and suggestions for subsequent action.) Does it? — Yes [ ] No [ ]

10. The report ordinarily should supply necessary appendix or attachment materials to document and amplify the discussion. Does it? — Yes [ ] No [ ]

Checklist 2.7.2. Final Checklist for Formal Reports

1. The title should be brief (10 words or fewer), should have an important noun in initial position, and should use substantive terms. The subtitle should suggest the purpose of the report. Do they? **Yes** [ ] **No** [ ]

2. The title page should provide sufficient information for any reader to understand the organizational roles and relationships of the source and audiences. It should also fix the report in time by date and complete references (file numbers, retrieval codes, project numbers, etc.). Does it? **Yes** [ ] **No** [ ]

3. The abstract should very briefly condense the discussion component of the report. It should provide substantive information about the problem, methodology, results, conclusions, and recommendations. Does it? **Yes** [ ] **No** [ ]

4. The table of contents should list, number, and give page references for the headings and major subheadings (in long reports usually to third-level headings) exactly as they appear in the report. Does it? **Yes** [ ] **No** [ ]

5. The foreword should begin on a separate page. It should identify the issue of organizational concern. It should state the technical questions or tasks in a manner accessible to all audiences. It should state the rhetorical purpose of the report. Does it? **Yes** [ ] **No** [ ]

6. The summary should begin on a separate page. It should provide a condensed statement of the conclusions and recommendations developed in the report. These should be detailed enough to be instrumentally useful to decision makers and general audiences who will read no further. Does it? **Yes** [ ] **No** [ ]

7. If there is a letter of transmittal bound with the report and functioning as the opening component, it should serve the functions of the heading, foreword, and summary segments. Does it? **Yes** [ ] **No** [ ]

8. The beginning of the discussion should begin on a new page. It should provide amplified explanation of the problem and technical questions or tasks for those audiences who need particulars. Usually it should state the rhetorical purpose. Does it? **Yes** [ ] **No** [ ]

9. The beginning of the discussion may provide background or amplification material. It usually provides a forecast of the structure to follow. Does it? **Yes** [ ] **No** [ ]

10. The discussion should be clearly arranged according to either rhetorical purpose, or the intellectual problem-solving process, or subject matter. Is it? **Yes** [ ] **No** [ ]

11. The discussion should contain all necessary—but only necessary—material. That is, it should be selective. Is it? **Yes** [ ] **No** [ ]

12. The discussion should be explicitly segmented down to unit divisions by format devices such as headings, numbering, and white space. Is it? **Yes** [ ] **No** [ ]

13. Each of the segments should have a clear rhetorical purpose, an evident core paragraph, and appropriate organization. Do they? **Yes** [ ] **No** [ ]

14. The discussion should incorporate simplified visual aids to clarify the text. Does it? **Yes** [ ] **No** [ ]

15. The discussion should be ended by generalizing. It can summarize the problem-solution process or reintroduce the rhetorical purpose and recommend subsequent organizational actions. Does it? **Yes** [ ] **No** [ ]

16. The report should supply necessary appendix materials to document and amplify the discussion. Does it? **Yes** **No**

Appendices

The appendices to this book consist of nine complete reports—all written on the job by professionals. Differing in length, complexity, type, and format, these reports range from one-page memoranda to fifty-page formal reports. In fundamental respects, however, the reports all embody rhetorical principles we have explained throughout the book. Individually these reports serve as useful models; together, despite their stylistic idiosyncracies, they illustrate the underlying similarity of reports designed to address audience needs.

CONTENTS

APPENDIX A

Memorandum Report:
Request for Exception to Midland Division Safety Standard S-641

THE DOW CHEMICAL COMPANY

BAY CITY PLANTS & HYDROCARBONS

April 6, 1973

E. R. Wegner
Manager
Petrochemicals Section

Subject: REQUEST FOR EXCEPTION TO MIDLAND DIVISION SAFETY STANDARD S-641, PAR. IV, SECTION H, REQUIRING TOP UNLOADING OF TANK CARS

Due to its high freeze point (158°F), biphenyl must be heated to near its flash point (223°F) before handling is possible. Even then its vapor pressure is low. Priming a pump to top unload a tank car under these conditions is very difficult. Consequently, I request that an exception to division safety standards be made and that biphenyl tank cars be bottom unloaded.

Since ease of handling is really the problem here, bottom unloading would be the most appropriate solution. The only probable objection to bottom unloading of biphenyl tank cars is that of spillage. However, biphenyl quickly freezes at ambient temperatures, and therefore spillage can be cleaned up easily.

The most logical alternate unloading method would be pressure unloading with air or nitrogen. This method is questionable because a broken line would mean substantial spillage before the car could be depressured.

Bottom unloading of biphenyl tank cars would be in exception to Midland Division Safety Standard S-641, Par. IV, Section H, requiring top unloading of tank cars.

Alan C. Lukes /wb

Alan C. Lukes, Superintendent
Unifiner Benzene Plants

wb

APPENDIX B

Job Letter and Resume

1 March 1975

Mr. Jeffrey J. Mundth
2754 Golfside Road #713
Ann Arbor, Michigan 48104
(313) 434-2940

Mr. Herbert B. Spence, President
Spence Bros. General Contractors
3245 Bay City Road
Saginaw, Michigan 48672

Dear Mr. Spence:

Professor Donald H. Gray, who has been a soils consultant for your firm, has informed me that you might have an opening for a civil engineer, and has suggested that I contact you. My experience and educational background in business, construction management, and structural design would be an asset to your company in maintaining a strong position in the construction industry.

The enclosed resume details my qualifications, work experience, and educational background. These include:

Summer employment at GMC Truck and Coach, where I had responsibility for administration of $100,000 asphalt repair contract and $60,000 roof repair contract, including specification preparation, bid requests, and field supervision.

Graduate teaching assistantship in Civil Engineering, where I assist in decisions concerning construction engineering course content and design.

Recipient of Outstanding Undergraduate Civil Engineer award; rank in top 15% of my class in engineering.

I will complete my Master of Science program in construction management at The University of Michigan in May, and would appreciate being considered for permanent employment starting anytime thereafter. I would like to call you to arrange to drive up for an interview, if that would be convenient for you.

Thank you for your time and cooperation; I am looking forward to hearing from you.

Sincerely yours,

Jeffrey J. Mundth

Jeffrey J. Mundth

JJM/amm
ENCLOSURE

PERSONAL DATA RESUME

Jeffrey John Mundth
2754 Golfside Rd. No. 713
Ann Arbor, Mich. 48104
Phone No. (313) 434-3940

Career Plans and Objectives:

I am seeking employment in the construction industry or a closely allied field. My experience and educational background in the areas of business, construction management, and structural design will enable me to make significant contributions to the growth and development of a construction or design and build firm. I will obtain my professional engineer's registration.

Qualifications and Experience:

May 1974 to Present	Master's Degree in Civil Engineering—Construction Management (Univ. of Michigan; expected completion: May 1975). Graduate teaching assistantship in the Civil Engineering Dept. I assist in decisions concerning construction engineering course content and design, and assist with individual student instruction and grading.
Summer 1973	Employed by G.M.C. Truck and Coach, Pontiac, Mich. in the Plant Engineering Dept. As a member of the field supervision group, I had responsibility for administration of $100,000 asphalt repair contract and $60,000 roof repair contract including specification preparation, bid requests, and field supervision.
Summer 1971 1972	Self-employed in home improvement construction business. Self-run operation including marketing, design estimating and field operation. Total volume averaging approximately $6,000 per summer. Enterprise resulted from prior experience gained as general laborer-carpenter for custom homes builder, Dumbroski Homes, Inc., Bel Air, Maryland.

Academic Summary:

Education	Rackham School of Graduate Studies, University of Michigan: Master's Degree in Civil Engineering–Construction Management. (Completion expected: May 1975.) College of Engineering, University of Michigan: Bachelor's Degree in Civil Engineering–Construction Option (May 1974). Class standing above the 85th percentile. Moore School of Electrical Engineering, Univ. of Pennsylvania: Preliminary courses in engineering (9/70–5/72).
Awards, Honors, Societies	Recipient of Outstanding Undergraduate Civil Engineer for 1974. Chi Epsilon, Univ. of Mich. Chapter; Officer (Marshall) 1974. Tau Beta Pi, Univ. of Mich. Chapter. A.S.C.E. Student Chapter, Univ. of Mich.; Officer 1973. A.S.C.E. (national); Associate Member 1974.

References and transcript will be supplied on request.

APPENDIX C

Progress Report:
Progress Achieved by the Petrochemical Department During First Half 1974

DOW CHEMICAL U.S.A.

BAY CITY PLANTS · MICHIGAN DIVISION
P. O. BOX 516
BAY CITY, MICHIGAN 48706
517 · 684-1330

BAY CITY PLANTS & HYDROCARBONS
August 7, 1974

TO: E. R. Wegner, Manager, Petrochemicals Section

FROM: E. C. Williams, Superintendent, Butadiene Plant
D. H. Smith, Superintendent, Ethylene Plant

SUBJECT: PROGRESS ACHIEVED BY THE PETROCHEMICAL DEPARTMENT DURING FIRST HALF 1974

ATTACHMENTS: Maintenance statements by foremen R. Jenkins and H. Chase

At your request we are submitting a semi-annual progress report for the Petrochemical Department. This report covers production, safety, and the status of the Expanded Day (5th Shift) Program for the first six months of 1974. Since there have been extensive job restructuring and maintenance preparation for the new 5th Shift, a report on the status of the department, though routine, is especially important now.

Brief

The maintenance and job restructuring required for the 5th Shift program have not harmed either productivity or safety in the department. Productivity during the first quarter exceeded prior records. Despite a slight decrease in the second quarter, we project a record year for production. Safety also has been good. We had no disabling injuries in the department during the reporting period.

The Expanded Day Program development continues to progress as scheduled. Necessary maintenance has been done because the men have adapted well despite the obstacles. The first step of job restructuring has already been accomplished. To continue this progress we now need to begin hiring, to provide equipment, and to provide an adequate building. The Unifiner also needs a minor maintenance building of 400–500 square feet.

Productivity

The 1974 productivity of the department has been excellent. During the first quarter, production exceeded previous production records by 7%. During the second quarter the production rate did decrease slightly (3% below the first quarter). The decrease was caused by steam plant boiler outages, a power failure, an exchanger leak, and warmer weather. Despite this downturn, if we continue our present rates of production, 1974 will be a record year.

Safety

The department continues to work safely. There have been no disabling injuries (DI's) and very few near-miss incidents during the reporting period. We attribute this good safety record to our on-going drive to produce a safer work climate. For example, more platforms have been installed, sight glasses on equipment have been eliminated, and housekeeping has improved. Safety discussions and instruction, of course, continue. As our new people are becoming more experienced, we can expect even better safety performance.

The 5th Shift (Expanded Day Program)

Preparing for the 5th shift by maintenance and by job restructuring is progressing well. We now need to begin hiring and to expand facilities.

What we have done so far

We have good reason to be proud of the maintenance work being done to prepare for the Fifth Shift. The foremen and production supervisors deserve a lot of credit for their enthusiasm and good work. It is especially gratifying to witness the same thing from the operating technicians. Their willingness to learn and show us what can be done if they are allowed and trusted to perform well is encouraging. (Please read the attached statements by two of the shift foremen regarding the maintenance performed by their crews.)

The maintenance items performed were necessary to the smooth operation of the plant, and they have been accomplished in the face of reduced manpower due to (1) job restructuring, (2) the Midland strike, (3) vacation, and (4) unusual sickness.

A second and very important part of the program is the job restructuring. The first step was accomplished two months ago. The furnace-dehydrator duties were combined into one job (and the pay level upgraded) and the ethylene loading, depentanizer and furnace assistance duties were added to the transfer and lab tech duties. Again, the operating people responded well and the results have been excellent. Not only have our people learned their own duties, but many are, on their own and under the direction of the shift foremen, learning other jobs as well. Again, a good deal of credit goes to the shift foremen and those people who are taking up the challenge and doing something about it.

We feel strongly that we should promote training and equipment, and provide supervisory encouragement so our people can show what they can do. The plant will run better, human relationships will be improved, and vacations etc. can be more easily accommodated.

Here is where we need some help

Our people have shown that they are very capable, and have responded over and above what we had proposed. To continue the progress and to make the 5th shift a solid success we need to do the following:

1. Begin Hiring: By September 1, 1974 we need 6 people. (A request for hiring dated 17 July 1974 has already been submitted to Personnel Department.)

2. Provide Equipment: We have provided small hand tools already. In addition, we need a drill press, a pipe threader, and two trucks.

3. Provide an Adequate Building: We have consulted an outside architect regarding the building proposed for Area 4. The architect's report indicates 3000 square feet would be required for shop area and offices. In addition, a minor maintenance building for the Unifiner (approximately 400–500 square feet) is necessary. The present Diphenyl building would do, but it is being used for research purposes.

Meeting these needs is crucial to the success of the Expanded Day Program.

Conclusion

As you can see, preparation for the Expanded Day Program is moving ahead on schedule. At the same time the department has been able to maintain high production and safety records. We therefore feel our progress during the first six months of 1974 has been excellent.

E C Williams

E. C. Williams, Superintendent,
Butadiene Plant

D. H. Smith

D. H. Smith, Superintendent
Ethylene Plant

APPENDIX D

Informal Report in Letter Form:
Adjudication of final estimate for modifications:
Take Home Motor Speed Control and Reversing

AMERICAN PRESIDENT LINES

INTERNATIONAL BUILDING

601 CALIFORNIA STREET • SAN FRANCISCO, CALIFORNIA 94108 U. S. A.

Trans-Pacific • Round-the-World

August 16, 1972 File: Change Order No. 16

TO: John Smith, Head
Office of Ship Construction
Maritime Administration
U.S. Department of Commerce
Washington, D.C. 20235

FROM: H. F. Monroe, Manager,
Engineering Department
American President Lines

SUBJECT: Adjudication of final estimate for modifications:
Take Home Motor Speed Control and Reversing
APL Seamasters, MARAD DESIGN C4-5-69a
CONTRACT NO. MA/H8B-46
MARAD HULLS 191-194 and 206
DOE HULLS 489-493
Change Inquiry No. 8—Change Order No. 16

DISTRIBUTION: Doe Shipbuilding, Doeville
Brown Naval Architects, New York
Marad-Doeville

Dear Mr. Smith:

On August 8, 1969, American President Lines authorized specification changes on five Seamasters through the issuance of Change Order No. 16 (ref. e). The Owner in making this economic decision relied upon a preliminary cost estimate given by Doe Shipbuilding, the Contractor. The Contractor now submits a final estimate exceeding the preliminary estimate by 600%. The Owner considers this final estimate unreasonable, and in accordance with standard contractual procedure asks Marad to establish a fair and reasonable cost. The Owner requests Marad to review and adjudicate this final estimate.

Modifications to the motor, controllers, and resistor bank drew a preliminary cost estimate by the Contractor of $42,221 for the five ships (ref. c). The final estimate is $279,866 (ref. g). The Owner feels that costs related to a change in gear ratio, structural changes to bulkheads, modification of gears and coupling, and related material costs are not chargeable to the Change Order and should be considered as a development of the contract. The Owner furthermore questions the costs due to disruption in man-hours (678). The Owner also observes an escalation in costs for successive vessels not provided for in the contract. The final estimate by the Contractor, therefore, is excessive, and should be made consistent with the preliminary estimate and with the specific changes requested in Change Order No. 16.

In examining the reasonableness of the final cost estimate submitted by the Contractor, the Owner thinks the following questions must be addressed:

1. Most important, do the charges relate specifically to the particular modifications required by the Change Order, or should the charges be considered as a development of the contract?

2. Are the charges realistic or inflated on the basis of typical construction procedures?

3. Are the charges provided for in the contract?

The disparity between the Contractor's preliminary and final cost estimates suggests that the final estimate must be scrutinized.

To answer these questions, the modifications to the original contract specifications required by Change Order No. 16 must be noted. Then specific items in the Contractor's final estimate can be examined in light of the original contract and modifications. Other charges in the estimate can be examined separately.

Disparity Between Preliminary and Final Charges

Final charges made by the Contractor exceed the preliminary estimate by 600%.

The preliminary estimate totaled $42,221. In response to Change Inquiry No. 8 (ref. b), the Contractor submitted a preliminary estimate of $8,502 increased cost for the first four ships (ref. c). This estimate was later supplemented by an additional cost of $8,213 for the fifth ship (ref. d). After a thorough study of the possible benefits to be gained, based on this estimate the Owner concluded that the increased cost was justified. He authorized the change (ref. e) on an unsubsidized basis following Marad approval (ref. f).

3

The final estimate submitted by the Contractor amounted to $279,866 for five ships (ref. g). Hulls 489 and 490 cost $53,719 each for modifications; Hulls 491, 492, and 493 cost $57,354 each for modifications.

Modifications Required by Change Order No. 16

Original contract specifications for the Seamasters called for a 750 HP non-reversing electric take-home motor of unspecified type. The motor chosen by the Contractor had to be modified to suit Change Inquiry No. 8 and Change Order No. 16.

The original contract required only that the motor be suitably geared and clutchable to the main turbine gears. No stipulation was made for propeller RPM or vessel speed. If an externally mounted resistor bank were to be required for the type of motor chosen, it was to have a screen-type enclosure for better ventilation.

The single-speed non-reversing wound rotor motor selected by the Contractor required an externally mounted resistor bank. The resistor bank was to be used during acceleration of the motor for start-up only. Heat generated in this resistor could be dissipated without forced ventilation.

Change Inquiry No. 8 was initiated to permit full motor loading under various conditions of draft, weather, and sea state. The change required two additional speed steps at 80% and 90% of full RPM as well as non-plugging reversing capability of the motor. The motor and controllers were modified to accommodate the change. Furthermore, the resistor bank had to be enlarged and arranged for continuous use when operating at the two reduced speed steps. Heat dissipation under these conditions required forced draft ventilation.

Final Cost Estimate for Modifications to Motor, Controller, and Resistors

On the basis of the original contract specifications and of the modifications required by Change Order No. 16, three items of charges in the Contractor's final cost estimate are unsupportable. These items should be considered as a development of the original contract rather than as required by the change order. Related material charges consequently are unjustified.

Item II (2). Gear and Coupling. The need for change in gear ratio to provide a 32.5 propeller shaft RPM is not clear. The change inquiry scope did not stipulate any propeller speed, only that provision be made for speed steps at 80% and 90% of maximum. In fact, Addendum No. 4 to the original purchase requisition (ref. h) set the propeller shaft RPM at 32.5

because the purchase requisition simply called for the reduction gear to provide "an appropriate output speed." Hence the choice of 32.5 as an "appropriate RPM" is considered as a development of the contract, and is not attributable to the change order.

Item II (7). Structural Changes. Structural changes were made to bulkheads at aft end of machinery space and at the tank top level. APL records show modifications made to these bulkheads were done as a development of the contract. The modifications resulted from discussions at Doeville on March 3 and 4, 1969. These discussions are documented in Doe Shipbuilding Inter-Office Memos dated March 9 and March 17, 1969. There is no evidence that the modifications had anything to do with Change Order No. 16.

Item IVA. It is not clear why any increased costs attributable to modification of gears and coupling should be charged to this change order. The gearing has been commented on above. The range of RPM under speed control is small, and the coupling size is influenced as much by starting torque as running torque. It thus seems to us that any change in this item was a development of the take-home propulsion mode design called for in the original contract.

Material Costs. From back-up material supplied, we are unable to confirm amount or cost of Davis 168 cable attributed to this change. We do question the increased costs of gears and couplings due to modifications, as explained above.

Final Cost Estimates for Labor and Cost Escalation

The final cost estimate includes other charges that the Owner finds unjustified. Charges for labor, especially for disruption, due to the modifications are not realistic on the basis of typical construction procedures. Escalation of costs for successive vessels is not provided for in the contract.

Labor. The entire number of disruption in man-hours (678) is questioned. This change was authorized nearly five months before the keel of the first ship was laid, and over seventeen months before the keel of Hull 493 was laid. The Owner has too much admiration for the yard's planning procedures to believe that such a minor modification could result in any disruption. Ample lead time was available.

The manpower charged to this change is unrealistically inflated. Particularly inflated are the charges for welders/burners/tackers, welders, cleaners, sheet metal crafts, and machinists.

Cost Escalation. While material costs for all five vessels are identical, the manufacturing and engineering labor, and overhead and insurance are higher for Hulls 491-492-493 than for Hulls 489-490. The contract does not provide for escalation of these costs for successive vessels.

The preliminary estimates indicated a lower cost for Hull 493 than for the first four ships, presumably based on the learning curve. Only upon review of the final estimate for this change did we note any cost escalation of significance.

It is unfortunate, to say the least, that an Owner in making an economic decision must rely on the engineering expertise of the shipyard to submit a reasonably accurate preliminary estimate only to find the final estimate to be in error by almost 600%. That the final estimate is in error seems clear for reasons explained above. The Owner therefore requests your review and adjudication of this estimate. Enclosed for your use are all of the references, as well as the Contractor's back-up material for this charge.

Very truly yours,

H. F. Munroe

H. F. Munroe
Manager, Engineering Department

Enclosures: Ref. a—Ref. h
Back-up material

APPENDIX E

Technical Paper:
An Educational Laboratory in Contemporary Digital Design

AN EDUCATIONAL LABORATORY IN CONTEMPORARY DIGITAL DESIGN

by

Janis Beitch Baron

and

D. E. Atkins

Department of Electrical and Computer Engineering
and
Program in Computer, Information and Control Engineering
The University of Michigan
Ann Arbor, Michigan 48104

Summary

Formal education in computer architecture rests upon the integration of numerous specialized courses. One such course, a new laboratory concerning contemporary, register-transfer level digital design, is described. Design in the course proceeds at the level of interconnection of memories, arithmetic logic units, I/O interfaces, and buses, rather than at the individual gate level associated with traditional combinational and sequential logic design. The principle items of hardware used in the laboratory are DEC Register Transfer Modules, the DEC PDP/16M, and the INTEL SIM8-01 Microcomputer. Numerous software packages have been developed to support activities within the lab including an RTM simulator, a graphical simulation of an RTM control sequencer, a PDP/16M assembler, a microcode generator, and SIM8-01 utility routine, assembler, and simulator. During one fourteen-week term, students are expected to complete six assigned lab projects and one special project. Examples of projects are described. A critique based upon the students' reactions and the technician's experience with the equipment is given together with proposals for future additions to the course.

Introduction

Computer architecture is a system-oriented design discipline requiring broad understanding within the hierarchy of computer system descriptions: the logic level, the instruction set level, the programming level, and the PMS level. (See reference [1] for a discussion of these description levels.) Formal education in computer architecture cannot be accomplished within a single course or two, but rather rests upon the integration of numerous specialized courses. This paper describes one such course in the computer engineering curriculum at The University of Michigan: a laboratory course concerning contemporary digital system design including, but not limited to, general purpose computers.

Laboratories in digital design ("logic labs") have existed within university curricula; several others exist within The University of Michigan. What justifies describing yet another? The answer to this question relies on our notion of "contemporary" digital system design which underlies the structure of this laboratory. Contemporary digital system design includes the following correlated properties:

Uses medium and large scale integration (MSI, LSI) technology.

Focuses on algorithms (not physical quantities such as voltages) and their alternate modes of implementation: hardware, software, microprogramming (firmware).

Stresses register-transfer level primitives (registers, adders, memories, etc.) rather than gate level (gates, flip-flops, inverters).

Employs a top-down approach (similar to structured programming) using flowchart specification of computation.

Laboratory projects are implemented using primarily two commercially available digital hardware products: the Digital Equipment Corporation (DEC) Register Transfer Modules (RTM)[1] and the INTEL SIM8-01 Microcomputer. A DEC PDP/8 minicomputer is also available for projects involving augmentation of general purpose machines with special purpose subsystems.

The course begins with several modest experiments to introduce the student to the Register Transfer Modules. Design proceeds at the level of interconnection of memories, arithmetic logic units, I/O interfaces, and buses rather than at the individual gate level associated with traditional combinational and sequential logic design. Design at this higher level permits the design and implementation of a moderately complex digital system, e.g. a mini-computer, within several laboratory periods. In later phases of the course, students select, design, implement, and analyze projects of their own choosing.

The course serves at least two broad purposes. It provides hands-on experience with digital hardware and immerses the student in the act of digital system design. Secondarily, the course develops skills in the application of two relatively new commercial products which exhibit an increasing influence on the realities of digital design.

The course also serves as an introduction to techniques for implementing special-purpose digital structures. Today there are many incentives to organize systems around a computer. However, with the present low cost of mini-computers, there would probably not be economic incentive to build a general purpose computer from modules used in this laboratory. The commercial value of register-transfer modules rests in the area of implementing special-purpose systems: either stand-alone or augmented general-purpose computers.

This paper will include a description of the hardware and software facilities established for the laboratory, a description of the content of the course, and a critique based upon experience to date.

The intent of the paper is not to sell the goals of the laboratory, but rather to offer our experience to other educators who basically agree with the goals and who are also attempting to implement them.

Facilities

Hardware

The principal items of hardware used in the laboratory are DEC Register Transfer Modules, the DEC PDP 16/M, and the INTEL SIM8-01 Microcomputer. Other auxiliary hardware required for interconnection will also be mentioned.

Register Transfer Modules (RTMs): Register Transfer Modules represent a commercial approach to the implementation of modular computer systems which permit design and construction at the register transfer level. A history of modular computer systems, beginning with Estrin's "fixed-plus-variable" computer systems, is traced in reference [2].

Medium and large scale integration circuits in themselves provide a loosely constrained modular approach to the design of the data flow section of a digital computer. Modular systems, in the spirit described in reference [2] and exemplified in the DEC RTM's, go further by also standardizing and simplifying the control part of the system. They represent attempts to provide simple alternatives to the complexities and inefficiencies of programming general purpose computers for applications which lend

[1] Register Transfer Module (RTM) is a registered trademark of Digital Equipment Corporation, Maynard, Massachusetts. The product is also known in a special form as the PDP-16.

themselves to specialized hardware organizations. The standardization implicit in the Register Transfer Modules enables students to design and construct several non-trivial digital systems within one term.

The RTMs are not projected as *the way* to implement digital systems. They do, however, provide a cost-effective solution to some design problems and are a pedagogical tool in studying solutions to many others.

The reader not familiar with a detailed description of Register Transfer Modules is referred to references [3]–[7]. Reference [3] contains the most correct description of the module pin assignments.

PDP-16/M Subminicomputer: In addition to the RTMs, the laboratory is equipped with one PDP-16/M, a specific configuration of RTMs sequenced by a microprogrammed controller. The unit is a functional computer with 256 words of read only control memory, 4 words of data read only memory, a general purpose register unit, 96 instructions, fully implemented 16 bit I/O enclosure, and power supplies. A detailed description of the unit is contained in reference [8].

The basic PDP-16/M sells for about $2000. The back plane is prewired to accept additional standard RTMs to extended memory and I/O capabilities.

The availability of the 16/M provides a specific contrast to the "chained evoke" control technique typically used with the RTMs. The economic advantages of a stored program approach to control become apparent for systems involving more than a few dozen control steps.

Both reprogrammable (PROM) and fusible link read only memories (ROMs) are available on the control memory module. DEC provides several options for programming ROMs: one using a PDP/8 and special interfacing hardware, one using the PDP-16/M itself. None of these is used in the U. M. laboratory. The PROMs on the PCS16-B control board have been removed and replaced by zero insertion force, dual-in-line sockets. With this modification the PROMs can be easily removed and replaced. They are programmed using the INTEL SIM8-01 prototype system described in the next section. An assembler to produce PDP-16/M code, including punched paper tape compatible with the PROM burning equipment, is described in the software portion of Section II.

INTEL SIM8-01—An MCS-8 Microcomputer: Microprocessors are exerting an increasing influence on digital design, not only as the basis for microcomputer systems, but also as components within other systems. Evidence is accumulating that they are cost effective replacements for even small amounts of random logic. Exposure to microprocessor technology is an essential ingredient in computer engineering education. The INTEL SIM8 and associated hardware have proved to be a satisfactory vehicle for such exposure.

The SIM8-01 is a specific configuration of components from the INTEL Microcomputer Set built around the 8008-01 central processor chip. It is a prototyping board which forms an operational microcomputer. It includes a central processor (INTEL 8008), 1024 x 8 bit read/write memory, six I/O ports (two in and four out), a clock generator, and sockets for 2048 words of read only memory.

The SIM8-01, a single 10″ x 12″ board, is used in conjunction with the MCB8-10 interconnect and control module. With the addition of an MP7-03 PROM Programmer, an ASR teletype, and a utility program stored in PROMs, the hardware becomes the MCS-8 PROM Programming System. Two such configurations are included in the laboratory. The purchase price of each is approximately $1200. The organization and operation of this equipment is well documented in reference [9].

Interfacing Equipment: As discussed in Section III, the later part of the course is devoted to a free choice design project. Many of the projects require the interconnection of various combinations of the following: RTM systems, PDP-16/M, SIM8, and PDP 8 minicomputer. The logic levels of all of these are compatible. However, some additional interfacing logic may be required to satisfy timing constraints. Suitcase type logic kits, developed for a long standing introductory course in gate-level design, have proved valuable in meeting the requirement. Ribbon cable terminated with 16 pin dual in line connectors (DIPs) are the primary means of interconnection.

A Microcomputer—RTM Student Project: Figure 1 shows a student built music synthesizer using RTMs, the INTEL equipment, and a suitcase logic kit.

Software

Several software packages have been developed to support activities within the laboratory. These packages, all but one of which are reasonably transportable, are briefly described in this section. References to more complete descriptions will be included. The exception cited above is a computer graphics package to simulate the operation of an RTM control sequence. A computer generated motion picture based upon the graphical simulation is scheduled for production within the coming year and will be available on a loan basis.

RTM Simulator: To reduce the amount of time that must be spent debugging once hardware assembly has begun, the RTM designer may choose to simulate his/her design on the central computing facility before building it. With SIM-16, a FORTRAN simulation of part of the RTM system, the digital hardware designer can create, test, and modify the logical structure of any single-bus RTM design. The program requires the user to describe the design as a sequence of FORTRAN CALL statements. These statements accomplish transfers of data between symbolic source and destination registers and perform control functions such as branching, merging, and decoding. The program allows close monitoring throughout the progress of the simulation. It is more fully described in Ref. [13].

Wiring Table Generation: To facilitate the assembly of an RTM device once its flowchart has been decided upon, a program called *WIREWRAP is used. *WIREWRAP, written in a combination of FORTRAN and IBM 360 Assembly Language, is similar to DEC's CHARTWARE, which runs only on the PDP-10. *WIREWRAP is a general-purpose logic design aid, extended to include RTM designs. Once a design is complete, the user need only describe it in a simple form that is derived directly from the flowchart. The program then decides which RTMs are needed, where each will be placed on the wire-wrap panel, and which pins should be connected together. It checks all gates and storage elements for overloading, calculates total cost and power requirements, minimizes wire length wherever possible, and generates complete wiring instructions and documentation. It is further described in Ref. [11].

Control Simulation on a CRT: As a pictorial aid for demonstrating the operation of the hard-wired control structure of the RTM system, The University of Michigan's PDP-9 Logic Simulation graphics system has been extended to provide an RTM control simulation. The simulation emphasizes the details of control timing rather than the trace of data flow provided by SIM-16. The control timing is graphically illustrated in slow motion on a DEC 339 display screen. The particular display, composed and controlled by the user by means of a light pen, simulates the operation of a specified CONTROL SEQUENCER. The CONTROL SEQUENCER may consist of EVOKE steps (mostly register transfers), two-way Boolean decisions, branches, merges, and subroutine calls. Figure 2 shows a student using the simulator.

PDP-16M Assembler: The PDP-16M Assembler, which runs on the central computing facility, translates symbolic programs for the PDP-16/M into binary object modules that are usually produced in the form of paper tapes. The contents of these paper tapes can be electronically burned into the Programmable Read Only Memories (PROMs) for use with the PDP-16M by means of the SIM8-01 microcomputer set and the Michigan INTEL Programmer and Loader Routines (described below). Ref. [16] is a user's guide for the Assembler.

Microcode Generator: As a special project, three students have developed a tool for implementing interpreters by means of microprogramming. The General Purpose Microcode Generator for Emulation on the PDP-16/M is an IBM 360 Assembly Language program which generates microcode for the control memory of the PDP-16/M. The program's input is a description of the instruction set of the emulated computer. This facility gives the students the opportunity to compare hardware and firmware implementations. It is further described in Ref. [8].

Utility Routines: To support software development for the INTEL Corporation SIM8-01 microcomputer, a set of utility programs has been developed. The Michigan INTEL Programmer and Loader Routines (MIM-9/PL) enable the user to load object code contained on paper tape into the writable memory of the SIM8-01, to debug the code interactively using a variety of breakpoint and status examination commands, and to electronically program the programmable read only memories (PROMs) once the code has been debugged. Although INTEL itself provides some utility programs, the MCS-8 PROM Programming System, the programs lack facilities for interactive debugging, are inflexible, and require an unnecessarily verbose format for object code. MIM-8/PL includes all of the facilities of the INTEL version plus others. The programs are available for general distribution. They are further described in Ref. [15].

MCS-8 Assembler: Object code for the SIM8-01 is typically produced by the Michigan INTEL MCS-8 Assembler (MIM-8/AL), an IBM 360 Assembly Language program. The assembler produces a binary-coded paper tape which is of the same format as the output of the PDP-16/M assembler. The contents of the paper tape can be electronically burned into a PROM by means of the SIM8-01 and MIM-8/PL (described above). MIM-8/AL can run under OS/VSI or the Michigan Terminal System and is available for general distribution. It is further described in Ref. [14].

MCS-8 Simulator: Students in courses without access to the MCS-8 hardware and students in the digital design lab who wish to debug their code before using the hardware make use of the MCS-8 Simulator. The simulator, written in IBM 360 Assembly Language, can execute about 300 simulated instructions per minute. It is designed to handle all of the MCS-8 instructions, including the input/output instructions. Its internal registers and I/O ports can be monitored throughout the simulation. In addition to providing a means for debugging code, the simulator generates a trace that is useful later when the code is being checked on the real machine. It is more completely described in Ref. [17].

Content of the Course

The content of the digital design laboratory course is built around class lectures, assigned laboratory projects, and special student projects, with little emphasis on written examinations. Each week of the two-credit hour, fourteen-week course contains a one-hour lecture and a three-hour lab. The students spend additional time outside of class preparing for and documenting each lab project. Written examinations are not used since the projects themselves are a comprehensive test of what the students have learned.

The Class Lectures

In addition to covering the various hardware and software facilities and the assigned projects, the class lectures discuss the following topics:

1. Levels of description of digital design, in particular, the register transfer (RT) and processor-memory-switch (PMS) levels.
2. Design choices behind the production of the Register Transfer Modules; for example, the logical design, the bus structure, the asynchronous timing.
3. Speed/cost, hardware/firmware/software, and special-purpose/general-purpose trade-offs.
4. Alternate forms of parallelism and synchronization of parallel activities.

Assigned Projects

During one fourteen-week term students are expected to complete six assigned digital design lab projects and one special project.

Table 1 lists the assigned lab projects and the number of lab periods allocated to complete each project.

PROJECT	ASSIGNED PROJECT	NUMBER OF 3-HOUR LAB PERIODS
1	1 to N Summer	1
2	Count the number of 1's or 0's in a word	1
3	Register I/O Utility	1-2
4	Single Precision, Integer Multiplication and Division	1-2
5	Simulated Serial Synchronous Receiver	2
6	MCS-8 Program	1-2

Table 1. Assigned Lab Projects

Projects 1 and 2 enable the students to gain familiarity with the RTMs and the design process for using them.

Project 3 is the design and implementation of a "Register I/O Utility Routine." The routine uses the lights and switches for depositing in and examining the registers within the scratch pad memory. It also produces an I/O facility that is used in later projects, for example in lab 4.

Project 4 gives the students their first experience with the design and implementation of a fairly complex RTM system and with multiplication and division algorithms.

Project 5 is the design of a serial synchronous receiver. It is not built in hardware, but is implemented with the SIM-16 RTM simulation package. The students must design the receiver to satisfy a set of requirements including transmission speed, maximum message length, and specified character length.

Project 6 enables the students to gain familiarity with the CPU architecture and the instruction set of the MCS-8 micro-computer. They write and debug a subroutine to sort a group of records stored in a buffer in an external random access memory.

Special Design Projects

The final six lab periods of the course are devoted to free choice individual or small-group special projects. One suggested approach to the projects is intended to illustrate "fixed plus variable structure" architecture.

The students are told to:

1. Define an application problem.
2. Implement a software solution on the SIM8-01.
3. Identify a time-critical portion of the software realization and design a special purpose RTM or PDP-16M system to implement this portion.

4. Interface the special purpose processor to the SIM8-01 software.
5. Modify the program to use the hardware processor instead of the software routines.
6. Conduct a performance comparison of the two approaches.

Projects following this approach give the students experience with interfacing and with analysis of hardware/software trade-offs.

The following are examples of some of the more ambitious special projects:

1. The implementation of part of an assembler on the SIM8-01 with an RTM content-addressable memory for the symbol table.
2. A SIM8-01 based desk calculator with a PDP-16M floating point arithmetic unit and elementary function generator (sin,cos, etc.).
3. A SIM8-01-controlled signal processor with a PDP-16M Fast Fourier Transform (FFT) routine.
4. A CRT line drawer. The SIM8-01 receives line-drawing instructions from the teletype consisting of the endpoints of a set of lines. It then sends the data to an RTM device which controls the CRT display.
5. An infix to postfix translator for arithmetic expressions. The project includes a speed and cost comparison of an RTM vs. a PDP-8 assembly language implementation.
6. A "Digital Music Machine." The SIM8-01 receives the musical score via teletype, then translates and forwards it to an RTM music synthesizer. (See Figure 1.)

Critique of the Course

The following critique will discuss the students' reactions to the course, the technician's reactions to working with equipment, and some accumulated ideas for possible future additions to the course facilities.

The Students' Reactions

Aside from a few complaints, most students claim that they find the course challenging, interesting, and rewarding. The hands-on experience, they feel, is clearly superior to classroom lecture learning. The special project, most agree, is particularly interesting. Perhaps more time should be allotted for the final project by eliminating one of the first six projects.

The students express only two significant complaints. The main complaint concerns the great amount of time that the course work requires. Many suggest that three hours of credit should be given instead of two. In addition, the students feel that the large amount of hand-wiring becomes tedious after about four projects. For this reason, work with the RTM simulator instead of the actual hardware by at least the fifth project is advisable.

The Technician's Reactions

The laboratory technician has mixed reactions about the desirability of working with the equipment used in the lab. His job consists of acquiring, assembling, and maintaining the equipment. These tasks, though not time-consuming by themselves, have been complicated by what the technician calls "growing pains." In other words, the equipment being used is new and still in the development stage. It is, therefore, insufficiently—and often incorrectly—documented. The problems of tracking down needed information often consume weeks. Fortunately, the availability of information from INTEL and DEC has subsequently improved.

Excluding all problems related to the "growing pains," the time requirements of the technician are not great. The assembly stage consists of a few weeks for the construction of four RTM assembly racks for the eight- to twelve-student labs, and a week or two for the remaining assembly details. Maintenance requirements are minimal. Burnt out PROMs constitute the only problem encountered with the MCS-8 system. In the past two years, we have found only two faulty RTM boards. No maintenance problems have arisen from use of the PDP-16M.

The technician has compiled lists of tips for use by others working with the same equipment. The lists, available from the authors, include tips on trouble-shooting, repair, facilities for ease of operation, purchase of the needed pieces of hardware, and proper care of the equipment.

Future Additions to the Course Facilities

Some possible future additions to the course facilities include:

1. *Movies of the RTM control simulation*—Students generally agree that the CRT display facility is a helpful, clear, pedagogic device. Hopefully, soon, movies of the display will be available for general distribution.
2. *An interface between the SIM-16 RTM simulator and the *WIREWRAP program*—After an RTM

design has been logically debugged by means of SIM-16, the next step is to input the design to *WIREWRAP for implementation details. A program to convert the input for SIM-16 to a form compatible with the input for *WIREWRAP would greatly simplify the user's job.

3. *An interface to enable the PDP-16M to substitute a minicomputer for the PROM during the micro-program development stage*—Since PROMs cannot be re-programmed very easily or very often, debugging PDP-16M programs using the PROM to store the program is clumsy. A micro-program stored in a minicomputer, however, is easily altered. Perhaps, while the micro-program is being debugged, the minicomputer could feed it the micro-instruction that will, after the debugging is complete, be read from the PROM.

Acknowledgments

Contributions to this paper have been made by the laboratory technician: Duane Haines; the instructors: K. B. Irani, S. Goldner, M. Bauer; those who developed the software facilities: S. Goldner, L. Uzcátegui, J. Gilbert, M. Ziegler, J. Blinn, K. D. Kanaby, D. R. Hanson, E. Berelian, J. Mulla, C. Zervos; and the students who designed and implemented the special projects described: T. Harkaway, P. Kostishak, D. Luther, J. J. Puttress, E. A. Berra, M. Shrader, R. Bryant.

This work was supported in part by the NSF Grant GY-8318.

References

[1] C. Gordon Bell and Allen Newell, Computer Structures: Readings and Examples, McGraw-Hill, New York, 1971, Chapter 1.

[2] Robert A. Ellis, "Modular Computer Systems," Computer, Vol. 6, No. 10 (Oct. 1973), p. 13.

[3] C. Gordon Bell, John Grason, and Allen Newell, Designing Computers and Digital Systems of Using PDP 16 Register Transfer Modules, Digital Press, 1972.

[4] PDP 16 Computer Designers Handbook, Digital Equipment Corporation, 1971.

[5] Register Transfer Modules . . . A Cost-effective, Easy-to-use, Method for Designing Logic Systems, Digital Equipment Corporation, 1973.

[6] C. G. Bell and J. Grason, "Register Transfer Modules (RTM) and their Designs," Computer Design, May 1971.

[7] C. G. Bell, J. L. Eggert, J. Grason, and P. Williams, "The Description and Use of Register Transfer Modules (RTMs)," IEEE Trans. Comput., Vol. C-21 (May 1972), pp. 495-500.

[8] PDP 16/M User's Guide, Digital Equipment Corporation, 1973.

[9] MCS-8 Micro Computer Set User's Manual, INTEL Corporation, March 1973.

[10] INTEL Data Catalog, INTEL Corporation, February 1973.

[11] Janis Beitch Baron, Using *WIREWRAP for PDP/16 Register Transfer Module Design, Department of Electrical and Computer Engineering, The University of Michigan, Ann Arbor, Michigan, July 1973.

[12] Daryl E. Knobloch, *WIREWRAP User's Guide, Computer Center Memo #M181, The University of Michigan Computing Center, July 1971.

[13] James G. Gilbert, User's Manual for SIM-16, Electrical and Computer Engineering, The University of Michigan, Ann Arbor, Michigan, September 1973.

[14] S. M. Goldner, L. A. Uzcátegui, D. E. Atkins, K. B. Irani, User's Guide for the Michigan INTEL MCS-8 Assembly Language, Electrical and Computer Engineering, The University of Michigan, Ann Arbor, Michigan, July 1973.

[15] S. M. Goldner, L. A. Uzcátegui, D. E. Atkins, and K. B. Irani, User's Guide for the Michigan INTEL MCS-8 Programmer and Loader System, Electrical and Computer Engineering, The University of Michigan, Ann Arbor, Michigan, July 1973.

[16] Leonardo A. Uzcátegui, PDP 16-M Assembler Description and User's Guide, Electrical and Computer Engineering, The University of Michigan, Ann Arbor, Michigan, July 1973.

[17] K. D. Kanaby and D. R. Hanson, MCS-8 Simulator, Electrical and Computer Engineering, The University of Michigan, Ann Arbor, Michigan, revised April 1973.

[18] Elias Berelian, Jamshed Mulla, and Christian Zervos, General Purpose Microcode Generator for Emulation on the PDP-16M, Electrical and Computer Engineering, The University of Michigan, Ann Arbor, Michigan, April 1974.

APPENDIX F

Short Formal Report with Memorandum of Transmittal: *Distribution Underbuild on 138 kV Circuits*

Consumers Power Company

INTERNAL CORRESPONDENCE

To: A. C. Fagerlund, P-13-213, Manager
Northwest Division

From: D. C. Tarsi, Principal Engineer, Equipment
and Materials Division
P. Ghose, Senior Supervisory Engineer,
Systems Analysis Section

Date: 26 October 1973

Subject: Distribution Underbuild on 138 kV lines.

cc: H. J. Jenson, Senior Engineer, Operations
C. R. Brown, Senior Engineer, Systems Protection
M. L. O'Brien, Principal Engineer, Planning
R. W. Wiarda, Senior Engineer, Transmission Lines

Attached is a copy of the final report on the proposed distribution underbuild on 138 kV lines. A preliminary copy of this report was circulated among the concerned divisions and departments, and their comments have been included in this final version.

After a comprehensive evaluation of the electrical performance characteristics resulting from the underbuild, we have concluded that no unusual problems would be encountered to jeopardize the performance of either the 138 kV or the distribution circuit. The following are the highlights of the report:

1. Several utilities which have similar distribution circuits suspended beneath transmission circuits have reported excellent results. Because the distance between the transmission and distribution circuits in our proposed construction is similar to those used by these utilities, we expect equally satisfactory results.
2. The electrostatically and electromagnetically induced voltages during normal, distribution de-energized, or minor fault conditions will have no significant effect on either the distribution or transmission circuits. Neither the reliability nor the performance of the system will be degraded by the proposed construction.
3. In rare situations, excessive fault currents during a single line-to-ground fault on the 138 kV relaying isolates the faulted line. This is of no concern since the transformer saturation in a wye grounded system provides the necessary current to trip the distribution circuit. The probability for distribution line tripping for this reason is once in 200 years.
   Under the same conditions in a delta connected system, it is possible for a distribution lightning arrester to fail. The distribution circuit would then continue to operate without interruption, but without adequate lightning protection.
4. Although significant induced voltages will exist on the distribution circuit when it is de-energized, proper safety procedures during circuit maintenance will eliminate any shock effects. Also, the distribution transformers and their connected load will drain off most of the electrostatically induced voltages that are of any concern.

A more detailed analysis of the electrical performance will be found in the attached report. If you have any questions, we will be happy to discuss them with you.

THE CONSUMERS POWER COMPANY
JACKSON, MICHIGAN

DISTRIBUTION UNDERBUILD ON 138 kV CIRCUITS

RE-316

N.G.G. Swamy, P.E.
General Engineer
Systems Analysis Section
Equipment and Materials Division
Electrical Engineering Department
October 26, 1973

DISTRIBUTION UNDERBUILD ON 138 kV CIRCUITS

RE-316

INTRODUCTION

Although Consumers Power presently has distribution underbuild on 46 kV lines, it does not have distribution underbuild on 138 kV lines. The Planning Division has recently proposed adding distribution circuit underbuilds to existing single and double circuit wood pole 138 kV transmission lines in the Northwest area. Therefore a study was conducted by the Systems Analysis Section of the Equipment and Materials Division to determine the feasibility of the proposal and to coordinate requirements relating to the proposed construction. The two lines studied were (1) the Boardman-Munson Junction double circuit 138 kV line with approximately two miles of 7.2/12.46 kV distribution underbuild and (2) the Emmet-McGulpin single circuit 138 kV line with approximately three miles of 13.2 kV and 0.3 miles of 2.4 kV distribution underbuild. (The locations of the distribution underbuilds are shown in Figures 1 and 2 attached.) This report discusses the analysis and conclusions from the study and makes recommendations for the construction, maintenance, and operation of the system.

SCOPE OF THE STUDY

The study considered the effect of electromagnetic and electrostatic induced voltages and currents from the transmission circuit on the distribution circuit and vice versa. Included were the effects due to (1) both circuits energized and (2) one circuit de-energized. The study also considered factors such as service performance and reliability, safety of operating personnel, transmission and distribution design requirements, relaying operation, and effects on distribution transformers.

CONCLUSIONS

The study indicates that no unusual problems should be encountered to jeopardize the performance of either the 138 kV or the distribution circuit. The reasons for this conclusion are:

1. We have encountered no special problems on the existing double circuit 46 kV distribution underbuild although significant induced voltages exist on these lines. In addition, several other utilities have distribution circuits under transmission circuits and have reported excellent operation from them. The distance between the 138 kV circuit(s) and the distribution circuit in our proposed construction is similar to the distances used by these utilities.

2. Because of the short line lengths in the proposed underbuild, the induced voltages during steady state and fault conditions do not significantly influence the customer voltage, relaying operations, or performance of distribution fusing or equipment.

3. Although significant induced voltages exist on the distribution circuit when it is de-energized, proper safety procedures during circuit maintenance will assist in eliminating shock effects. In addition, the distribution transformers and their connected load will drain off most of the electrostatically induced voltages.

I. PREVIOUS DISTRIBUTION UNDERBUILD

The induced voltages and currents are determined primarily by the voltages of each circuit and the distance between them. The proposed underbuild is similar in both these respects to underbuilds presently operated by Consumers Power and other utilities. None of these existing systems has experienced any degradation of reliability.

A. *Effect on our Existing 46 kV Distribution Underbuild Lines*

The electromagnetic and electrostatic induced voltages and currents were calculated for the proposed 138 kV distribution underbuild construction in order to determine their relative magnitudes with respect to the existing double circuit 46 kV distribution underbuild. Presently, induced voltages and currents are highest when the distribution circuit is de-energized and the two 46 kV circuits are energized, especially for fault conditions. Even though these induced voltages and currents are lower than those of the 138 kV distribution underbuild, severe shocks can be experienced during circuit maintenance if proper safety procedures are not followed. A detailed listing of various induced voltages and currents on the distribution circuit when it is de-energized is given in Table I. It was concluded from this analysis that the induced effects on the proposed 138 kV distribution underbuild are not significantly different from those of the existing 46 kV distribution underbuild.

B. *Effects on the Distribution Underbuild of Other Utilities*

Several utilities were contacted to find out about their experiences with distribution underbuild on their transmission circuits. Eight utilities indicated that they have distribution underbuild on single lines for transmission voltages ranging from 115 kV to 345 kV. Five of these eight utilities informed us that they have underbuild on either 115 kV or 138 kV double circuit lines for a distance of one to twelve miles. The distribution voltages ranged between 4.8 kV and 34.5 kV for these underbuild constructions. The clearance between distribution circuit(s) and transmission circuit(s) was found to be in the range of 13 to 20 feet. All of these utilities have reported excellent performance and have experienced no problems so far. There appears to be no reason why our lines should not operate equally well.

II. CALCULATION OF THE INDUCED EFFECTS

The electrostatic and electromagnetic induced voltages on the transmission and distribution circuits of the proposed underbuild were calculated for normal, one circuit de-energized, and fault conditions. Our analysis assumed that the placement of the lines on the poles would be as is shown in Figure 3. The method used for calculating these effects is described in Table I.

The conclusion drawn from these calculations is that induced voltages are highest when the distribution circuit is de-energized and the 138 kV circuit(s) is energized. This is especially true during fault conditions. Open circuiting the distribution circuit under the two 138 kV circuits results in a maximum electrostatic voltage of 14 kV on one of the de-energized phases for normal 138 kV operation and 19.5 kV for a double line-to-ground fault on the top two phases of the 138 kV circuits. The maximum electromagnetic induced voltages for the same situations were found to be .25 V/A/Mi for normal operations and 1.0 V/A/Mi for a single line-to-ground fault on the bottom phase of the two 138 kV circuits. The 1.0 V/A/Mi corresponds to approximately 10 kV for the distribution circuit under the proposed double circuit Boardman-Munson Junction line if the fault occurs at the end of the underbuild. A more detailed listing of induced voltages and currents is given in Table I.

III. SYSTEM CONSIDERATIONS

The effect of induced voltage or current on the hardware of the system appears to be minimal. All of the various components studied were shown to be able to function normally, even under major fault conditions. The normal protection built into the system appears to be quite adequate for use with the proposed distribution underbuild.

A. *Effect of Induced Voltage and Current on the Transmission and Distribution Circuits*

Due to the short line lengths in the proposed underbuild, no significant increase in operating voltage of either the 138 kV or the distribution circuits should be experienced during normal or fault conditions. In rare situations, excessive current during a single line-to-ground fault on the 138 kV line(s) can cause an increase in the distribution operating voltage until 138 kV relaying isolates the faulted line (usually within approximately 30 cycles). This is of no concern in a wye grounded system since transformer saturation due to higher operating voltage provides the necessary current to trip the distribution circuit protection. For the proposed Boardman-Munson Junction double circuit distribution underbuild, we calculate a 50% probability of distribution tripping for this reason once in 200 years.

B. *Effect of Induced Quantities on the Lightning Arrestors*

Distribution underbuild should not cause any severe problems with the lightning arrestors. In a delta connected distribution system, the excessive induced voltage during a 138 kV fault is discharged by the line-to-ground connected lightning arrestors rather than by transformer saturation. In a very rare situation, it is possible to fail a distribution lightning arrestor on a delta connected system due to short circuit ground faults occuring on the nearest 138 kV phase conductor. Under this situation the distribution circuit will continue to operate without interruption of service, but without adequate lightning protection.

C. *Effect of Induced Quantities on Fusing*

The proposed underbuild should allow fusing on the distribution system to be maintained with no problems. This conclusion is based on reasonable projections of system conditions to the year 2000. However, if additional underbuilds are undertaken in the future, each must be considered by its own merits. Longer underbuilds and/or higher 138 kV currents would cause distribution fusing problems that would require separate resolution. Both the probabilities of 138 kV faults and of possible failure of distribution fuses and cutouts due to their inability to clear at high induced voltages would have to be re-evaluated under the new conditions.

D. *Effect of Induced Quantities on Relaying*

No adverse effects on either transmission or distribution relaying should be encountered from the proposed underbuild construction. As with the fusing, the System Protection Department should be consulted if any future underbuilds are undertaken to make sure that no relaying problems would exist.

E. *Effect of Induced Quantities on Distribution Transformers and Loads*

In a wye grounded system, the majority of the induced voltage on the de-energized distribution circuit from the transmission circuit(s) will be drained off by the transformers and their secondary loads connected to the distribution circuit. This is because even a small load acts as a low resistance path for draining off these induced voltages compared to the de-energized line capacitive reactance. Further, if induced voltages higher than the nominal distribution voltage should exist during some rare situation, the transformer saturation lowers the magnetizing reactance, thus providing the required draining of the induced voltage. Therefore, the effect of induced voltages on the distribution transformers or on loads connected to the secondary of the transformers is considered to be insignificant.

F. *Design Requirements for Reducing Induced Quantities During Normal Operation*

The final conclusion the study made in examining system considerations was that changes in sag, circuit clearances, and circuit height within the design limits will not affect the induced voltages significantly. However, reverse phase sequencing of the second circuit with respect to the first circuit, as shown in Figure 4, will reduce induced voltages. Previously it was indicated that the probability of distribution tripping is once in 200 years during steady-state operation of the proposed double circuit distribution underbuild. This probability can be somewhat improved by using the reverse phase sequence construction. Therefore, this configuration is recommended for all such distribution underbuild.

IV. SAFETY CONSIDERATIONS

Adherence to the following guidelines will eliminate or minimize the hazards associated with electrostatically or electromagnetically induced energy on the conductor. These procedures are in line with the rules and regulations specified by OSHA.

A. *The Proper Method for Grounding Work Areas*

Even though high induced voltages can be encountered on a de-energized circuit, they can be eliminated by proper grounding. A ground anywhere on a de-energized conductor is sufficient to drain off any electrostatically induced charge. However, a second ground is necessary to provide a path for an electromagnetically induced charge to drain off. By using two grounds and keeping near them, maintenance personnel can eliminate most of the danger from induced voltage or current.

B. *Guidelines for Safe Maintenance*

Field personnel should routinely follow these work procedures:

1. Low resistance safety grounds should always be applied on either side of the working personnel before workmen touch the conductor. The two grounds should be as close to each other as possible (20 feet or less). Use of the tower footing ground or driving a temporary grounding rod will provide satisfactory grounding for the safety grounds.

2. All conductors should be cleaned using hot-stick techniques before ground or jumper clamps are attached.

3. When more than one phase is grounded, jumpering from phase to phase with minimum cable sag and employing a single ground lead will minimize cable stress and assist in providing effective grounds.

4. Gloves should be used by personnel while working on the de-energized distribution line. Personnel should also keep their hands as dry as possible.

5. All conducting objects within a work area should be wired together so that it is not possible to receive a shock through contact with two different grounds present in the work area.

SUMMARY

Induced voltage and current in distribution underbuild affects many areas in the operation of a system. Special methods of design and maintenance can alleviate most of the difficulties arising from these induced effects. The problems with the proposed construction of 138 kV single and double circuit underbuild can be solved by the use of the methods outlined in this report. In the future, if a distribution line has to be built under a single or double circuit 138 kV line for different system conditions, the effect of induced voltage or current should be determined by conducting a separate study. The System Analysis Section should be consulted in such a case to determine the effect of induced voltage on the distribution system. In this way, we will be able to provide continuous, low cost electricity to our customers in these rural areas for many years to come.

REFERENCES

1. "Electrostatic Effect of Overhead Transmission Lines: Part I—Effects and Safeguards," by IEEE Working Group on E/S and E/M Effects, IEEE Transactions on Power Apparatus and Systems, Vol. PAS-91, pp. 422–426, 1972.

2. "Electrostatic Effect of Overhead Transmission Lines: Part II—Method of Calculation," by IEEE Working Group on E/S and E/M Effects, IEEE Transactions on Power Apparatus and Systems, Vol. PAS-91, pp. 426–433, 1972.

3. "Electromagnetic Effects of Overhead Transmission Lines—Practical Problems, Safeguards, and Methods of Calculation," by IEEE Working Group on E/S and E/M Effects, presented at the 1973 Summer Power Meeting.

4. "ACSR: Aluminum Conductor, Steel Reinforced Wire," by Alcoa, American Electricians Handbook, Table 120, pp. 2–64, 1972.

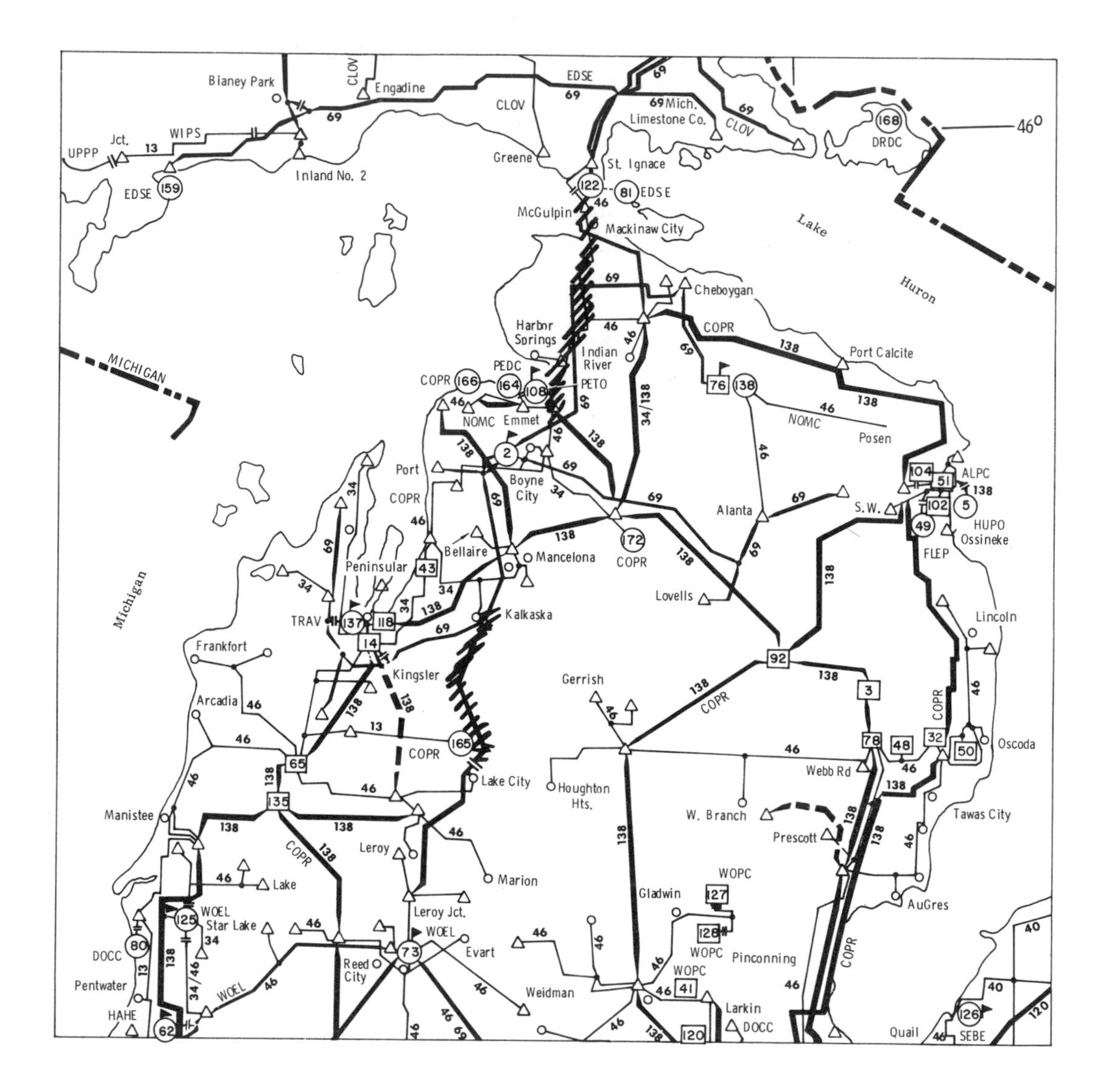

Bianey Park
Engadine
CLOV
EDSE
UPPP
Jct.
WIPS
Inland No. 2
Greene
Mich. Limestone Co.
St. Ignace
McGulpin
Mackinaw City
DRDC
46°
Lake
Huron
Cheboygan
COPR
Port Calcite
Harbor Springs
Indian River
PEDC
PETO
NOMC
Emmet
Posen
MICHIGAN
Michigan
Port
Boyne City
Alanta
ALPC
HUPO
Ossineke
S.W.
FLEP
Bellaire
Mancelona
Peninsular
Lovells
TRAV
Kalkaska
Lincoln
Frankfort
Kingsler
Gerrish
Arcadia
Oscoda
Lake City
Houghton Hts.
Webb Rd
Manistee
W. Branch
Prescott
Tawas City
Leroy
Marion
Lake
WOEL
Star Lake
Leroy Jct.
Gladwin
WOPC
AuGres
DOCC
Reed City
Evart
Pinconning
Pentwater
HAHE
Weidman
Larkin
Quail
SEBE

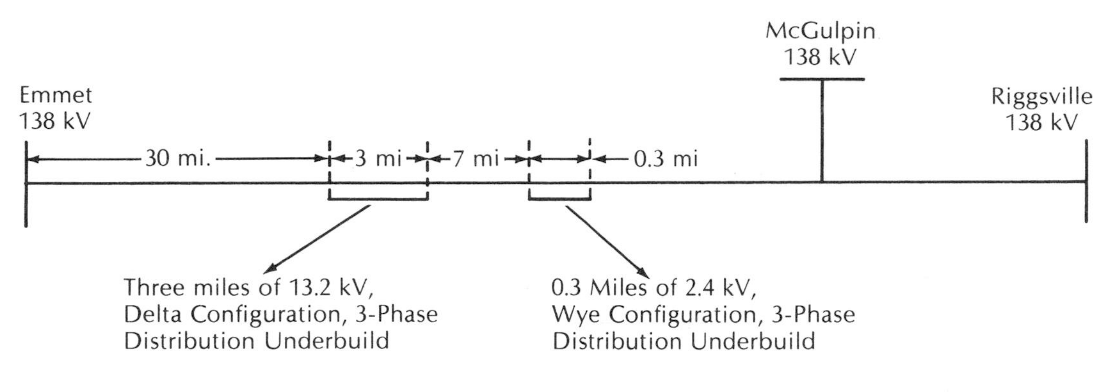

Figure 2a

Location of Single Circuit Underbuild Construction

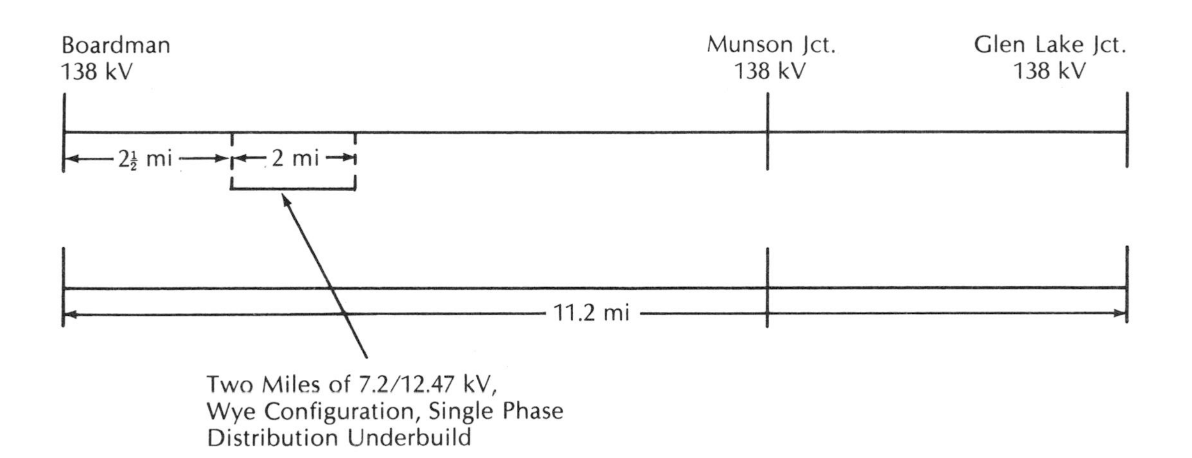

Figure 2b

Location of Double Circuit Underbuild Construction

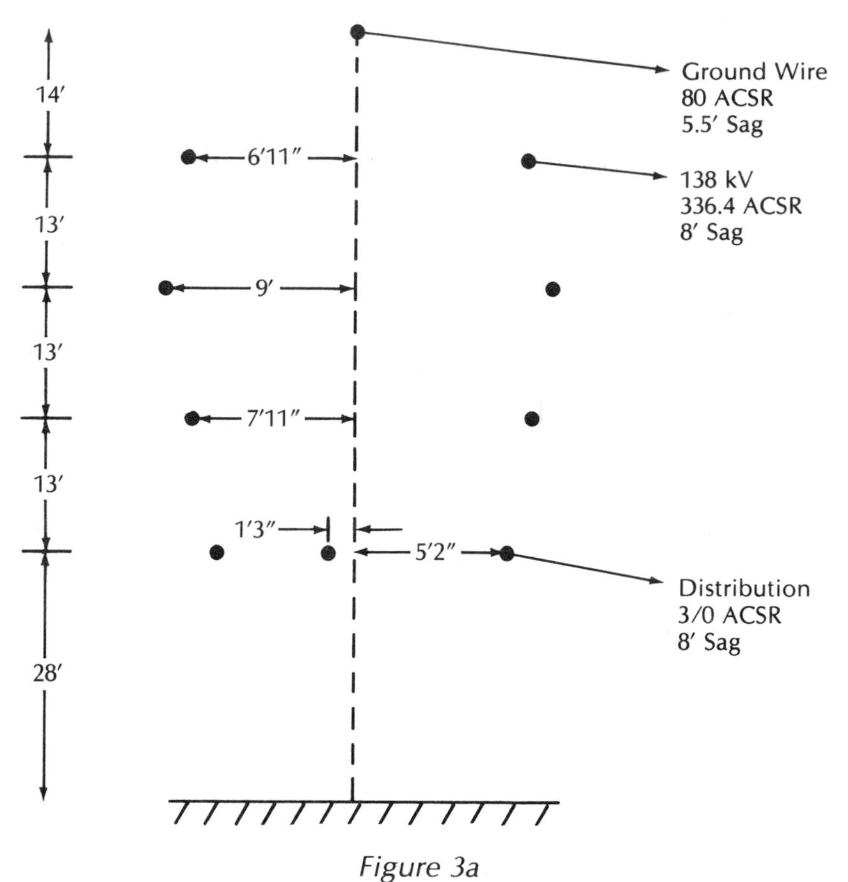

Figure 3a

Distribution Underbuild on Double Circuit 138 kV Lines

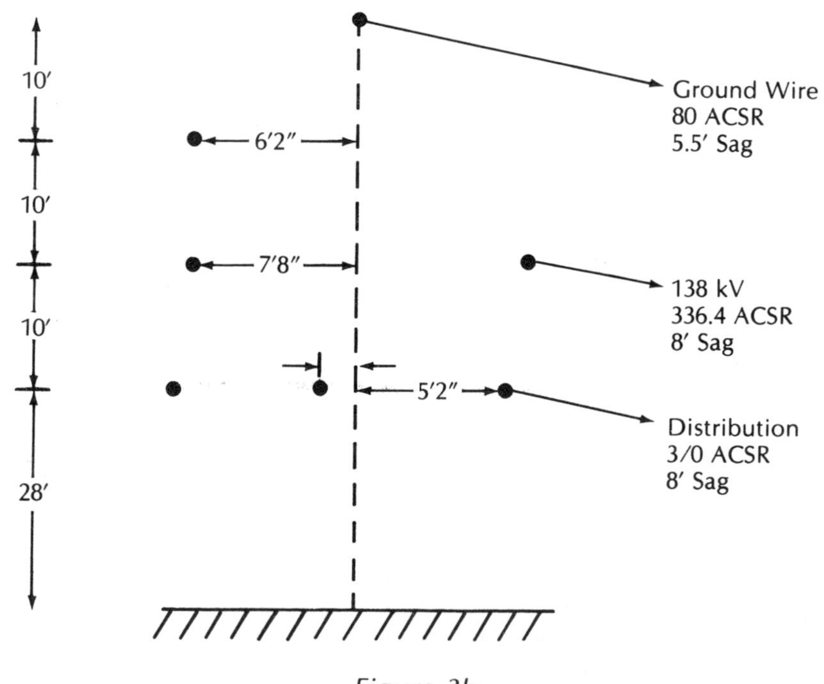

Figure 3b

Distribution Underbuild on Single Circuit 138 kV Lines

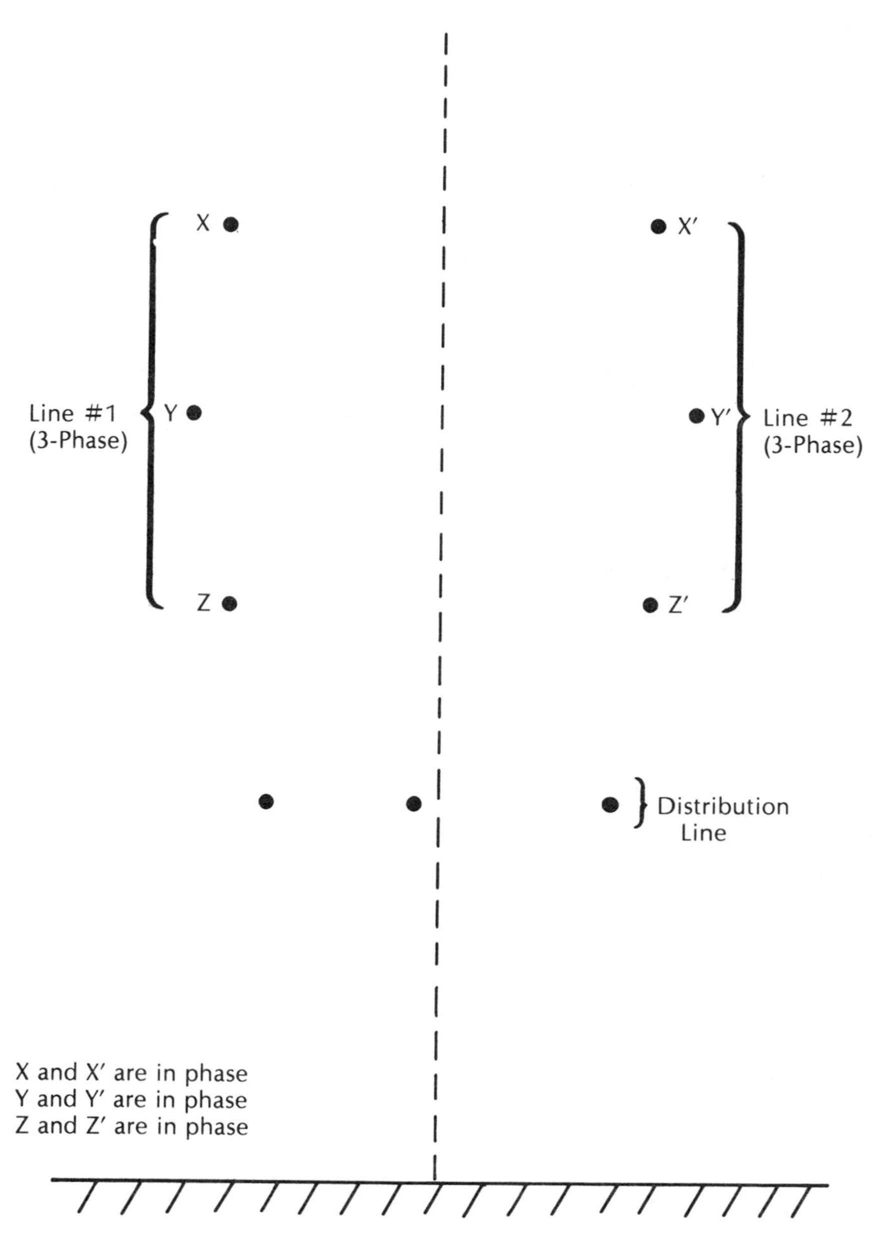

Figure 4

Reverse Phase Sequence Arrangement for Two 3-Phase Lines (to Reduce Induced Voltage)

TABLE I.

INDUCED VOLTAGES AND CURRENTS ON A DE-ENERGIZED DISTRIBUTION CIRCUIT DUE TO 138 kV CIRCUIT(s)

	ENERGIZED CIRCUIT (138 kV)		
DESCRIPTION	138 kV Single Circuit	138 kV Double Circuit	46 kV Double Circuit
Electrostatic Effects			
Maximum Induced Voltage During Normal Operation	8.0 kV	14.0 kV	6.0 kV
Maximum Induced Voltage During Faults on the 138 kV Circuit	14.5 kV	19.5 kV	8.5 kV
Maximum Normal Charging Current Through a Man (1000 Ohms Resistance) in Direct Contact with an Ungrounded Line	.38 A/Mi	.48 A/Mi	.24 A/Mi
Electromagnetic Effects			
Maximum Induced Voltage During Normal Operation	.12 V/A/Mi	.25 V/A/Mi	.27 V/A/Mi
Maximum Induced Voltage During a Single Phase Fault	.50 V/A/Mi	1.00 V/A/Mi	1.15 V/A/Mi

APPENDIX G

Manual:
Cyphernet FFIO Routines

Network Services, Inc.
Cyphernetics Division
175 Jackson Plaza
Ann Arbor, Michigan 48106
(313) 769-6800

Dear Cyphernet System User:

Cyphernet Fast FORTRAN Input/Output (FFIO) is now available to all users of the Cyphernet System. FFIO is a set of FORTRAN-callable subroutines for operating on files which cannot be read or written using standard FORTRAN input/output.

The primary purpose of the FFIO routines is to enable FORTRAN programs to read and write magnetic tapes in formats compatible with other computer systems. The FFIO routines will, however, operate on any device (including disk and DECtape), and are useful for reading and writing binary files which do not conform to Cyphernet System conventions.

For a complete description of Cyphernet FFIO, order the new FFIO User's Manual. This manual includes the following information:

Accessing FFIO Subroutines

Basic FFIO Subroutines—FOPEN, FREAD, FWRITE, FSPACR, FSPACE, FREWND, FCLOSE

Character Conversion Subroutines—CONVRD, ICONV, CONVER

Sample FFIO Program

To order your copy of the FFIO User's Manual, call your Cyphernetics representative or simply mail the enclosed coupon.

Sincerely yours,

Stephanie Rosenbaum

Stephanie Rosenbaum
Publications Manager

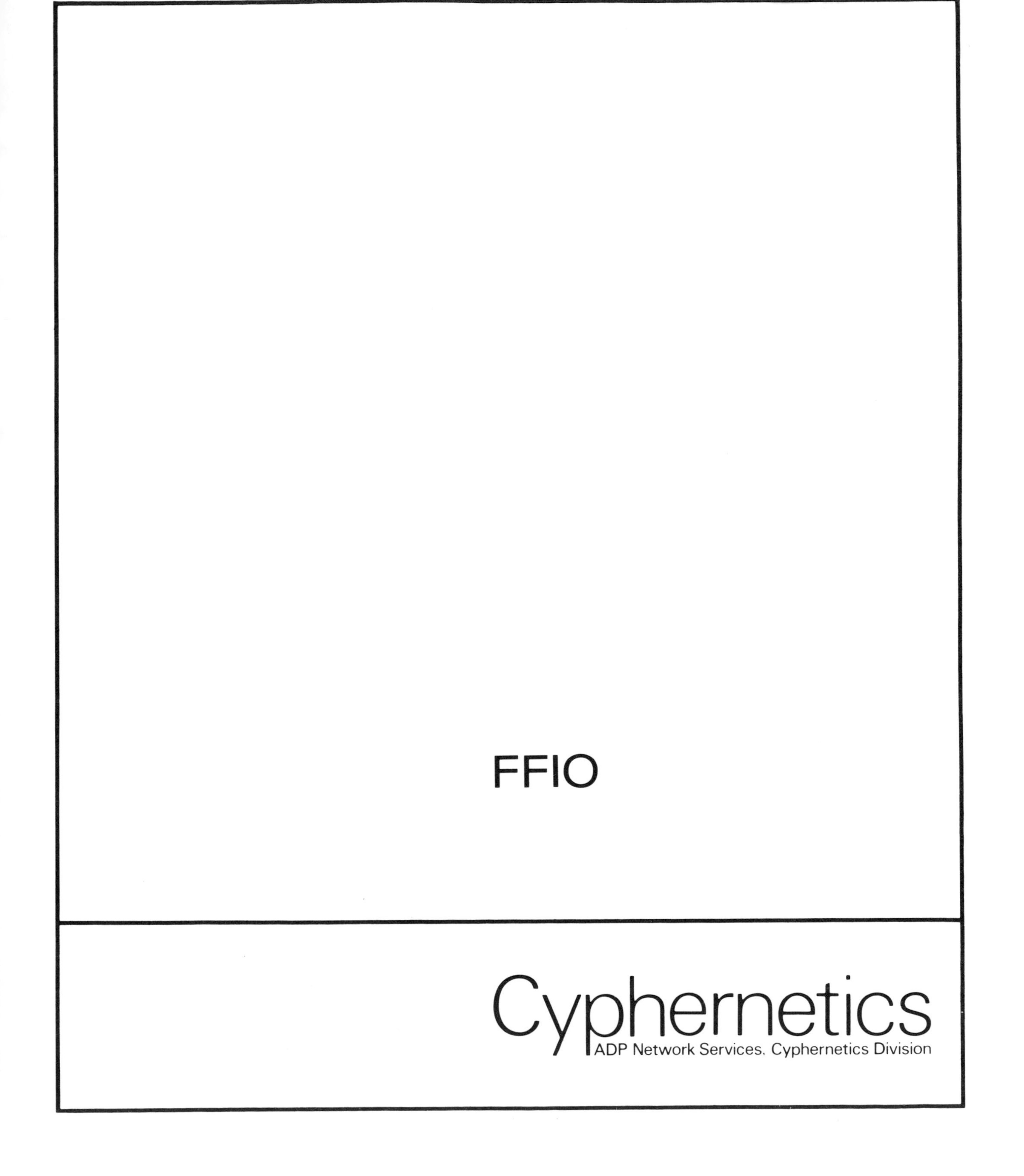
FFIO
Cyphernetics
ADP Network Services. Cyphernetics Division

Cyphernetics makes no warranty, expressed or implied, as to the documentation, function, or performance of this system; and the user of the system is expected to make the final evaluation as to the usefulness of the system in his own environment.

Cyphernetics Publication 120-5
First Edition
September, 1975

This manual was edited, formatted, and photocomposed using CypherText.

Table of Contents

Introduction

Cyphernet FFIO (Fast Fortran Input/Output) is a set of FORTRAN-callable subroutines for operating on files which cannot be read and written using standard FORTRAN input and output statements.

The primary purpose of the FFIO routines is to enable FORTRAN programs to read and write magnetic tapes in formats compatible with other computer systems. The FFIO routines will, however, operate on any device (including disk and DECtape), and are useful for reading and writing binary files which do not conform to Cyphernet System conventions.

The FFIO subroutines are divided into two groups. One group is a set of basic input/output routines. The other is a set of very flexible routines for doing translations from one set of character codes to another.

Accessing FFIO Subroutines

The FFIO subroutines are contained in the subroutine library file, FOR:FFIO.REL. To request the FFIO subroutines when running your program, type the Command Language command RUN, the name of your FORTRAN program, and the library option (LIB) followed by FOR:FFIO. The library option retrieves from the file FOR:FFIO only those FFIO subroutines which your program calls.

For example, if the program MYPROG calls FFIO subroutines, type:

RUN MYPROG,(LIB)FOR:FFIO

Here the system compiled, loaded into core, and executed MYPROG. The user specified the library option (LIB) followed by the file name FOR:FFIO, to retrieve the subroutines called by MYPROG.

To request subroutines from the FOR:FFIO file when you create and store a core image version of your program, use the same procedure as described above, with the Command Language CREATE command instead of the RUN command. For example:

CREATE MYPROG,(LIB)FOR:FFIO

For complete information on program execution and compilation commands, refer to the Cyphernet Command Language Manual (Publication Number 100-1).

Basic FFIO Subroutines

This section describes the basic FFIO input/output subroutines contained in the library file, FOR:FFIO.REL. A standard nomenclature is used to show the general form for calling each subroutine.

In the general form of each subroutine call:

> Parameters shown in upper-case letters, special characters, and punctuation marks are *literal* parameters. When you use these parameters, you must enter them exactly as shown in the general form.
>
> Parameters shown in lower-case letters are *variable* parameters. When they are used with the subroutine, you must supply a suitable substitute in place of the parameter name used in the general form.
>
> Parameters shown in square brackets (i.e., [ ]) are *optional* parameters. You may supply these parameters or not, depending on the way you use the subroutine.
>
> Parameters not enclosed in square brackets are *mandatory* parameters — you must supply the parameter when you use the subroutine.

The following pages present descriptions of all the basic FFIO subroutines.

FOPEN

PURPOSE

The FOPEN subroutine initializes a device or file on a FORTRAN logical unit for operation by the FFIO subroutines.

GENERAL FORM

CALL FOPEN (unit,name,direction,status [,max,nbufs,density,parity,mode])

Where:

unit is the FORTRAN unit number.

name is a standard device/file name (e.g., 'M305:' or 'DSK:XYZ.DAT').

direction specifies the direction of transfer: 'I' for input, 'O' for output.

status is an integer variable in which the subroutine returns a status code as follows:

Status Code	Code Description
0	The device or file has been opened successfully.
1	Bad parameter.
2	Device or file not available or does not exist.

The optional parameters must appear in the order shown in the general form; if one is specified, all preceding it must also be specified. If you specify a value of −1 for any of these parameters, the default value for that parameter will be used.

max is the maximum physical record size (in DECsystem-10 words). The default is 128. For 7-track tapes, allow one word for every 6 tape frames. For 9-track tapes, use one word for every 4 frames on the tape in 'I' mode, and one for every 5 frames in 'D' mode.

nbufs is the number of buffers to use. The default is 2.

density is the tape density, coded as 2=200 bpi, 5=556 bpi, 8=800 bpi. The default is 5.

parity is the tape parity, coded as 'O'=odd, 'E'=even. The default is 'O'.

mode specifies the recording mode for 9-track tapes. 'D' specifies DEC mode, 'I' specifies industry standard mode. The default is 'I'. 'I' is the appropriate specification for all tapes for non-DEC computer systems.

EXAMPLE

CALL FOPEN (22,'M305','I',IS,500,2,8,'O','I')

Here the user called FOPEN to open M305 on unit 22 for input. M305 has records up to 500 words long, and is an 800 bpi, odd parity, industry standard 9-track tape. FFIO will use 2 input buffers internally.

FREAD

PURPOSE

The FREAD subroutine reads a selected number of words from a FORTRAN logical unit into a specified array.

GENERAL FORM

CALL FREAD (unit,buff,nwords,status)

Where:

unit is the FORTRAN unit number.

buff specifies an array.

nwords is the number of words to be read from **unit** into the array **buff**.

status is an integer variable in which the subroutine returns a status code as follows:

Status Code	Code Description
0	Data read successfully.
1	Bad parameter.
3	Device error (hardware).
4	Record too large (i.e., a physical block was larger than **max** as specified in FOPEN).
5	End of file.
6	End of tape.
7	Beginning of tape.

NOTES

The array **buff** must be at least **nwords** long. If **nwords** is initially zero, FREAD will read one physical block and return its size in **nwords**.

If you wish to read another file on magnetic tape after encountering an end-of-file, you must call FCLOSE and call FOPEN again.

FWRITE

PURPOSE

The FWRITE subroutine writes a selected number of words from a specified array as one physical record.

GENERAL FORM

CALL FWRITE (unit,buff,nwords,status)

Where:

unit is the FORTRAN unit number.

buff specifies an array.

nwords is the number of words to be written from the array **buff** as one physical record.

status is an integer variable in which the subroutine returns a status code as follows:

Status Code	Code Description
0	Data written successfully.
1	Bad parameter.
3	Device error (hardware).
4	Record too large (i.e., a physical block was larger than **max** as specified in FOPEN).
5	End of file.
6	End of tape.
7	Beginning of tape.

NOTES

FREAD and FWRITE neither rearrange data nor insert and delete control words. Thus, if you are reading or writing a binary data file created in FORTRAN, BASIC, or COBOL, consider that the first word of each 128-word block is an internal control word and not a data value.

FSPACR

PURPOSE

The FSPACR subroutine spaces forward or backward a specified number of physical blocks.

GENERAL FORM

CALL FSPACR (unit,nrecs,status)

Where:

unit is the FORTRAN unit number.

nrecs is the number of physical blocks to be spaced forward. A negative value of **nrecs** will cause FSPACR to space backward.

status is an integer variable in which the subroutine returns a status code as follows:

Status Code	Code Description
0	Spacing operation successful.
1	Bad parameter.
3	Device error (hardware).
4	Record too large (i.e., a physical block was larger than **max** as specified in FOPEN).
5	End of file.
6	End of tape.
7	Beginning of tape.

NOTES

FSPACR is effective only on disk and magnetic tape. On disk, a physical block is 128 words.

When **nrecs** is negative, a status return of 7 may or may not indicate an error condition.

When **nrecs** is positive, a status return of 5 or 6 may or may not indicate an error condition.

FSPACF

PURPOSE

The FSPACF subroutine spaces forward or backward a specified number of files.

GENERAL FORM

CALL FSPACF (unit,nfiles,status)

Where:

unit is the FORTRAN unit number.

nfiles is the number of files to be spaced forward. A negative value of **nfiles** will cause FSPACF to space backward.

status is an integer variable in which the subroutine returns a status code as follows:

Status Code	Code Description
0	Spacing operation successful.
1	Bad parameter.
3	Device error (hardware).
4	Record too large (i.e., a physical block was larger than **max** as specified in FOPEN).
5	End of file.
6	End of tape.
7	Beginning of tape.

NOTES

FSPACF is effective only on magnetic tape.

When **nfiles** is negative, a status return of 7 may or may not indicate an error condition.

When **nfiles** is positive, a status return of 5 or 6 may or may not indicate an error condition.

FREWND

PURPOSE

The FREWND subroutine rewinds a magnetic tape or positions a disk file to the beginning.

GENERAL FORM

CALL FREWND (unit,status)

Where:

unit is the FORTRAN unit number.

status is an integer variable in which the subroutine returns a status code as follows:

Status Code	**Code Description**
0	Operation successful.
1	Bad parameter.
3	Device error (hardware).
4	Record too large (i.e., a physical block was larger than **max** as specified in FOPEN).
5	End of file.
6	End of tape.
7	Beginning of tape.

NOTES

FREWND is effective only on disk and magnetic tape. On disk, FREWND positions the file to its first block.

A status return of 7 should not be considered as an error.

FCLOSE

PURPOSE

The FCLOSE subroutine closes files opened with FOPEN.

GENERAL FORM

CALL FCLOSE (unit)

Where:

unit is the FORTRAN unit number.

NOTES

If you wish to read another file on magnetic tape after encountering an end-of-file, you must call FCLOSE and call FOPEN again.

Character Conversion Subroutines

This section describes the FFIO subroutines which enable FORTRAN programs to perform character code conversions of the type done by the CODE program. They are typically used in conjunction with the basic FFIO input/output subroutines to read or write magnetic tapes compatible with other computer systems. The character conversion subroutines operate in characters of any size up to 36 bits.

The conversion occurs in memory by translating a string of characters stored in an input array to a string of characters stored in an output array, according to the contents of a translation table stored in a third array.

Conversion is done by table lookup; the numeric value of each character to be converted is used as an index into the translation table. Translation table entries are stored one character per word, left-justified. Null entries (for characters which are to be skipped rather than converted) are specified by setting the value of the corresponding word in the table to −1.

To specify end of each input record being converted, either specify the record length in characters, or specify an end-of-record character. One or two end-of-record characters can also be specified for insertion at the end of each output record.

If you wish any input character to generate two output characters, you can do so by specifying a second translation table having the same format as the first.

The following pages present descriptions of the FFIO character conversion subroutines.

CONVRD

PURPOSE

The CONVRD subroutine sets up a conversion table by reading a CODE conversion file. This is the usual way to set the values in a conversion table, since CODE conversion files are available for most of the commonly used character sets.

GENERAL FORM

CALL CONVRD (fname,table,nchars,isiz)

Where:

fname is the name of a CODE conversion file (see NOTES).

table specifies the array which will contain the conversion table.

nchars is the number of entries in the table.

isiz is the size (in *bits*) of the *converted* characters.

NOTES

The following CODE conversion files are available for your use. Of course, you may create your own CODE files if required.

Filename	For Converting
SYS:BTOA.IBM	IBM BCD to ASCII
SYS:ATOB.IBM	ASCII to IBM BCD
SYS:ETOA.IBM	IBM EBCDIC to ASCII
SYS:ATOE.IBM	ASCII to IBM EBCDIC
SYS:BTOA.GE	GE BCD to ASCII
SYS:ATOB.GE	ASCII to GE BCD
SYS:BTOA.UNI	UNIVAC BCD to ASCII
SYS:ATOB.UNI	ASCII to UNIVAC BCD

Conversion files are read using a FORTRAN "8O" format (8 Octal values per line) A file entry of the form 777777 signifies that no output character is to be produced and is thus stored as −1 in the conversion table.

ICONV

PURPOSE

The ICONV subroutine is called to specify the information necessary for the conversion process.

GENERAL FORM

CALL ICONV (table1,insize,outsize
[,brkchr,eorchr,eorchr2,table2])

Where:

table1 is the array containing the conversion table as described for the CONVRD subroutine.

insize is the input character size in bits (1 to 36 bits).

outsize is the output character size in bits (1 to 36 bits). The value of **outsize** is thus the same as the value of **isiz** in the call to CONVRD.

The optional parameters must appear in the order shown in the general form; if one is specified, all preceding it must also be specified. If you specify a value of −1 for any of these parameters, the parameter will be ignored.

brkchr is the input end-of-record character.

eorchr is the first character to output at the end of a logical record.

eorchr2 is the second character to output at the end of a logical record.

table2 is the second conversion table, required only if you wish to output two characters for each input character.

CONVER

PURPOSE

The subroutine CONVER converts one logical record.

GENERAL FORM

CALL CONVER (ibuff,ipos,obuff,opos,ichars,ochars)

Where:

ibuff is the array containing the record to be converted.

ipos is the first character position in **ibuff** to input. Character numbering begins with 1.

obuff is the integer array to contain the output record.

opos is an integer constant or integer variable specifying the first character position in **obuff** to receive output. Character numbering begins with 1.

ichars is an integer variable whose value is the number of input characters to be processed. Do not use a constant. Note that processing continues until **ichars** characters have been processed or **brkchr** (if specified in the call to ICONV) is encountered, whichever comes first. In either case, **ichars** returns the number of characters in the input record.

ochars is an integer variable in which the number of converted characters stored in the output record is returned.

NOTES

ichars and **ochars** will generally differ due to skipped characters and end-of-record characters.

FFIO Sample Program

The following sample program illustrates the use of FFIO subroutines. This program reads M305:, a 7-track, 800-bpi, odd-parity BCD tape with 80-character (card image) logical blocks blocked into physical blocks which are variable length, up to 800 characters long. The output is written into a standard ASCII file, OUT.DAT.

Sample Program

The text on the right-hand side of the page explains the program flow.

```
00100      DIMENSION TABLE(64),IBUFF(134),OBUFF(16)
```

TABLE will contain the conversion table, word per character in input character sequence.

```
00110      INTEGER OCHRS,TABLE,OBUFF
00120  C
00130      CALL CONVRD('SYS:BTOA.IBM',TABLE,64,7)
```

CONVRD reads the file SYS:BTOA.IBM, the BCD to ASCII conversion table.

```
00140      CALL ICONV(TABLE,6,7)
```

ICONV initializes the conversion routines. TABLE is the conversion table, 6 bits is the input byte size, 7 bits is the output byte size. There is no input break character (this application has fixed-length records).

```
00150      CALL FOPEN(1,'M305:','I',ISTAT,134,2,8)
```

*FOPEN establishes M305: as the input tape with 804 byte records (134*6), 2 buffers, and 800 bpi. Since parity and mode are not specified, parity is odd and mode is industry compatible.*

```
00160      IF (ISTAT.NE.O) GOTO 4
```

A check for a successful open.

```
00170      CALL OPEN(2,'OUT.DAT','ASO',ISTAT)
```

Opens the output disk file which is a standard ASCII sequential output file.

```
00180      IF (ISTAT.NE.O) GOTO 5
```

A check for successful open.

```
00190  C
00200   1       NWORDS=0
```

This is the start of the loop which reads a physical record, breaks it into logical records, then converts and writes each to the output file.

```
00210        CALL FREAD(1,IBUFF,NWORDS,ISTAT)
00220        IF (ISTAT.NE.O) GOTO 6
```

A check for a successful read.

```
00230        NREC=NWORDS*6/80
```

Gives the number of records: 6 characters/word, 80 words/record.

```
00240  C
00250        DO 2 I=1,NREC
```

This DO loop converts the input block and writes the records to the output file.

```
00260        IPOS=I*80-79
```

Position to next logical record.

```
00270        ICHARS=80
```

Sets the number of characters in the input record.

```
00280        CALL CONVER(IBUFF,IPOS,OBUFF,1,ICHRS,OCHRS)
```

Converts one record.

```
00290        WRITE(2,3) OBUFF
00300   3       FORMAT(16A5)
00310   2       CONTINUE
00320        GOTO 1
```

Reads another block.

```
00330  C
00340   4       TYPE,'?INPUT TAPE NOT ASSIGNED'
00350        STOP
00360  C
00370   5       TYPE,'CANNOT CREATE OUTPUT FILE'
00380        STOP
00390  C
00400   6       IF (ISTAT.NE.5) GOTO 7
```

Checks for end of file.

```
00410        CALL CLOSE(2)
00420        CALL FCLOSE(1)
00430        STOP
```

End of input file, close the files and stop.

```
00440  C
00450   7       TYPE,'?INPUT ERROR'
00460        STOP
00470  C
00480        END
```

Cyphernetics Division
175 Jackson Plaza
Ann Arbor, Michigan 48106
(313) 769-6800

Atlanta	(404) 252-1033	The Hague	070-948866	Morristown	(201) 267-1600
Boston	(617) 227-8707	Hartford	(203) 677-8584	New York	(212) 757-0322
Brussels	02/538.89.62	Honolulu	(808) 947-1267	New York Financial	(212) 349-4800
Chicago	(312) 346-1044	Houston	(713) 236-1632	Norfolk	(804) 461-1903
Cincinnati	(513) 621-2074	Indianapolis	(317) 632-1161	Philadelphia	(215) 985-1144
Cleveland	(216) 531-2214	Kalamazoo	(616) 381-9173	Pittsburgh	(412) 355-2867
Cologne	0221-731029	London	01-836-5081	San Diego	(714) 280-8100
Dallas	(214) 651-9351	Los Angeles	(213) 480-3122	San Francisco	(415) 986-4810
Dayton	(513) 228-0375	Miami	(305) 577-3981	Stockholm	(08) 23.51.81
Denver	(303) 534-2824	Milan	(02) 225310	Toledo	(419) 255-8076
Detroit	(313) 271-7300	Milwaukee	(414) 224-9797	Washington, D.C.	(202) 872-0480
Frankfurt	0611-664081	Minneapolis	(612) 338-8753	Zurich	01-470337

APPENDIX H

Formal Report:
Charlevoix, Michigan Water Supply System Improvements

CHARLEVOIX, MICHIGAN WATER

SUPPLY SYSTEM IMPROVEMENTS

Recommendations to Meet
Projected Water Demands

McNAMEE, PORTER AND SEELEY
CONSULTING ENGINEERS
ANN ARBOR, MICHIGAN

McNamee, Porter and Seeley
Consulting Engineers
Ann Arbor, Michigan

CHARLEVOIX, MICHIGAN WATER SUPPLY SYSTEM IMPROVEMENTS

Recommendations to Meet
Projected Water Demands

December 1972

by

J. M. Holland, P.E.
Project Supervisor

J. P. Oyer, P.E.
Project Engineer

Submitted to
Mayor and City Council
Charlevoix, Michigan

This study was authorized by the Charlevoix
City Council on December 6, 1971

ACTIVE PARTNERS
J. C. SEELEY R. M. BACHTEAL
J. M. HOLLAND D. H. NOLAND, JR
M. R. VAN EYCK S. C. WARTINBEE
RETIRED PARTNERS
R. L. McNAMEE W. S. HERBERT

McNamee, Porter and Seeley

2223 PACKARD ROAD · ANN ARBOR, MICH. 48104 · AREA CODE 313 769-9220

Consulting Engineers

December 27, 1972

Honorable Mayor
Councilmen and
Mr. Thomas G. Hanna, City Administrator
Charlevoix, Michigan 49720

Gentlemen:

On December 6, 1971, the City Council of Charlevoix, Michigan, authorized McNamee, Porter and Seeley to study the Charlevoix Water Supply System and make recommendations for improvements to meet future needs. Since the installation of the present system in the early 1920's, the population of Charlevoix has doubled. Recent projections indicate that the population will double again by the year 2000.

Our study to review the adequacy of the present system began in May 1972 and has just recently been completed. This report, being submitted to you for your approval, makes recommendations for improvements to the water supply system to meet projected water demands for the year 2000. The report includes detailed cost estimates for these improvements.

The study used population projections to compare the capacity of the present water supply system with future water supply requirements. Charlevoix should anticipate a significant increase in population by the year 2000. The permanent population should increase from 3519 to 4800, and seasonal population from 7000 to 10000.

Population to be served	Permanent	Seasonal
Present (Based on 1970 Census)	3519	7000
1980	3940	8000
1990	4370	9000
2000	4800	10000

The present water supply system cannot now provide the theoretical maximum day flow or maximum potential fire flow demand during summer population peaks. The maximum safe capacity of the present water system is 1.89 million gallons per day. Population projections indicate that in the year 2000 maximum day flow can be expected to be 3.5 million gallons per day. Without improvements, the present water system would have a deficiency of 1.61 million gallons per day.

Therefore, we recommend the following improvements in the Charlevoix Water Supply System. Our cost estimates are based on current prices.

Recommended Improvements and Estimated Costs

1. Additional supply, chlorination, pumping station revisions	$ 720,000
2. Improvements to distribution system	
A. Northeast feeder main and river crossing	285,000
B. Southeast feeder main	150,000
3. Elevated Storage Tank	300,000
Total Estimated Cost of Recommended Improvements	$1,455,000

We wish to express our appreciation to Mr. Thomas G. Hanna, City Administrator, and other City Officials who have assisted us in the preparation of this report.

Respectfully submitted,

McNamee, Porter & Seeley

By J M Holland
J. M. Holland, P.E.
Project Supervisor

J. P. Oyer
J. P. Oyer, P.E.
Project Engineer

TABLE OF CONTENTS

INDEX TO MAPS AND FIGURES

INTRODUCTION

The purpose of this investigation was to review the adequacy of the present Charlevoix water supply system to meet present and anticipated future water demands of the community. The present system was installed in 1922, when the permanent population of Charlevoix was about 2200. The population since then has doubled, and recent projections indicate that the population will double again by the year 2000.

The following questions therefore arise:

1. Does the present system have a maximum safe capacity to meet projected water demands?
2. Are the pumping, distribution, and elevated storage facilities capable of meeting projected demands?
3. Are these facilities capable of meeting recommended fire flow for the projected population?

The present system also lacks the disinfection facilities now required by the State for all municipal water supplies.

To evaluate the adequacy of the present water supply system in the face of future demands, the study uses population projections to establish future requirements and analyzes the capacity, equipment, and facilities of the present system. A comparison of future requirements with the abilities of the present system to meet those requirements yields the recommendations for improvement of the Charlevoix water supply system.

We use existing records, field studies, and earlier engineering reports to determine the needs of the water system. We review all general areas of the present water system. This report recommends several improvements in the present system plus some additional construction to bring the total system up to levels sufficient to meet anticipated requirements. The report presents the estimated total cost of the projects and a breakdown of costs for each part of the project.

The report is divided into the following main sections:

I. WATER SUPPLY REQUIREMENTS. This section analyzes the trends of population growth, pumping patterns, and per capita consumption in order to project future design flow requirements.

II. THE PRESENT WATER SYSTEM AND RECOMMENDED IMPROVEMENTS. This section compares the future requirements with the abilities of the present system and recommends system improvements.

III. COST ESTIMATES. This section itemizes the estimated costs for the improvements recommended.

I. WATER SUPPLY REQUIREMENTS

Water supply requirements in general are measured by per capita consumption in terms of annual average rate in millions of gallons per day (mgd) of flow. Specific design parameters are defined by average rates of flow for the maximum hour, maximum day, and maximum month. Significant variations from average requirements also have to be determined.

In order to make projections of future water supply requirements, it is essential to analyze the trends of population growth, pumping pattern, and per capita consumption. In addition, major improvements should be planned to serve the demands of the service area for a period of time at least equal to the number of years required to finance the project. For this report, the year 2000 is used as the basis for design.

It is also necessary to define the design parameters for this project. When a water supply system is designed, certain demand rates have to be used to define sufficient capacity. The maximum flow in any one day is used to size treatment facilities and intake facilities. It is also used to size major transmission mains. The maximum hour flows and fire flows, which are based on population, are used to size pumping facilities, feeder mains, and water storage facilities.

Finally, it is necessary to look at a city to see if water consumption at certain times varies significantly from average levels. As Charlevoix is a community in which population increases substantially in the summer months due to tourist trade, the water supply system must be designed to handle the increased load during the summer months.

Population Trends

Past population data has been analyzed in order to project future trends. We project a growth in permanent population for the city of Charlevoix from 3519 in 1970 to approximately 4800 in the year 2000. This means a projected summer population of about 10,000 in the year 2000.

Past population data for the city of Charlevoix was obtained from United States census figures and is tabulated in Figure 1.

Figure 1

Population Data for Charlevoix

Year of Census	1910	1920	1930	1940	1950	1960	1970
Population	2420	2218	2247	2295	2695	2751	3519

These population figures reveal that the City experienced a substantial growth of 768 persons from 1960 to 1970. This contrasts with the total growth of the City from 1910 to 1960 of only 331 persons. Although a growth rate similar to that from 1960 to 1970 is not likely to continue unabated through the design year 2000, it is reasonable to project a growth rate higher than that prior to 1960. A reasonable estimate of the future growth rate would be a rate between the rate prior to 1960 and the rate since 1960. Therefore we project an increase in population of 430 persons per decade. From 1970 to 2000, then, we project the population to grow to 4800 persons. We use 4800 persons for the permanent design population in our calculations of future water supply requirements. This projection is illustrated in Figure 2.

In addition to permanent population, the projected summer population must be calculated. According to the best estimates obtained by your City Administrator, the summer population of Charlevoix is about double the permanent population. With a permanent population of approximately 4800 persons in year 2000, then, we can project a summer seasonal population of 10,000 persons.

Of course, several factors which cannot be foreseen might have unexpected effects on the actual growth rate. Industrial and commercial decisions as well as development of the transportation system could cause the growth rate to be higher than that we assume. Another factor which could affect the growth rate would be expansion of the Service Area. The area now served by the present Charlevoix water system is generally within the City limits. Few areas outside the City presently are served with City water. A need exists for extending the Charlevoix water system into adjacent areas to promote growth of these areas and to provide a safe water supply in those areas where the underground water supply is poor. This extended service area should be considered in future expansion of the present water system.

Figure 2

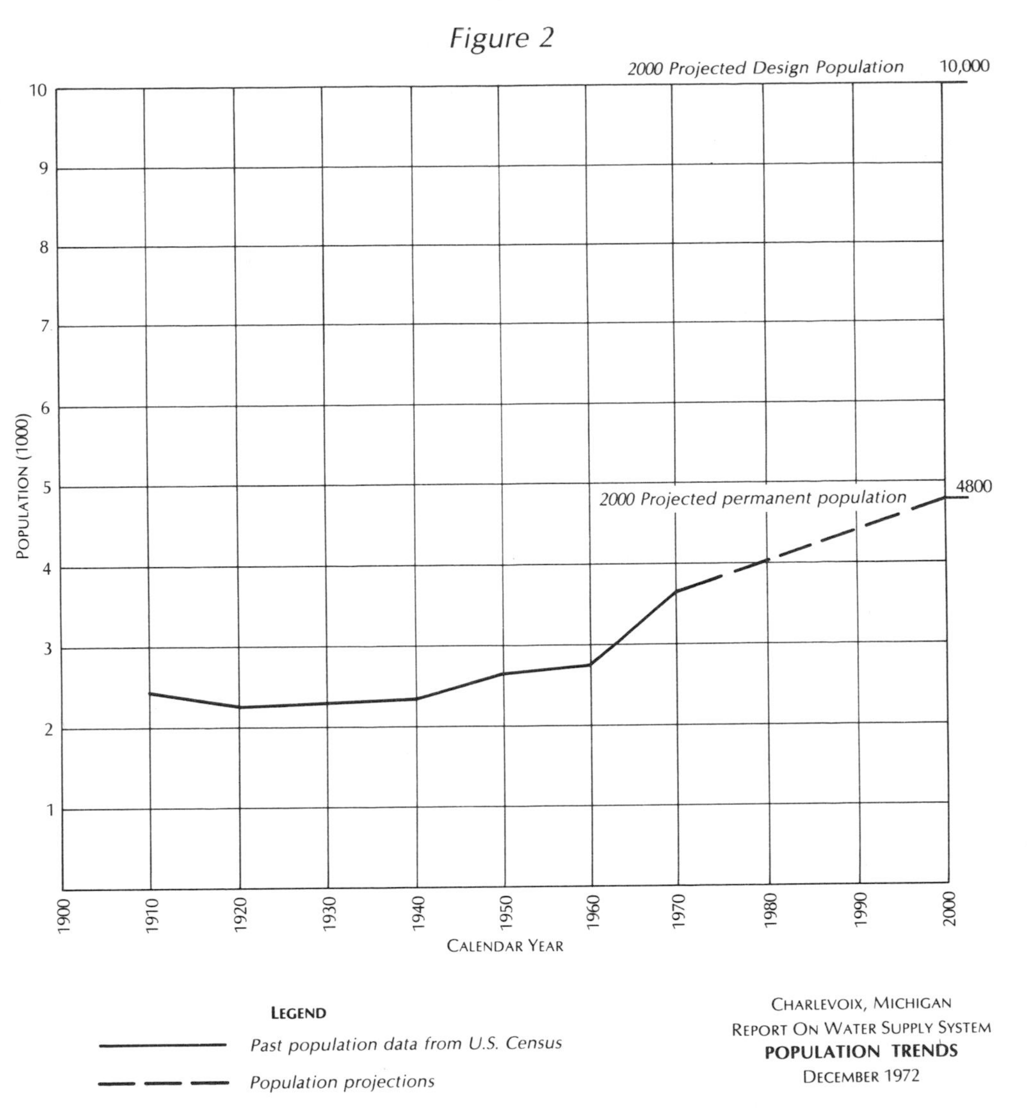

You will notice that population projections found in "Charlevoix County, Comprehensive Area-Wide Plan for Water and Sewer Utilities," Williams and Works, 1971, are somewhat higher than the projections presented in this study. However, the population figures in that area-wide plan include "population equivalents" for industrial and commercial establishments. In this report, we use recorded population and past water consumption data to establish per capita consumption, which thus includes an allowance for industrial and commercial users.

Per Capita Consumption

Per capita consumption has increased from 122 gallons per day (gpd) in 1950 to

157 gpd in 1970. It probably will continue to increase to 180 gpd by year 2000. With the projected population, this results in an annual average rate of 1.17 million gallons per day (mgd).

Per capita consumption is obtained from water pumpage records and past population data. Average per capita flows are determined and then future per capita consumption calculated on the basis of projected population trends. Changes in living style which will increase per capita consumption are anticipated.

Available water pumpage records since 1945 are used. Pumpage rates for 1950 and 1960 through 1971 are tabulated in Appendix A. These figures show a general increase in consumption from 1950 to 1970, as shown in Figure 3.

Figure 3

Pumpage Rates (in mgd)

	Annual Average	Winter Average	Summer Average
1950	.39	.33	.53
1960	.52	.44	.70
1970	.67	.59	.86

Although there is considerable fluctuation on a year-to-year basis, the trend of increase in consumption is apparent.

On the basis of these pumpage rates, the per capita winter consumption has increased from 122 gallons per day (gpd) in 1950 to 157 gpd in 1970. Present average per capita consumption is determined by dividing the winter average flows for the entire city by the permanent population. (See Figure 4.)

Figure 4

Per Capita Consumption

	Winter Average (mgd)	Permanent Population	Per Capita Consumption (gpd)
1950	.33	2695	122
1970	.59	3519	157

This increase in per capita consumption is to be expected because of the increasing use of water-consuming household appliances such as automatic dishwashers, garbage grinders, and lawn sprinklers.

Due to this trend of increased use of water-consuming appliances, the average per capita consumption may continue to increase to an estimated 180 gpd in the design year 2000. With a projected permanent population of 4800, this will result in an average per capita winter consumption rate of .86 mgd.

The per capita summer consumption has increased in the same manner. Based on a projected summer population of 10,000, it can be expected to increase to an estimated 1.80 gpd in the design year 2000. Combining these figures in a weighted average (Appendix A shows the annual average to be closer to the winter than the summer average) results in an annual average rate of 1.17 mgd.

Design Flows

In addition to per capita consumption and annual average consumption rates, the average rates of flow for the maximum hour, maximum day, and maximum month must be determined. These design flow rates are used to define the specific parameters of the water supply system. Figure 5 shows the design flows projected for the future.

Figure 5

Design Flows (in mgd)

Year	Annual Average	Winter Average	Summer Average	Maximum Month	Maximum Day	Maximum Hour
1980	0.86	0.63	1.29	1.68	2.52	3.34
1990	1.00	0.74	1.54	2.00	3.00	4.00
2000	1.17	0.86	1.80	2.35	3.50	4.70

The water supply system should be designed to meet the maximum hour, maximum day and maximum month flow rates projected for the year 2000.

These rates are determined from past records (see Appendix A) by establishing ratios of the various rates to past annual average rates. These ratios then are applied to the design year annual average in order to determine design flows. (See Appendix B for ratios used to calculate design flows.) Since the annual average rate for 1970 is .67 mgd and that projected for the year 2000 is 1.17 mgd, the projected design flows are significantly higher than present flows.

One other design flow must be determined: fire flow. Based on a projected summer population of 10,000 persons, the fire flow required for the City of Charlevoix is 3120 gallons per minute (gpm) or 4.46 mgd (required duration of 10 hours). These figures are determined by the National Board of Fire

Underwriters' formula (see Appendix C). The total quantity of water used for fire protection is usually a small percentage of the total annual water consumption. However, the flow rate at which water is required for extinguishing a fire is much larger than average flow rates. Thus, fire flow is a factor in the design of pumping equipment, storage facilities, and distribution mains. Determination of fire flow demands is based on population. Fire flow demands usually are met by a combination of elevated storage and pumping capacity.

Summary of Design Year Water Supply Requirements

In the design year 2000, Charlevoix will have a permanent population of approximately 4800 persons and a summer seasonal population of approximately 10,000 persons. At that time the per capita winter consumption rate will be about .86 mgd, and the summer rate about 1.80 gpd. This will result in an annual average rate of 1.17 mgd. The design flows will be:

Maximum Month	2.35 mgd
Maximum Day	3.50 mgd
Maximum Hour	4.70 mgd

In addition, fire flow requirements will be 4.46 mgd for a duration of 10 hours.

II. THE PRESENT WATER SYSTEM AND RECOMMENDED IMPROVEMENTS

Charlevoix obtains its municipal water supply from a unique collection system along the eastern shore of Lake Michigan. After collection, the water is flouridated and pumped directly to the distribution system for use. This system will be evaluated in terms of the design year water supply requirements established in the previous section. The condition of the facilities and equipment also must be examined. Finally, the water quality must be evaluated by Michigan Department of Health standards.

Specifically, we must determine the following:

1. The maximum safe capacity of the existing raw water collection system.
2. The capacity and the condition of the pumping station.
3. The adequacy of the present distribution system.
4. The function and fire flow capacity of the elevated storage.
5. The adequacy of the water treatment facilities.

Each of these is examined in turn in the following sections, and recommendations for improvements made as necessary.

The Raw Water Collection System

The maximum safe capacity of the existing raw water collection system is calculated to be 1.89 mgd. Based on the year 2000 maximum day flow of 3.50 mgd, the existing system would have a deficiency of 1.61 mgd.

The Present System

The present raw water collection system, located near the intersection of Park and Clinton Streets, consists of a long wooden underground flume paralleling the shoreline of Lake Michigan. Ground water enters the flume through slotted openings in the bottom and the flume discharges to a 16 foot diameter receiving well. The receiving well serves as a suction well for the high service pumping units.

The collection system was installed in the early 1920s and the design included two flumes, each 225 feet long, laid north and south from the receiving well. The City uses only the northern flume and it is believed that the other flume does not exist. An effort was made in June 1972 to locate the southern flume with City owned equipment, but the second flume was not found.

In June 1972 field tests were made to determine the capacity of the existing collection system. Based on the results of those tests, the safe capacity of the system is calculated to be 1315 gpm or 1.89 mgd. The safe capacity was based on low lake levels and the present pump suction piping and conditions. At the time field tests were made, lake levels were at near record highs and an additional six feet of head was available on the flume. At low lake levels the available drawdown is reduced, thus limiting the capacity of the system.

Recommended Improvement

To increase the raw water supply, we recommend the construction of a new collection system similar in design to the present system. The proposed collection system would be located south of the present pumping station and would have a design capacity of 5 mgd.

The proposed collection system differs from the present system in that the receiving well will be designed as a low service pumping station to pump raw water from the collection system to a proposed chlorine contact chamber (see section on Water Treatment Facilities below). The station would be designed to accommodate a minimum of two vertical pumps of the mixed flow or propeller type.

We further recommend that the existing receiving well be converted for use as a low service pumping station. If a vertical pump is installed in the existing receiving well, the capacity of the existing collection system may be increased to 3.84 mgd. The increased capacity is a result of setting the pump impeller closer to the bottom of the well and making available an additional 3.2 feet of drawdown.

We recommend that the low service pumping units be of the axial flow or propeller type, which will provide a large capacity of water at low pumping heads. The proposed low service pumping units are shown in Figure 6.

Figure 6

Proposed Low Service Pumping Units

Pump No.	Location	Capacity (gpm)
1	Existing receiving well	2100
2	Proposed receiving well	2100
3	Proposed receiving well	1400

The firm capacity of the low service pumping units would be 3500 gpm.

Another alternative to increase the raw water supply would be to construct an intake line extending into Lake Michigan. However, this alternative would require additional pumping facilities and a filtration plant. The City should consider this alternative to meet future water demands beyond the design year (Appendix D).

The Pumping Station

The present pumping station, although in good physical condition, will require a few alterations. We recommend that three of the four suction lines from the existing receiving well be plugged and a new suction line from the proposed ground storage reservoir to the present pumping station be constructed. The fourth suction line should be left in place for emergency operation. Installation of a pump suction header to the pumping units will require modifications to both the present pump suction and discharge piping. Check valves in the existing suction piping will be relocated to the discharge side of the pumps. Shut off valves will be provided at each pump suction and discharge end to isolate each pumping unit for maintenance purposes.

The Present Pumping Station

The present pumping station, located just east of the receiving well, was constructed in the early 1920s. The station is a masonry and concrete structure designed to accommodate four pumping units with related piping and electrical equipment. Separate rooms also were provided for an office and laboratory, with one of the rooms presently being used to house fluoridation equipment.

The present pumping units and year of installation are tabulated in Figure 7.

Figure 7

Existing Pumping Units

Pump No.	Capacity (gpm)	Head-Feet	Motor HP	Year Installed
1*	500	260	75	1922
2	1200	260	100	1931
3	400	235	40	1931
4**	1200	260	125	1964

* Pump No. 1 is equipped with an 85 HP gasoline engine.
** Pump No. 4 is installed with vertical shaft and motor mounted on an 11 foot high platform.

Recommended Improvements

We recommend the following improvements in the present pumping units in order to meet the maximum day flow of 3.50 mgd.

Pump No. 1. This pump was installed in 1922 and has outlived its useful life. A standby engine connected to the pumping unit has become a safety hazard. Personnel working in the pumping station, while the standby engine is running, have been overcome by the fumes. The pumping unit, electric motor, and standby engine are also located below the high lake level and are subject to flooding. Therefore, we recommend that Pump No. 1 be replaced with a new 900 gpm pumping unit and standby engine. The motor and engine should be mounted above high lake level on a platform and connected to the pumping unit by way of a vertical shaft and right angle gear drive, respectively.

Pump No. 2. This pump, installed in 1931 and rated at 1200 gpm, has provided adequate service. The pumping unit and motor are also located below the high lake level. Field tests in June 1972 indicate that this pumping unit has decreased slightly in capacity. We recommend that Pump No. 2 be reconditioned and modified for installation with a vertical shaft. A platform should be provided to elevate the motor above expected high water levels.

Pump No. 3. Pump No. 3 is installed below high water levels and field tests indicate its capacity has been reduced to 300 gpm when operated with Pump Nos. 2 and 4. We recommend that Pump No. 3 be replaced with a 700 gpm vertical shaft unit with its motor supported on a platform.

Pump No. 4. Pump No. 4 was installed in 1964 and is rated for 1200 gpm at 260 feet head. This pump has a vertical shaft and its motor is mounted on an 11 foot high platform, clearly above high water levels. Field tests indicate the unit is pumping at its rated capacity.

In addition to the above pumping unit modifications, we recommend that a suitable ventilation system be installed in the pumping station.

Recent discussions between the City of Charlevoix and the Michigan Department of State Highways indicate that the cost of protecting the present pumping station from wave action and pumping units from flooding may be paid for by the State. We estimate that raising the three pumping units now below flood level and protecting the exterior of the pumping station could be done at a cost of $100,000 plus contingency allowances and engineering fees. We recommend that the necessary applications be submitted to the State to perform this work.

With the recommended improvements to the pumping station, the total pumping capacity will be 4000 gpm; with the largest unit out of service, the firm capacity of the pump station will be 2800 gpm or 4 mgd. The firm capacity of the station would then exceed the maximum day flow of 3.5 mgd. (A drawing of the existing and proposed improvements to the collection system and pumping station is in Appendix E.)

The Distribution System

The distribution system should be improved primarily to anticipate future expansion of the system into adjacent townships. The ultimate configuration of the distribution system should be a peripheral-type feeder main system generally following the present City limits. Water service could then be extended to areas either within the City or to adjacent townships as development occurs. The improvements would also improve the reliability of the present system and serve the proposed improvements in elevated storage. A map (Map 1) of the present system and recommended improvements as well as future extensions is attached to this report.

Cost estimates for the recommended immediate improvements have been made and may be found in the Cost Estimates section of this report. Cost estimates for improvements labeled "future" have not been made due to unforeseen development which may require a slightly altered configuration than that shown on Map 1.

The Present Distribution System

The present distribution system consists of an underground network of 4- to 12-inch water mains. The system consists of three major transmission subsystems: the Sherman Street System, the Park Street System, and the Division Street System.

The Sherman Street System. The Sherman Street System, consisting of 10-inch cast iron pipes, begins at the pumping station and follows Sherman Street south to Carpenter and east on Carpenter to a point east of Bridge Street. This system provides an adequate water supply to the southwestern portion of the city.

The Park Street System. The Park Street System, also constructed of 10-inch cast iron pipe, begins at the pumping station and follows Park Street to the intersection of Park and Bridge Streets. This system serves as the feeder main to the existing elevated storage tank.

The Division Street System. This system, constructed in 1964, consists of 12-inch cast iron pipe. It serves as the major transmission main in the northern part of the City. When this system was constructed, numerous connections were made to existing 4- and 6-inch mains to provide better pressure distribution north of the River.

Two river crossings supply water to the north side of the City. One crossing is a 10-inch line at Bridge Street, and the other is a 6-inch line near the U.S. Coast Guard station.

Recommended Improvements

We recommend the following improvements to the present distribution system in order to serve the anticipated population growth of the City and area as well as to improve the present system.

First, construction of a 12-inch river crossing near the United States Coast Guard Station. This crossing is required to provide adequate quantities of water to the area north of the river. The importance of this crossing would be even more evident in the case of the failure of the 10-inch crossing at Bridge Street.

Second, construction of a 12-inch feeder main on the City's north and east sides. The feeder main would begin at the proposed river crossing at the Coast Guard Station and would follow Mercer Avenue north to US-31, northeast along US-31 to Martian Road, north along Martian Road approximately 500 feet and west through the industrial park area to North Point Road. The feeder main would then follow North Point Road south to Division Street, where it would connect with the existing Division Street feeder main system. This feeder system is recommended to supplement the existing 4- and 6-inch mains in the north area of the City and to serve as the feeder main for a proposed elevated storage tank in the industrial park area. Existing 4- and 6-inch mains would be connected to the proposed feeder to eliminate dead ends and provide better pressures in the distribution system.

Third, construction of a 12-inch feeder main system on the east side of the City in Ferry Street. This system would begin at the proposed Pine Channel crossing and would follow Ferry Street south to Stover Road and connect with the existing 6-inch main at Stover Road and Ferry Street. A branch of this feeder main system would be constructed in Eaton Street and proceed westerly to the existing 10-inch feeder main in Carpenter Street. This system is recommended to boost the distribution system on the City's southeast side.

These recommended improvements are compatible with earlier discussions in the report that improvements to the City of Charlevoix water supply system be designed to anticipate future extensions of the system into adjacent townships.

Elevated Storage

The City presently maintains a 100,000 gallon elevated water storage tank. The tank is located on Park Street between Grant and State Streets. The high capacity line of the tank is 146 feet above its foundation. The present 100,000 gallon storage tank provides inadequate volume to meet expected water demands. We recommend that a 500,000 gallon elevated reservoir be erected in the industrial park area on the north side of the City. At this location the reservoir would serve primarily as a storage and pressure equalizing reservoir.

Present Elevated Storage

Storage reservoirs on a water distribution system are classified according to the material of which they are built and their position, whether elevated or surface. In addition, a reservoir may also be classified according to its function within the distribution system. Typical functions are for the storage of water, equalizing rates of flow, and equalizing distribution system pressures.

The existing 100,000 gallon elevated steel reservoir on Park Street between Grant and Sherman serves all of the above three functions. The reservoir has recently been repaired and repainted and should continue to provide its intended service for a few more years. The existing elevated tank should be inspected annually to evaluate its condition.

Recommended Improvement

We recommend that a 500,000 gallon elevated storage tank be erected. The size of the proposed reservoir is based on the volume of stored water required to meet the recommended fire flows in conjunction with the proposed high service pumping capacity.

The recommended fire flow for a population of 10,000 persons is 3120 gpm, which must be provided for a duration of 10 hours. The difference between the proposed high service pumping capacity of 4000 gmp and the maximum day demand of 2450 gpm is 1550 gpm and would be available for a fire demand. The volume of water that should be made available from storage is then 1570 gpm for 10 hours. A total storage capacity of 935,000 gallons would be required.

Together with the present 100,000 gallon tank, the 500,000 gallon tank recommended would make available a total volume of 600,000 gallons of storage. This volume, in conjunction with the proposed high service pumping capacity, would sustain a 3120 gpm fire flow for 6.4 hours. The additional 335,000 gallons of elevated storage recommended by the National Board of Fire Underwriters does not appear justifiable on the basis that the fire flow is based on a seasonal population. With the proposed pumping capacity and 600,000 gallons of elevated storage, a fire flow of 2550 gpm could be maintained for 10 hours. According to the National Board of Fire Underwriters, a 2550 gpm fire flow would be adequate for a population of 6000 persons, which exceeds the projected permanent population of 4800 persons.

We recommend that a site approximately 200 feet by 200 feet in the Industrial Park area be made available for the proposed elevated storage tank.

Water Treatment Facilities

The analysis of the water supply indicates that treatment in the form of clarification and filtration is not necessary. In general, the present supply meets or exceeds drinking water standards of the Public Health Service (see Water Quality Tables in Appendix F). Presently, the only treatment applied to the water supply is fluoridation. This is accomplished by two small chemical feed pumps which apply fluoride chemicals to the receiving well. The feed rate is adjusted according to which pump is running.

However, the Michigan Department of Public Health presently requires that all municipal water supplies be disinfected prior to distribution to insure a safe water supply. It is for this reason that facilities and equipment for disinfection are included in the recommended improvements. A water supply, with a total hardness of over 200 mg/1 as $CaCO_3$, is generally considered "hard" water. Water hardness has no sanitary significance other than requiring greater amounts of soap and detergents. As Charlevoix residents are accustomed to this quality in the water supply, facilities to soften the water either by base exchange or the lime-soda process are not recommended at this time. It is to be noted, however, that the recommended improvements would be compatible for future softening facilities.

Recommended Improvements

We recommend a chlorine contact chamber to properly disinfect the raw water. This will require a contact basin to provide sufficient chlorine contact time. However, injecting chlorine at the pumping station with the present piping configuration will not provide sufficient contact time before distribution. Therefore, in conjunction with the chlorine contact chamber, a ground storage reservoir is proposed to serve as a high service pump suction reservoir.

The chlorine contact tank will have a minimum volume of 73,500 gallons, which will provide 30 minutes of contact time at the maximum day flow of 3.5 mgd. We recommend that the chamber be of the baffle wall design to prevent short circuiting and enhance mixing.

The ground storage reservoir is necessitated due to piping revisions required at the present pumping station. It will provide a relatively constant source of water to which the high service pumps are connected. The volume of the ground storage reservoir will be 100,000 gallons. This volume will be adequate for proper pump operation.

A small service building should be built over a portion of the proposed chlorine contact chamber and ground storage reservoir. The service building would contain the chlorine feed room, control equipment, and pipe gallery.

The chlorine feed room would contain two chlorine dispensers, each with a capacity of 150 pounds per day, chlorine scales, space for storing additional chlorine cylinders, and other related equipment. The chlorine room would be provided with an exterior door, and access to the chlorine room would not be permitted from the interior of the service building.

Metering and control equipment would be provided to meter the raw water flow, control the feed rate of chlorine in proportion to flow, and control the starting and stopping of the low service pumping units.

A pipe gallery would be provided adjacent to the chlorine contact chamber and would contain the raw water flow element and the necessary pipe and valve work to direct the raw water flow to the chlorine chamber or to the ground storage reservoir if it is required to bypass the chlorine contact chamber.

Summary of Recommended Improvements

To meet the anticipated increase in permanent and seasonal population by the design year 2000, we recommend that the Charlevoix water supply system be improved. We recommend that the raw water supply capacity be increased, the pumping units modified, the distribution system improved, the elevated storage capacity increased, and the water treatment facilities improved to meet Michigan Department of Public Health standards.

To increase the raw water supply, we recommend the construction of a new collection system similar in design to the present system. Both receiving wells should be low service pumping stations.

To increase the capacity of the pumping station, we recommend modifications to three of the four existing pumping units. With these modifications, the firm capacity of the pumping station will exceed the design year maximum day flow of 3.5 mgd.

To improve the distribution system, we recommend construction of a 12-inch river crossing and construction of two 12-inch feeder mains. These improvements will anticipate development of adjacent townships.

To increase storage capacity to meet future fire flow demands, we recommend that a 500,000 gallon elevated reservoir be erected. The increased volume of storage will sustain a fire flow adequate for population exceeding the projected permanent population.

To meet Michigan Department of Public Health standards we recommend the installation of a chlorine contact chamber with associated facilities, including a ground storage reservoir.

Detailed cost estimates for these recommended improvements are presented in the next section.

III. COST ESTIMATES

Cost estimates have been prepared for the recommended improvements. These costs are based on current prices for industrial type buildings and high quality equipment. Land costs have not been included.

Cost Estimates

A summary of proposed improvements to the water supply system is as follows:

Source, Treatment and Pumping Equipment	$ 720,000
Distribution System	
A. Northeast Feeder Main and River Crossing	285,000
B. Southeast Feeder Main	150,000
Elevated Storage Tank	300,000
Total Estimated Project Cost	$1,455,000

Note: The above Total Estimated Project Cost includes contingency allowances and engineering fees but does not include land costs.

Source, Treatment and Pumping Equipment

Item	Description	Estimated Amount
1	5 mgd raw water collection system	$100,000
2	Low service pumping station at new source	100,000
3	Convert existing receiving well to low service pumping station	50,000
4	Chlorine contact chamber and ground storage reservoir	250,000
5	Revisions to existing pumping station	100,000
	Total Estimated Construction Cost	$600,000
	Contingencies and Engineering	120,000
	Total Estimated Project Cost	$720,000

Distribution System

Improvements to the distribution system are given in a descending order of priority.

Northeast Feeder Main System and River Crossing at U.S. Coast Guard Station

Quantity	Unit	Description	Estimated Unit Price	Estimated Amount
12,350	LF	12-inch cast iron water main	$ 14.00	$172,900
100	LF	6-inch cast iron water main	8.00	800
17	Ea.	12-inch valves and manholes	700.00	11,900
5	Ea.	6-inch valves and valve box	300.00	1,500
200	LF	12-inch cast iron river crossing	200.00	40,000
14	Ea.	6-inch fire hydrant assemblies	700.00	9,800
		Total Estimated Construction Cost		$236,900
		Contingencies and Engineering		48,000
		Total Estimated Project Cost		284,900
		Rounded Off		$285,000

Southeast Feeder Main System

Quantity	Unit	Description	Estimated Unit Price	Estimated Amount
7,450	LF	12-inch cast iron water main	$ 14.00	$104,300
400	LF	6-inch cast iron water main	8.00	3,200
11	Ea.	12-inch valve and valve manhole	700.00	7,700
5	Ea.	6-inch valve and valve box	300.00	1,500
8	Ea.	6-inch fire hydrant assembly	700.00	5,600
65	LF	12-inch casing pipe construction at railroad	100.00	6,500
		Total Estimated Construction Cost		$128,800
		Contingencies and Engineering		26,000
		Total Estimated Project Cost		$144,800
		Rounded Off		$150,000

Elevated Storage Tank

Item	Description	Estimated Amount
1	500,000 gallon elevated steel storage tank	$200,000
2	Foundations for elevated steel storage tank	50,000
	Total Estimated Construction Cost	$250,000
	Contingencies and Engineering	50,000
	Total Estimated Project Cost	$300,000

APPENDIX A

Pumpage Rates

Year	Average Rates in MGD				
	Annual Average	Winter Average	Summer Average	Maximum Month	Maximum Day
1950	.39	.33	.53	.59	
1960	.52	.44	.70	.87	
1962	.54	.48	.69	.95	1.59
1963	.67	.54	.95	1.14	1.89
1964	.47				
1965	.52	.46	.66	.93	
1966	.66	.48	1.05	1.27	
1967	.55				
1968	.64	.59	.73	.87	
1969	.58	.54	.67	.96	
1970	.67	.59	.86	.95	
1971	.59	.44	.82	.98	

Note: Summer months are June through September

Rates calculated from water pumpage records.

APPENDIX B

Ratios Used to Determine Design Flows

The demand rates are determined from past records by establishing ratios of the various rates to past annual average rates. The ratios thus determined are then applied to the design year annual average to determine design flows. Where past records do not indicate a particular rate of flow, such as the maximum hour, an estimate is made based on engineering experience or comparison with similar communities where that information is available. The following ratios are used for estimating the future water requirements:

$$\frac{\text{Average rate for max. hour}}{\text{Annual Average}} = 4.0$$

$$\frac{\text{Average rate for max. day}}{\text{Annual Average}} = 3.0$$

$$\frac{\text{Average rate for max. month}}{\text{Annual Average}} = 2.0$$

$$\frac{\text{Annual rate for summer months}}{\text{Annual Average}} = 1.54$$

$$\frac{\text{Annual rate for winter months}}{\text{Annual Average}} = .74$$

APPENDIX C

Fire Flow Calculations

The total quantity of water used for fire protection is usually a small percentage of the total water used in a year's time. However, the rate at which water is required for extinguishing a fire is generally so great as to be a factor in the design of pumping equipment, storage facilities, and distribution mains. Water for fire demands is based on population. The National Board of Fire Underwriters' formula for determining fire flow is:

$$Q = 1020\sqrt{P}\,(1-0.01\sqrt{P})$$

Where Q = fire demand, gallons per minute
P = population, thousands

Based on a summer population of 10,000 persons, the required fire flow for the City of Charlevoix is 3120 gpm or 4.46 mgd. The required duration for fire demand is 10 hours.

Fire demands are generally met by a combination of elevated storage and pumping capacity.

APPENDIX D

Meeting Future Water Demands Beyond the Design Year

Improvements to the Charlevoix water supply system are based on the estimated water requirements for the year 2000. In order to meet future water demands beyond the design year and to aid City officials in future land use planning it is recommended that the City consider an off-shore intake system and filtration plant to be constructed when water demands require more than the present system can furnish. A suitable location for an intake appears to be along the Lake Michigan shoreline on the north side of the City. A site north of Pine River is more favorable for an off-shore intake than is the location of the present collection system due to the proximity of the sewage treatment plant outfall sewer to the latter and due to the discharge from the channel.

A suitable location for the filtration plant would be near the intake system to reduce pumping and operating requirements. The filtration plant could easily be designed for softening if desired by the citizenry. It is recommended that the City retain possession of or make available approximately 6 acres of land in the north part of the City, preferably along the Lake Michigan shoreline, for future construction of water treatment facilities.

It should be noted that proposed improvements to the present water system are considered adequate to meet the expected water supply demands to the year 2000 both qualitatively and quantitatively. The above recommendation concerning future expansion of the water system beyond the design year is mentioned primarily to alert City officials that consideration should be given to future water supply requirements and to assist City officials in planning for these future requirements.

APPENDIX E

Proposed Collection and Treatment Facilities

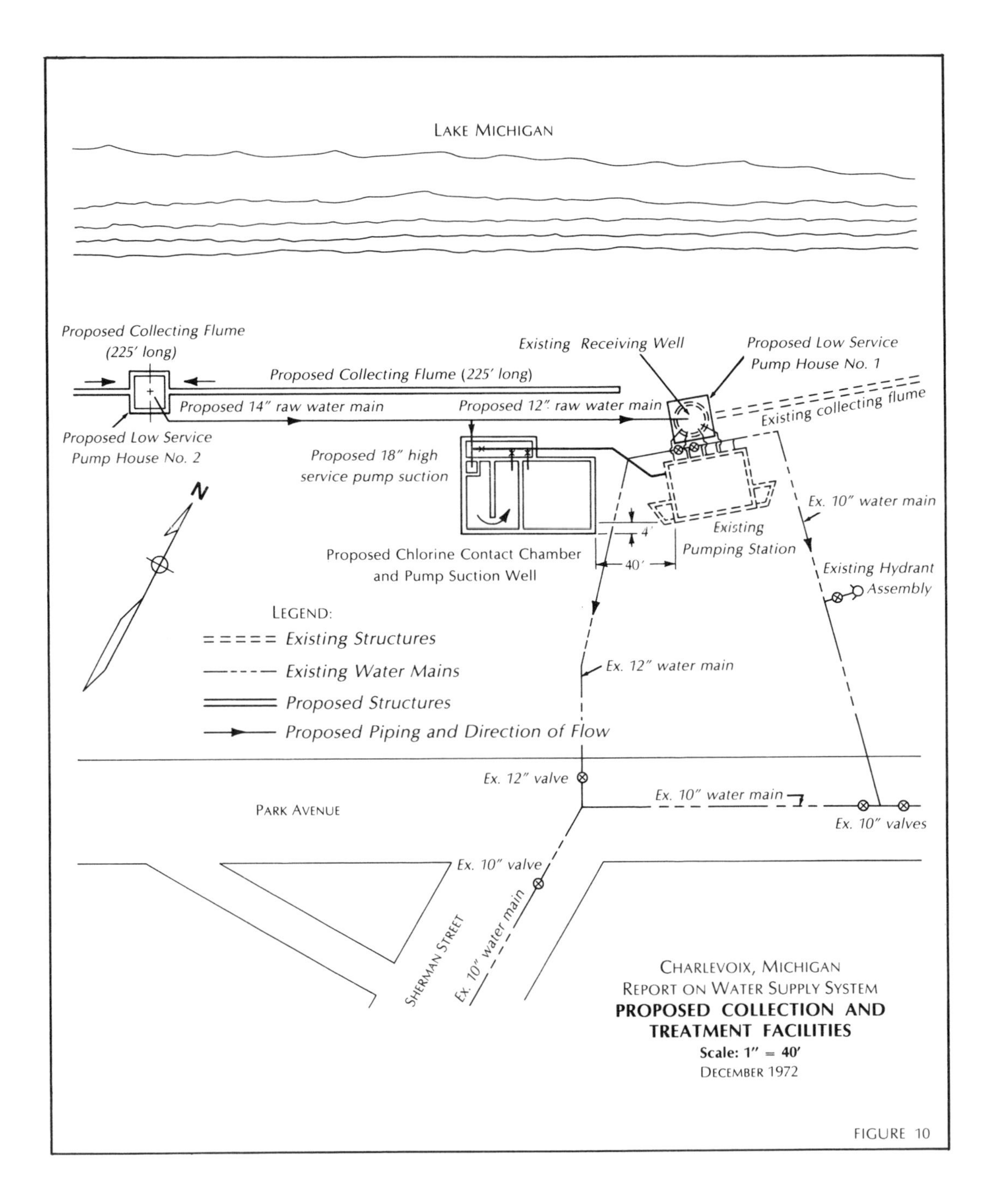

30

APPENDIX F

Water Quality Tables

A complete chemical analysis of the Charlevoix water supply, after fluoridation, was performed by the Michigan Department of Public Health. The results of the analysis are:

Finished Water Quality

Substance	Concentration-mg/1 unless noted
Color	0
Odor	0
Turbidity	1 Jackson Turbidity Unit
Total solids	242
Silica	5
Iron	less than 0.1
Manganese	less than 0.05
Calcium	56
Magnesium	17
Sodium	7
Potassium	1.2
Nitrate	0.7
Chloride	9
Sulphate	36
Bicarbonate ($CaCo_3$)	160
Hardness ($CaCo_3$)	210
Fluoride	0.78
PH	7.8 PH units
Conductance	385 micromhos
	Note: Sample taken August 30, 1972

In addition, a partial chemical analysis of a water sample taken directly from Lake Michigan on August 1, 1972, was made by the Michigan Department of Public Health. The results are:

Raw Water Quality

Substance	Concentration-mg/1 unless noted
Iron	0.0
Chloride	6
Hardness as $CaCo_3$	140
Conductance	250 micromhos
Fluoride	0.1

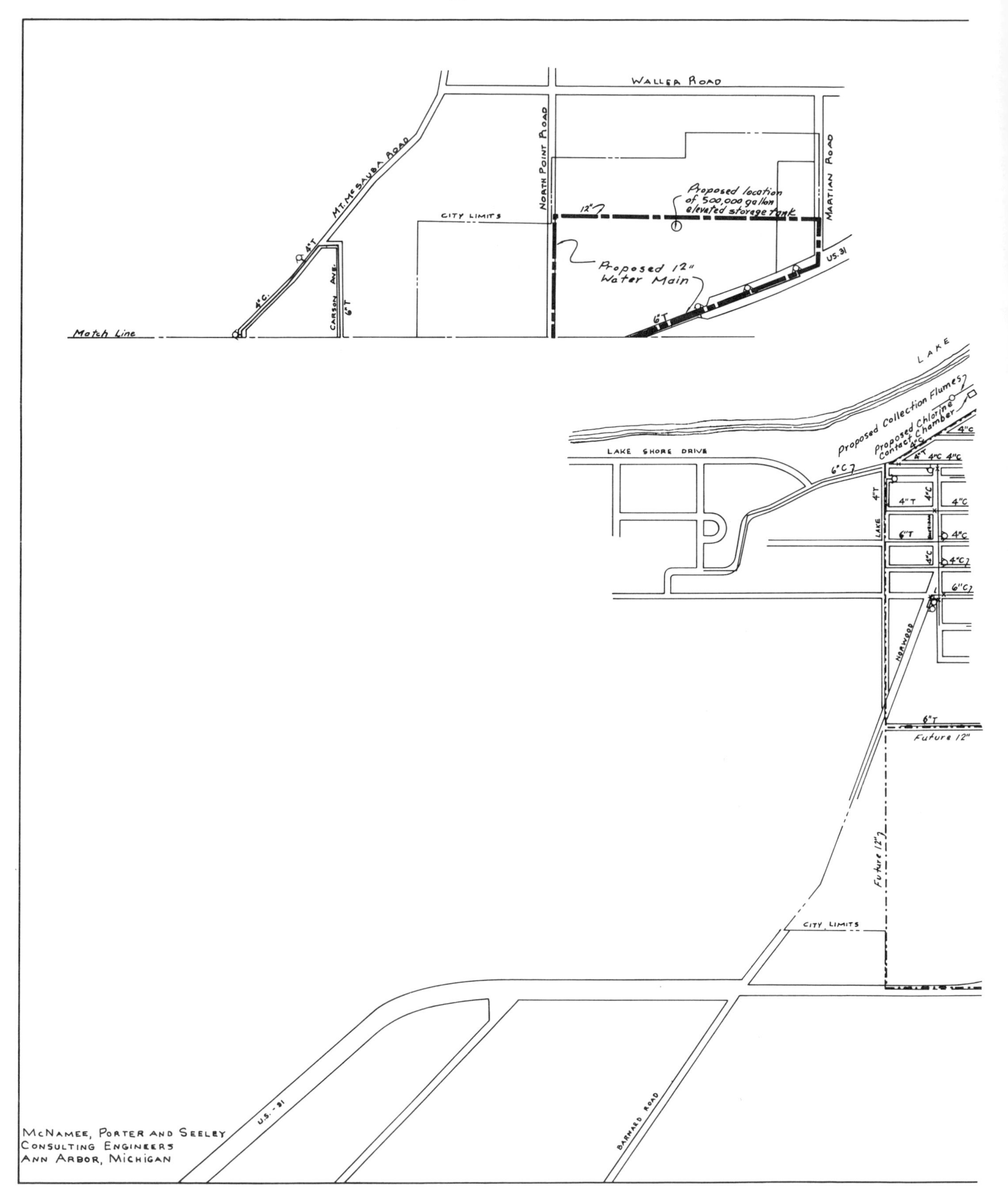

WALLER ROAD
NORTH POINT ROAD
MARTIAN ROAD
MT. MESAUBA ROAD
CITY LIMITS
Proposed location of 500,000 gallon elevated storage tank
Proposed 12" Water Main
US. 31
CARSON AVE.
Match Line
LAKE
LAKE SHORE DRIVE
Proposed Collection Flumes
Proposed Chlorine Contact Chamber
NORWOOD
Future 12"
CITY LIMITS
U.S.-31
BARNARD ROAD
McNAMEE, PORTER AND SEELEY
CONSULTING ENGINEERS
ANN ARBOR, MICHIGAN

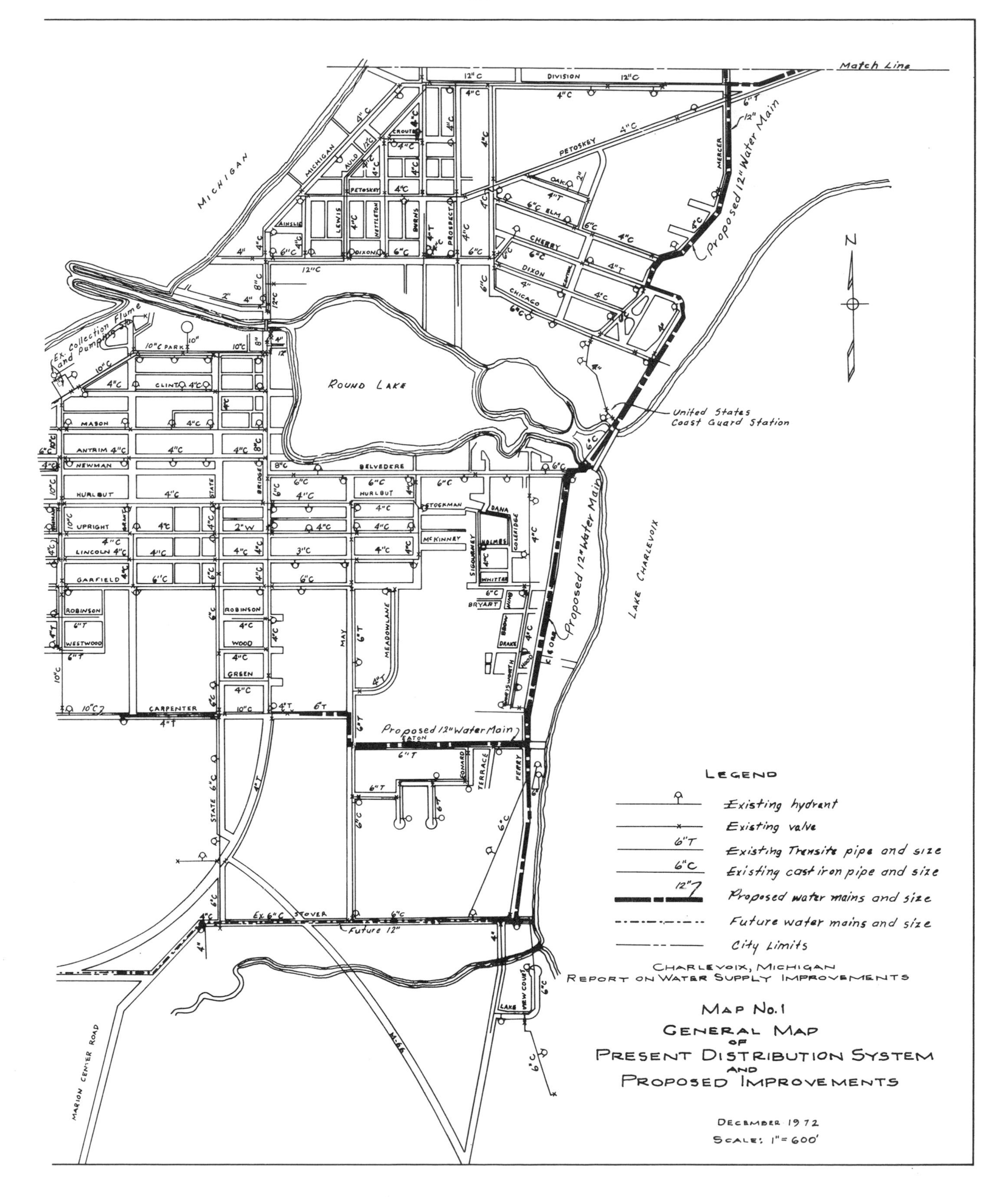
Legend
Existing hydrant
Existing valve
Existing Transite pipe and size
Existing cast iron pipe and size
Proposed water mains and size
Future water mains and size
City Limits
Charlevoix, Michigan
Report on Water Supply Improvements
Map No. 1
General Map of Present Distribution System and Proposed Improvements
December 1972
Scale: 1" = 600'
Round Lake
Lake Charlevoix
Lake Michigan
Proposed 12" Water Main
United States Coast Guard Station
Ex. Collection Flume and Pumping Sta.
Match Line

APPENDIX I

Formal Report: *Centralization of Medical Records Transcription Services*

CENTRALIZATION OF MEDICAL RECORDS TRANSCRIPTION SERVICES

Sunnybrook Medical Centre
Toronto, Ontario

Community Systems Foundation

2200 FULLER ROAD • ANN ARBOR, MICHIGAN 48105 • 313/761-1846

CORPORATE HEADQUARTERS

CENTRALIZATION OF MEDICAL

RECORDS TRANSCRIPTION SERVICES

Sunnybrook Medical Centre
Toronto, Ontario

ANN ARBOR | ATLANTA | AUGUSTA, ME. | BALTIMORE | CONCORD | MINNEAPOLIS | NEW HAVEN | SEATTLE | TORONTO | WASHINGTON, D.C.

September 20, 1974

Mr. J. R. Hamilton
Assistant Executive Director
General Services
Sunnybrook Medical Centre
2075 Bayview Avenue
Toronto, Ontario Canada

Dear Mr. Hamilton:

CSF is pleased to submit our analysis of the Medical Records transcription function at Sunnybrook Medical Centre.

The study primarily addressed the issue of centralizing this function in the first floor Steno-Pool. Centralization of the transcription function was deemed advisable as a result of the analysis. A staffing reduction from 26.0 full-time-equivalent positions to 14.3 positions is one advantage of the centralization. This reduction will result in an annual savings of over $82,000 including fringe benefits. It will also result in several non-economic advantages.

We would like to take this opportunity to acknowledge the excellent cooperation and assistance received throughout the study from Mr. McCandless, Ms. Evans, and Ms. Johnson of Medical Records. Their efforts significantly contributed to the success of this analysis and are sincerely appreciated by CSF Ltd.

We are looking forward to assisting the hospital in implementing the approved recommendations.

Respectfully submitted,

G.F. Braidwood

G. F. Braidwood,
Associate

GFB/dr

CENTRALIZATION OF MEDICAL RECORDS TRANSCRIPTION SERVICES:

Reducing Staff and Increasing Efficiency

Sunnybrook Medical Centre
Toronto, Ontario

Submitted by: CSF Ltd.
Project Engineer: G. F. Braidwood

Project Number: AA-SB-18
Date Submitted: 20 September 1974

Submitted to:
Mr. J. R. Hamilton, Assistant Executive Director, General Services
Mr. M. M. McCandless, Head, Medical Records
Ms. E. Johnson, R.R.L., Medical Records Supervisor, Coding and Indexing
Ms. J. Evans, Medical Records Supervisor, Stenographic Services

TABLE OF CONTENTS

LIST OF TABLES

THE PURPOSE OF THE STUDY

At the request of the hospital administration, CSF Ltd. has analyzed the medical transcription function of the Medical Records Department at Sunnybrook Medical Centre. The service, which presently consists of twenty-six (26) full-time Medical Dicta-Typists distributed throughout the hospital, appeared to be both costly and inefficient. Miscellaneous clerical tasks which should be performed by the forty-seven (47) cost-shared secretaries[1], have by habit and tradition become the responsibility of the Medical Dicta-Typists. Making appointments for patients, collecting x-ray and laboratory reports, filing admission and discharge cards, xeroxing, and running errands, the Medical Dicta-Typists are being used in a way which significantly departs from stated policy.

The purpose of this investigation was to determine if centralization of the medical transcription function of the Medical Records Department would make the service more efficient and more economical than it is presently. The study documents present activities, determines workload frequencies of present activities, develops required staffing projections for alternative organizations (both centralized and decentralized), considers the qualitative advantages and disadvantages of centralization, and recommends operating policies and procedures to implement centralization. The remainder of this report shows that by centralization it is possible to reduce staff and to improve the quality of the medical transcription function. The next section summarizes our major recommendations.

[1] In addition to the 26 Medical Records personnel, there are 47 other secretaries located throughout the hospital. These secretaries provide secretarial service to the Medical Staff. They are either hospital employees (cost-shared with Sunnybrook Hospital University of Toronto Clinic, SHUTC, and The University of Toronto) or SHUTC employees (cost-shared with the hospital). This report does not discuss the 47 cost-shared secretaries. See "Secretarial/Typing Support Service Provided by Sunnybrook Hospital," W. J. Kribs, December 15, 1972.

SYNOPSIS OF RECOMMENDATIONS

On the basis of our investigation we recommend centralization of the medical transcription function of the Medical Records Department. Not only is this consistent with recommendations of the Canadian Council on Hospital Accreditation, but it would achieve an annual savings of approximately $82,000 by staffing reduction. It would also eliminate performance of non-medical records work by Medical Dicta-Typists, afford better supervision and control, more efficient scheduling, and more even distribution of workloads. In addition, centralization would free one area on each of the ten floors and provide hospital-wide knowledge of work backlogs. Our specific recommendations are as follows:

1. Centralize the Medical Records transcription function in the Steno Pool located on D-1. The required staffing is 11.3 full-time-equivalent employees. This staffing level includes discharge summary and operating room note transcription, administrative typing at the current level, supervision, transportation, and ample allowances for fatigue and unavoidable delays. Implementation should be accomplished via attrition. When coupled with the three positions described in the next two recommendations, the resulting 14.3 positions represent a savings of 11.7 positions, or $82,080 annually.
2. Maintain the present staffing of 1 full-time position on the Ground floor assigned to the Medical Records Department (as distinguished in terms of location from the Steno-Pool).
3. Transfer 2 employees permanently to the Pharmacy and to Neuro-Psychology (C-8). This is essentially a budget transfer since 2 employees are already working in these areas.
4. Establish a regular messenger service within Medical Records to provide adequate pickup and delivery of charts, discs, and reports for all areas of the hospital.

viii

5. Do *not* implement a centralized dictating system. Further analysis is needed before this concept can be recommended.
6. Use the Hospital Messenger Service and the Print Shop (under the Materials Management Department) for all photocopying work.
7. Implement an index card file to control inpatient charts and, after discharge, to monitor transcription backlogs and physician reviews and signatures.
8. Implement regular Management Information Reports from the Steno-Pool supervisor to the Medical Records Department Head.
9. Review the design and number of copies of the Discharge Summary form, presently a 6-part carbon set.
10. Implement a well-communicated and administered secretarial replacement function which features a "first-come, first-served" policy until the maximums of two per day (or four per day for summer months) are reached.

CENTRALIZATION OF MEDICAL

RECORDS TRANSCRIPTION SERVICES

I. Inefficiencies in Present Staffing Workloads

The present Medical Records transcription service at Sunnybrook Medical Centre is both inefficient and costly. The service employs 10 of the total 26 Medical Dicta-Typists on the floors of the hospital. This distribution of personnel results in non-medical records work often being performed by Medical Records personnel, despite existing policies. It also creates overlaps with ward clerks and cost-shared secretaries. It makes supervision of records work difficult, and causes scheduling of work, workload distribution, and control of records to suffer because the records personnel are distributed throughout the hospital. In addition, the service employs 13 Steno-Pool Typists who presently perform the work of 8.2 Medical Dicta-Typists. (Along with these thirteen typists there are two supervisors in the Steno Pool and one typist located in the Medical Records Department on the ground floor.) The present service of 26 Medical Dicta-Typists represents a total staffing cost of $182,380 per year. Of this $161,400 are wages (average of $6,207 per employee) and $20,980 are fringe benefits (average of 13% of wages). Appendix A describes the normal work day and defines a full-time equivalent employee.

An examination of the activities and workload frequencies of both the Steno-Pool Medical-Dicta Typists and the Floor Medical Dicta-Typists reveals the inefficiency and therefore costliness of the present procedure.

A. Steno-Pool Medical Dicta-Typists

The Steno-Pool Medical Dicta-Typists presently perform three general categories of work: medical typing, administrative and other typing, and secretarial relief coverage for virtually all areas of the hospital. The Steno-Pool medical typing consists of discharge summaries, which by policy should be done exclusively by the Floor Medical Dictas; consultations, which should be done by the cost-shared secretaries; and operating room notes and miscellaneous medical reports, which are the appropriate responsibility of Steno-Pool typists. The administrative typing consists largely of letters, reports, theses, Form 106's, Social Service case reports, and a variety of miscellaneous typing tasks. Again, many of these tasks are the theoretical responsibility of the cost-shared secretaries; however, convention has made these the work of the Steno-Pool Medical Dictas.

These categories of work performed by Steno-Pool Medical Dicta-Typists are illustrated by the workload frequencies for Steno-Pool Medical Dicta-Typists during a recent twelve month period (Table 1). As one can see, the frequency of non-medical typing and activities which should in theory be done by other secretaries is very high.

TABLE 1.

STENO-POOL MEDICAL DICTA-TYPISTS
WORKLOAD FREQUENCIES

Medical Typing	Average No./Working Day
Discharge Summaries	18.604
Consultations	0.432
Operating Room Notes	20.212
Misc. Medical Reports	0.660
Administrative and Other Typing	
Letters	3.745
Reports	1.283
Theses, Stencils	0.528
Form 106's	48.692
Social Service Case Reports	2.112
Administration of Typing Tests	1.104
Requisitions, Invoices, Bills	3.297
Labels, Envelopes, Notices	4.153
Other	
File Assembly	14.256

In addition to the items in Table 1, other work performed by these Steno-Pool dictas includes transportation (5 round trips per day—4 of which are to the operating room), telephone, and special projects such as data reduction.

Along with the medical typing and administrative typing, a major part of the work of Steno-Pool Medical Dicta-Typists is to provide secretarial relief coverage. During a representative twelve-month period, for example (see Appendix B), these secretaries furnished 949 man-days worth of relief service. Administrative replacements were 267 days. Medical/Surgical replacements were 682 days. In all, replacement man-days averaged 3.8 per working day during the period examined; however, this activity is highly dependent upon seasonal variations. The peak demand for replacements comes during the summer months. (Included in the replacement days are those for the two positions which have effectively become "permanent replacements." One Medical Dicta has been assigned to the Pharmacy and one to Neuro-Psychology (C-8) on a full-time basis. Since neither the Pharmacy nor Neuro-Psychology has a budget position for a Medical Dicta, the positions are filled from the Steno-Pool.)

B. Floor Medical Dicta-Typists

The work of the Floor Medical Dicta-Typists consists primarily of transcribing discharge summaries. Of the 54.40 discharges per working day in May-June, 1973, for example, the Floor Medical Dicta-Typists transcribed 38.56. (See Appendix C for a summary of workloads.) Theoretically *all* discharge summaries should be done by the Floor Medical Dictas, but as the number of discharges actually done by Floor Medical Dicta-Typists indicates, transcription of 17.84 discharges per working day is performed by Steno-Pool personnel.

Although most of their time is given to transcribing discharge summaries, Floor Medical Dictas spend a significant portion of their time at activities other than those they should perform. In 1967, for example, the Sunnybrook Hospital industrial engineer surveyed twenty-five secretary/medical dicta positions located throughout the hospital. From this study, he computed an average of 0.71 hours/day/position for non-transcription. 0.32 hours/day—telephone; 0.12 hours/day—mail; 0.27 hours/day—filing. This study also indicated an average of 0.85 hours per day per position for errands. Data from the 1973 Materials Management Analysis documents the amount of this transportation work as 0.54 hours/day (see Appendix D).

Other activities performed by the Floor Medical Dictas include consultation transcription (which should be done by the cost-shared secretaries), making appointments for patients, collecting x-ray and laboratory reports, filing admission and discharge cards, xeroxing, and other typing (reports, letters, theses). Very few, if any, of these activities are the responsibility of the Floor Medical Dictas. In practice, however, work habits have resulted in these tasks being done by Floor Medical Dictas. (We have not sought to document the workload data for these activities. It would be expensive to obtain and of questionable usefulness.) Our investigation makes clear that Floor Medical Dictas do a significant amount of non-Medical Records work. This is both inefficient and costly.

II. Proposed Staffing: Two Alternatives

On the basis of our investigation we conclude that by revision of the present Medical Records transcription procedures it is possible to increase efficiency and to reduce costs by cutting the size of the staff. This can be accomplished by either of two methods, a centralized organization or a decentralized organization. Both would reduce the staff. The centralized organization would require 14.3 full-time equivalent employees rather than the present 26. The decentralized organization would require 21.2 full-time equivalents. Thus the cost of either of the two alternatives would be substantially lower than the cost of the present system. A centralized system would cost $100,300 per year (an annual cost reduction of $82,080 from the present $182,380). A decentralized system would cost $148,690 (an annual cost reduction of $33,690).

From a staffing (i.e., economic) viewpoint, the centralized Medical Records transcription service certainly appears to be the most desirable of the alternatives. In the following sections, the requirements for staffing of the two organizations are explained. (As a preliminary step in considering alternatives to the present system it was necessary to compute the time required to transcribe discharge summaries. An explanation of this computation will be found in Appendix E.)

A. Staffing: Centralized Organization

Centralizing the Medical Records transcription service would consist of locating all Medical Records personnel except one in a central area such as the present Steno-Pool (D-1). Medical Dicta-Typists would not be located on the Floors as they presently are; however, one position located in the Medical Records Department itself (Ground Floor) would not be affected by centralized organization. This position would be the assignment of one Medical Dicta, and will not be analyzed in detail.

A summary of the staffing necessary under a centralized system is presented in the table on p. 5. As the table indicates, 14.3 full-time equivalents are required under centralized organization.

The table of required staffing under a centralized organization results from the following calculations:

1. Determining workloads for former Steno-Pool production activities.
2. Determining the workload for transcription of discharge summaries formerly done by Floor Medical Dicta Typists.
3. Determining the workloads for supervisors.
4. Developing workloads for transportation functions.
5. Developing workloads for miscellaneous clerical work.
6. Calculating the replacement function.

TABLE 2.

STAFFING—CENTRALIZED ALTERNATIVE

Function	Hours per Working Day	Hours Per Year (250 Working Days)
1. Formerly Steno-Pool Production Activities	27.4	
2. Formerly Floor Medical Dicta Discharge Summary Transcriptions	14.8	
3. Supervision	4.0	
4. Transportation	6.4	
5. Miscellaneous Clerical	3.3	
Subtotal	55.9	13975.0
6. Replacement Function		5325.0
TOTAL		19300.00
7. Full-Time Equivalent Employees: Line 9 ÷ 1704 working hrs/employee		11.3
8. Medical Records (constant position)		1.0
Total: Medical Records		12.3
9. Transfer of "replacements" to Pharmacy and Neuro-Psychology Budgets		2.0
Total Staffing (FTE)		14.3

Staffing requirements for all of these activities were determined by using the workload frequencies previously shown in Table 1, p. 2, consulting the *Hospital Staffing Methodology Manual,* and consulting with Steno-Pool supervisors to estimate the time required for functions whose nature could not be determined in other ways. The following discussion explains the staffing needs for each of the activities described above.

1. Time required for former Steno-Pool production activities

Multiplying workloads presented in Table 1 by the standard time per activity yields the workload itemized in Table 3, "Steno-Pool Staffing Requirements: Production Activities." The daily times computed in the table above result in 1645.8 minutes per day, or 27.4 hours per working day. These figures include all allowances for personal time, fatigue, and unavoidable delays.

TABLE 3.

STENO-POOL STAFFING REQUIREMENTS
PRODUCTION ACTIVITIES

Activity	Frequency/ Working Day (Table 1)	Standard Time in minutes (see footnotes)	Total Time/ Working Day in minutes
Medical Typing			
Discharge Summaries	18.604	—[1]	449.249[1]
Consultations	0.432	25.533[2]	11.030
Operating Room Notes	20.212	26.677[3]	539.196
Misc. Medical Reports	0.660	25.533[2]	16.852
Chronic, Domiciliary Discharge Summary	—	—	30.000[4]
Administrative and Other Typing			
Letters	3.745	25.533[2]	95.621
Reports	1.283	120.000[5]	153.960
Theses, Stencils	0.528	120.000[5]	63.360
Form 106's	48.692	0.918[6]	44.699
Social Service Case Reports	2.112	20.000[5]	42.240
Typing Tests	1.104	15.000[5]	16.560
Requisitions, Invoices, Bills	3.297	11.554[7]	38.094
Labels, Envelopes, Notices	4.153	0.572[8]	2.376
Other			
File Assembly	14.256	10.000[4]	142.560
	TOTAL		1645.797

[1] Refer to Appendix E, Steno-Pool Discharge Summary Transcription

[2] *Hospital Staffing Methodology Manual*: Assume 40 lines/report, 2 pages/report, 1 disc/report. (See formula on page 15.)

[3] Refer to Appendix F, Steno-Pool Operating Room Note Transcription

[4] Estimated

[5] Estimated by Steno-Pool Supervisor

[6] *Hospital Staffing Methodology Manual*: 6 items/form; two alphabetical sorts; one matching operation

[7] *Hospital Staffing Methodology Manual*: Assume 20 lines/item, 1 page/item

[8] *Hospital Staffing Methodology Manual*: 6 items/label or notice

2. Time required for transcription of discharge summaries formerly done by Floor Medical Dicta-Typists

Table 4 (below), "Floors Discharge Summary Transcription," was determined in the same manner as Table 3 showing staffing requirements for former Steno-Pool production activities. The daily times computed show 886.8 minutes per day, or 14.8 hours per working day.

TABLE 4.

FLOORS DISCHARGE SUMMARY TRANSCRIPTION

	Floors Discharge Summary Transcription	
Hospital Area	Number Done[1] By Floors	Time/Working Day[2] (in minutes)
B-3	4.35	134.119
B-4 (I)	1.70	42.498
B-4 (N)	1.87	54.021
B-5 (H)	1.45	50.353
B-5 (R)	1.90	64.013
B-6	2.92	55.077
C-2 (G,M,N)	0.66	21.909
C-2 (D)	8.00	88.032
C-3 (D)	1.08	35.645
C-3 (R)	1.34	41.264
C-3 (FCM)	0.78	23.157
C-4	0.38	9.876
C-5	3.76	70.778
C-6 (G)	0.38	8.124
C-6 (T)	0.21	4.465
C-6 (C)	0.03	0.704
C-6 (P)	0.06	1.063
A-4 (Gyn)	1.72	35.392
A-4 (Ot)	2.64	44.748
A-4 (Op)	0.63	11.667
F-1	2.30	78.103
Rehabilitation	0.20	6.037
Dental	0.20	5.793
Totals	38.56	886.838 minutes/day
		or 14.8 hours/day

[1] From Appendix B (Hospital data for May–June 1973)
[2] Number done by Floors multiplied by standard time (from Appendix E)

3. Time required for supervision

At the present time there are two supervisory positions in the Steno-Pool. Both of these are working-supervisors. An estimated 4.0 hours per working day are allocated to supervisory duties only. This staffing is not expected to change under the centralized alternative.

4. Time required for the transportation

Transportation time under the centralized alternative will result in eliminating some of the present travel by Floor Medical Dictas and adding time to properly service dictating areas throughout the hospital from the centralized location. The total required time is 6.4 hours per working day. (Procedural recommendations for transportation tasks under the centralized plan are discussed later in this report.) This calculation of the transportation requirement results from identifying the time required for existing Steno-Pool transportation tasks (1.0 hours per day), estimating the time for additional travel required because the centralized location (D1) is located further from the dictating areas throughout the hospital than the Floor Dicta Typists presently are (1.0 hour per day), and determining the time for Floor transportation which will continue under centralization (4.37 hours per day).

The time required for the existing Steno-Pool transportation tasks is computed from data collected during the current Materials Management Study. Five round trips per day are made by Steno-Pool personnel. The average trip is 12.4 minutes. Multiplication results in 62.0 minutes per day, or 1.0 hour per working day.

The time required for additional transportation by the new centralized location (1.0 hour per day) is an estimate derived from interviews with supervisors, present Steno-Pool personnel, and present Floor Medical Dictas.

The time required for transportation (presently done by Floor personnel) which will continue under centralization (4.37 hours per day) is based upon an analysis of trips presently done by Floor personnel (total 5.43 hours per day) qualified as being either high (+) or low (−) in comparison to the time required for the same trips from the new centralized location. Table 5, "Distribution of Floor Medical Dicta Transportation," identifies those destinations which will require trips under the centralized plan and shows whether the times required will be either greater or less than present.

5. Time required for miscellaneous clerical work (phone, filing, mail, etc.)

The present miscellaneous clerical work required by Steno-Pool personnel was estimated to be 0.5 hours per day per position. (This is a conservative estimate. A 1967 study showed 0.71 hours per day were required for miscellaneous clerical work by Floor Medical Dictas.) For the 6.5 positions required to do the former Steno-Pool Production Activities, the Discharge Summaries and Transportation Functions, this totals 3.3 hours per working day. Since the

TABLE 5.

DISTRIBUTION OF FLOOR MEDICAL
DICTA TRANSPORTATION

Destination	%	Average Time (% 5.43 hrs)	Present Requirement
1. Nursing Station	35.3	1.92 hrs/day	+
2. Residents' Offices	17.1	.93	+
3. Doctors' Offices	15.7	.85	+
4. Print Shop	7.0	.38	
5. Medical Records	4.9	.27	–
6. Miscellaneous	4.7	.26	
7. Operating Room	4.0	.22	+
8. Outpatient (A-Wing)	3.4	.18	+
9. Steno-Pool	2.7	.15	
10. Medical Library	2.2	.12	
11. X-Ray	1.6	.09	
12. Medical Photography	1.3	.07	
	99.9	5.44	

Steno-Pool figure is conservative and the Floor figure should, if anything, decrease under a centralized plan, it is safe to assume these same figures for all positions under centralization.

6. Time required for the replacement function

The task of determining standard staffing levels for the secretarial relief replacement function is difficult due to the monthly demand variations (see Appendix B). The average number of replacements per working day throughout the year is 3.8. However, some months have a replacement demand much higher. August, for example, requires 7.5 replacements per day. Other big months are May (6.4), July (5.8), and September (4.6). Therefore, 710 replacement days per year, or 5325.0 hours per year, are needed.

The recommended staffing levels for the replacement function are highly dependent upon successful scheduling of personnel. For example, if personnel were not permitted to take their vacations during high demand months, the staffing could be kept to a minimum. However, this may not be advisable from an employee relations viewpoint. Table 6, "Staffing Requirements—Replacement Function," presents recommended staffing levels. It represents an attempt to balance demand with economy. (Procedural recommendations for administering replacement functions are presented later in this report.)

The recommended staffing levels in Table 6 do *not* include provisions for the two relatively permanent replacements which currently exist for the Pharmacy and for Neuro-Psychology. If these two departments have

10

TABLE 6.

STAFFING REQUIREMENTS—REPLACEMENT FUNCTION

Month	Present Demand (from Table 2)	Recommended Staffing Level	Number of Days Needed Per Year[1]
Jan.	2.8	2.0	42
Feb.	0.9	2.0	40
Mar.	3.1	2.0	42
Apr.	2.4	2.0	42
May	6.4	4.0	84
Jun.	3.5	4.0	84
Jul.	5.8	4.0	84
Aug.	7.5	4.0	84
Sep.	4.6	4.0	84
Oct.	3.6	2.0	42
Nov.	2.4	2.0	42
Dec.	1.2	2.0	40
TOTAL			710

[1] Obtained by multiplying the recommended staffing level (replacements per working day) by an estimated 21 working days per month (20 days used for February and December).

justified need for this secretarial service, we recommend adjustments be made in their respective budgets.

These six calculations for the activities that remain unchanged and for those that change under a centralized organization yield a total staffing requirement for centralized organization of 11.3 full-time equivalent employees. To that number add the one position located in the Medical Records Department which will not be affected by centralized organization. Also add the two "replacements" transferred to the Pharmacy and Neuro-Psychology budgets. Therefore, 14.3 full-time equivalents are required under centralized organization.

B. Staffing: Decentralized Organization

A decentralized medical records transcription service with an arrangement similar to the present organization is an alternative to centralized organization. Computation of workloads and staffing requirements for the Steno-Pool and for Floor Medical Dicta typists suggests that without radical revision of the present system it is possible to reduce staffing to 21.2 full-time equivalents from the present staffing of 26 employees. (As before, the one dicta typist located in the Medical Records Department will be considered constant and will not be analyzed.)

A summary of the staffing requirements under a decentralized system is presented in Table 7, "Staffing—Decentralized Alternative." This table of required staffing results from analysis of workloads for Steno-Pool personnel and Floor personnel. As before, standard times for these workload frequencies were determined by consulting the *Hospital Staffing Methodology Manual* and by estimates provided by supervisors. Since an explanation of staffing requirements for Steno-Pool functions and Floor functions has been presented in the preceding section, in this section we present only a summary of staffing required for decentralized organization.

TABLE 7.

STAFFING—DECENTRALIZED ALTERNATIVE

	Full-Time-Equivalents
1. Medical Records (constant position)	1.0
2. Steno-Pool (includes supervision)	8.2
3. Floors (includes overstaffing due to coverage requirements)	10.0
Sub-Total	19.2
4. Transfer of "replacements" to Pharmacy and Neuro-Psychology Budgets	2.0
Total Staffing (FTE)	21.2

1. Summary of staffing required for Steno Pool: Decentralized Alternative

The staffing required for Steno-Pool production activities under the decentralized alternative remains the same as that for the centralized alternative. Production activities (1645.797 minutes per working day, Table 3) require 27.4 hours per working day. Transportation, miscellaneous clerical, and supervision, when added, bring the total to 34.4 hours per working day. When the replacement function is calculated, the total becomes 8.2 full-time equivalents for Steno-Pool staffing as shown in Table 8, "Steno-Pool Staffing Summary—Decentralized."

2. Summary of staffing required for Floors: Decentralized Alternative

It is assumed in the decentralized organization that all ten existing locations on the floors are to be retained. The required hours are much less than ten full-time positions, so the average hours are also indicated in Table 9, "Floors Staffing Summary—Decentralized." These extra hours are more than enough to support those activities discussed earlier (pp. 3–4) which are not the stated responsibility of the Floor Medical Dictas.

TABLE 8.

STENO-POOL STAFFING SUMMARY—DECENTRALIZED

Function	Hours Per Working Day	Hours Per Year (250 Working Days)
1. Production	27.4	
2. Transportation	1.0	
3. Sub-Total	28.4	
4. Phone, Filing, Misc: 28.4 ÷ 7.5 ≅ 4.0 positions; 4.0 × 0.5/position =	2.0	
5. Supervision	4.0	
6. Sub-Total	34.4	8600.0
7. Replacement Function (Table 6): 710 days × 7.5 hours/day =		5325.0
8. TOTAL Hours/Year		13925.0
9. Full-Time-Equivalent Employees: Line 8 ÷ 1704 working hours/employee		8.2

As Table 9 indicates, 27.3 productive hours per day are required for the 10 Floor Medical Dicta positions. However, the present decentralized organization must have a staff of 10 people, which requires 75 hours per day. In other words, the present decentralized organization is costing the hospital 47.7 non-productive hours per day.

III. Policies and Procedures

A. Qualitative Advantages and Disadvantages of Centralization

The previous section explained two alternatives for staffing Medical Records transcription services. Of these two, the centralized alternative was clearly the more economical, with an annual savings of $82,200. However, before one can recommend the centralized transcription system, qualitative, non-economic aspects must be considered. The following section identifies the advantages and disadvantages of centralization. It shows that when the points are combined with economic considerations, the centralized service is definitely feasible and is, in fact, the more advantageous of the alternatives.

TABLE 9.

FLOORS STAFFING SUMMARY—DECENTRALIZED

Present Floor Position	Discharge[1] Summary Transcription	Phone, Mail,[2] Filing, etc.	Transportation[3]	Total Req'd Hours/Day	Overstaffed Hours/Day[4]
1. B-3	2.24	0.71	0.45	3.40	4.10
2. B-4	1.61	0.71	0.51	2.83	4.67
3. B-5	1.91	0.71	0.34	2.96	4.54
4. B-6	0.92	0.71	0.50	2.13	5.37
5. C-2	1.83	0.71	0.50	3.04	4.46
6. C-3	1.67	0.71	1.20	3.58	3.92
7. C-5	1.18	0.71	0.51	2.40	5.10
8. C-6	0.24	0.71	0.89	1.84	5.66
9. A-4	1.53	0.71	0.45	2.69	4.81
10. F-1	1.30	0.71	0.08	2.09	5.41
Other	0.36	—	—	0.36	(0.36)
TOTALS	14.79	7.10	5.43	27.32	47.68

[1] From Table 4, converted to hours per working day; nursing areas are combined where appropriate, i.e., where one Floor medical dicta transcribes summaries from more than one unit
[2] From page 5
[3] From Appendix D.
[4] 7.5 hours/day minus Total Required Hours/Day

It is important to note that centralized *dictation* is *not* being recommended. The feasibility of a central dictating area for physicians was not analyzed in detail in this study. While on the surface there may appear to be advantages (reduced chart movement, reduced transportation work, better chart control, more efficient administration of deficiencies, etc.), these must be weighed against the possible disadvantages: doctor inconvenience, privacy, space, noise, installation costs including equipment, etc. If centralized dictating is to be considered further, a detailed study is recommended.

1. Advantages of Centralization

The main advantages of centralization of records transcription are that it would eliminate the performance of non-medical records work by medical dictas. By putting all Medical Dicta personnel in one place it would also promote better supervision, scheduling, workload distribution, and record control. In addition, it would free the ten floor areas presently occupied. And finally it would be consistent with the recommendations of the Canadian Council on Hospital Accreditation. The advantages, then, are as follows:

a. Elimination of responsibility redundancies and inconsistencies. Non-Medical Records work is now being performed on the Floors, resulting in overlaps with ward clerks and cost-shared secretaries. A much clearer job definition for all positions will result.

b. Better supervision. This will result in increased control over quantity and quality of work, discipline, and work habits.

c. More efficient scheduling. It is always easier to arrange work schedules and time-off coverage when dealing with a group than individually.

d. Even workload distribution. With centralization, more equitable work assignments can be made (via the direct supervision) such that all medical dictas are contributing equally. Specialization is also possible, further increasing productivity.

e. More effective record deficiency control. One centralized transcription area will enable an efficient administration of all record deficiencies.

f. Centralized record control. The location control of medical records for inpatients would become more efficient and effective by the reduction of at least ten locations now present. Centralization should also improve retrieval time when responding to a record request.

g. Hospital-wide knowledge of workload backlogs. Meaningful information will be available daily concerning work yet to be done so that effective management decisions can be made.

h. Ten areas on the floors would be freed for alternative uses.

i. Consistent with recommendations from the Canadian Council on Hospital Administration. A centralized transcription function would be consistent with these recommendations.

2. Disadvantages of Centralization

Centralization of records transcription services does have some disadvantages. Chiefly, however, these are matters of communication and job reclarification with other hospital personnel, both medical staff and secretarial staff. The evident disadvantages are:

a. Adjustment by cost-shared secretaries and ward clerks. This disadvantage relates to these positions being required to perform some of the miscellaneous tasks now done by the Floor Medical Dictas. These jobs of making patient appointments, filing x-ray reports, typing consultations, etc. are not supposed to be done even now by the Floor dictas, so the elimination of the Floor dicta is really a compliance with stated hospital policy.

b. Messenger Service necessity. The centralization of the transcription function will require the establishment of regular pickups and deliveries to and from the floors. These trips will be necessary for charts, discs, reports, etc. This disadvantage may not be too serious if the existing Medical Records messengers (2 employees) can absorb the work.

c. Doctor deficiency reminders. Reminding physicians to complete charts will require more formality (lists, letters, etc.) under the centralized plan. While the central control concept is an improvement (advantage e. above), the ability to verbally remind doctors in person on the floors will be lost.

d. Record control. While advantage f. above discusses the elimination of ten areas to worry about chart location control, there does remain the disadvantage of not having a Medical Records employee on the floor to physically monitor charts in other locations, e.g., doctors' offices, dictating areas, etc.

e. Noise and space in the Steno-Pool. This disadvantage is only temporary. When implementation is complete, the proposed Steno-Pool staffing of 11.3 full-time equivalents (from Table 2, page 5) will definitely "fit" in the existing D-1 room.

B. Procedures for Implementing Centralization

The analysis has thus far shown that the centralized approach is feasible. In this section, operating policies and procedures are recommended to implement the centralized organization. These include recommendations on Messenger Service, Xeroxing, Inpatient Index File, Transcription and Signature control, Management Information System, Discharge Summary Form, and Replacement Days Service. In the event that the hospital administration prefers the decentralized alternative, several of these policies and procedures will nonetheless be capable of adaptation.

1. Messenger Service

Presently two Medical Records employees provide a messenger service for the department. With the centralization, the following tasks will need to be added to their responsibilities:

a. Pick up new charts after discharge from the nursing station.

b. Pick up old charts after discharge from doctors' offices.

c. Insure that all dictating stations are supplied with discs (blank) and the assembled charts are ready for dictating.

d. Pick up completed dictation (discs) and the corresponding charts.

e. Deliver reports and charts to doctors' offices for signatures.

f. Pick up reports and charts after signatures.

Obviously the above pickups and deliveries can be done simultaneously. Regular routes should be established to insure complete and dependable coverage of this function. Initially, one complete trip (to all nursing areas, operating room, offices, etc.) should be made every three hours at approximately 7:45 a.m., 10:45 a.m., 1:45 p.m., and 4:45 p.m. After some experience is gained, more or less frequent trips may be needed.

In the staffing calculations contained in this report, the allowance of 6.4 hours per day is included for transportation. If the two existing messengers are able to perform these tasks, the Medical Dicta staffing can be reduced by about 1 person. If not, then the staffing recommendations contain the provision for 6.4 hours per day for a Medical Dicta to perform the necessary travel. In the latter case, it would be desirable to hire a less-expensive employee to do this work and not pay a qualified Medical Dicta to walk.

2. Xeroxing

All xeroxing is to be accomplished by sending the material via the hospital Messenger Service to the Print Shop for photocopying and return via the Messenger Service. Unless dictated by emergencies, this entire function (transportation and photocopying) should *not* be performed by Medical Records personnel.

3. Inpatient Index File

To effectively control charts and other items, an inpatient index file must be maintained in the centralized Steno-Pool. This file (perhaps 3x5 index cards) will indicate all inpatients at any given time. Therefore, the file must be updated daily to reflect admissions and discharges. The cards should be filed alphabetically and should indicate the nursing unit for chart location control purposes.

4. Transcription and Signature Control

After a patient is discharged, the index card (see previous procedure) is pulled from the Inpatient File and placed in a Transcription File. This card can then be used to indicate the status of discharge summary dictation, discharge summary transcription, doctor's signature and perhaps other information. By the use of different files, tabs or other means, these various categories can be monitored. The information can then be used to provide signature deficiency reports, workload backlog reports, and other management status reports (see next procedure).

5. Management Information System

A regular Management Information report should be prepared by the Steno-Pool supervisor and submitted to the Medical Records Department Head.

While the exact format of the report can be left to the discretion of the supervisor, the following items should be included:

a. Workload volumes completed—discharge summaries, operating room reports, administration typing (by type), and number of replacement days provided.

b. Number of discharge summaries not dictated.

c. Number of discharge summaries dictated but not transcribed.

d. Administrative typing backlog.

e. Number of reports transcribed but not signed by the physician.

f. Description of current problems and/or unusual events.

This report should be submitted weekly unless desired more frequently by the Department Head. Another alternative is to break the report into two parts, a monthly report covering items (a) and (f) above and a weekly report covering the more variable and timely items (b), (c), (d), and (e).

The doctor deficiency items, (b) and (d), may require further detail such as a listing by physician so followup action can be taken by the Department Head, the Medical Records Committee, and/or Administration.

6. Discharge Summary Form

The present discharge summary form consists of the first page and then continuation pages. Both forms are a six-part carbon set. For two reasons, this form should be reviewed and perhaps revised.

First, only about three or four copies are now used for the chart and doctors. Reducing the carbon set to only four copies would produce a significant monetary savings in forms costs (a minimum of $500 per year). If more than three copies are needed, using the Xerox is more economical if the frequency of these instances is low.

Second, there are a great number of discharge summaries which are between 20 and 30 lines in length. If the discharge summary first sheet could be re-designed to accommodate more than the present 20 lines, many continuation pages would be unnecessary.

7. Replacement Days Service

In earlier sections of this report, the present secretarial replacement service was discussed. The recommended level for replacements is shown by Table 6, and consists of 2 replacements available for every working day during January-April and October-December. The other five months, May through September, have 4 replacements available per day.

A "first come-first served" policy is recommended. After sufficient communication to all potential users of this service, the Steno-Pool supervisor should maintain an "appointment-like" book, making reservations in advance for replacement dates. All requests for replacements must be submitted *in writing* to the supervisor to minimize conflict over the request date. The supervisor will acknowledge these requests, in writing, indicating whether a replacement is available (form letters can be used for this purpose). Replacements will be provided until the maximum of two or four per day is reached, depending on the month. This type of upper-limit control is mandatory if any degree of fairness and managerial planning is expected.

IV. Conclusion

The present system for Medical Records transcription services is clearly inefficient and costly. Either of the two alternatives discussed above would be an improvement. The centralized system, though requiring somewhat more revision and adjustment in the total hospital routine, appears the most effective solution. It would reduce costs by $82,080 per year by cutting the staff from 26 to 14.3 full-time equivalent personnel. At the same time it would promote more efficient service by affording better supervision, scheduling, and control of records.

The second alternative—a decentralized alternative much like the present situation—is also an improvement over the present system. It would reduce costs by $33,690 per year, and at the same time eliminate at least some of the inefficiencies in the present system. However, although it is clear that the second alternative is the easiest of the two and is even justified by a substantial economic gain, we believe the administration should adopt the centralized alternative.

APPENDIX A

Computation of Full-Time Equivalent Employee

The normal working day for a medical dicta-typist is from 8:30 a.m. to 4:45 p.m. with a 45 minute lunch break. This results in a 7.5 hour work day per position.

The average number of days off per year was next calculated. Vacation time data was collected, resulting in an average of 12.6 vacation days per year per employee. The average number of sick days per year was determined as 10.2 per employee. Combining these figures with the 10 statutory holidays, the average Medical Dicta will have 32.8 days off per year.

Using the above information, a full-time equivalent employee can be defined as 1704 working hours per year:

$$52 \times 5 = 260 - 32.8 = 227.2 \times 7.5 \text{ hours/day} = 1704$$

APPENDIX B

Replacement Days Variations by Month

The following table shows the monthly variations in demand for replacement secretarial services.

Month	Replacement Man-Days Admin.	Med./Surg.	Total	Net Working Days/Month	Replacements Per Working Day
Jan. '73	20	42	62	22	2.8
Feb.	15	3	18	20	0.9
Mar.	37	31	68	22	3.1
Apr.	23	24	47	20	2.4
May	31	117	148	23	6.4
Jun. '72	12	62	74	21	3.5
Jul.	20	95	115	20	5.8
Aug.	26	139	165	22	7.5
Sep.	53	44	97	21	4.6
Oct.	9	71	80	22	3.6
Nov.	14	39	53	22	2.4
Dec.	7	15	22	19	1.2

APPENDIX C

Discharge Summary Workloads of Floor Medical Dicta-Typists

Floor Medical Dictas	Hospital Area	Total Discharges Per Working Day (May–June 1973)	No. Done/Working Day by Floor Medical Dictas
1.	B-3	4.40	4.35
2.	B-4	4.10	3.57
3.	B-5	4.20	3.35
4.	B-6	3.30	2.92
5.	C-2 (incl. dialysis)	11.40	8.66
6.	C-3	3.90	3.20
7.	C-5	4.80	3.76
8.	C-6	5.40	0.68
9.	A-4	8.70	4.99
10.	F-1	2.30	2.30
Other:	C-4	1.50	0.38
	Rehabilitation	0.20	0.20
	Dental	0.20	0.20
	TOTALS	54.40	38.56

APPENDIX D

Floor Medical Dicta-Typist
Transportation Work

Floor Medical Dicta	Hospital Area	Trips Per Day	Hours Per Day
1	B-3	3.2	0.45
2	B-4	2.8	0.51
3	B-5	4.0	0.34
4	B-6	(No data)	0.50 (Assumed)
5	C-2	3.5	0.50
6	C-3	5.5	1.20
7	C-5	6.5	0.51
8	C-6	11.1	0.89
9	A-4	2.9	0.45
10	F-1	0.3	0.08
	TOTALS		5.43

APPENDIX E

Time Required for Discharge Summary Transcription

The total time required per working day for transcription of discharge summaries is 1269.3 minutes, or 21.2 hours per working day. An explanation of how that figure was computed follows.

The standard times described in the *Hospital Staffing Methodology Manual, MM-7 Medical Records* (The University of Michigan, USPHS) enable us to determine staffing requirements. The activities included in the standard times are listed below:

1. Removing recording from machine, inserting new disc, clip note; one disc per discharge summary is used.
2. Inserting disc in machine, putting on earphones, adjusting speed/volume, removing after typing, clipping note to disc.
3. Listening to dictation, making note of patient's name, case number, doctor, etc.
4. Paper and carbon assembled in proper holder, assembly inserted and aligned in typewriter, removed; includes listening to determine which "form" to use.
5. Same as #4 except for multiple pages of same report.
6. Typing patient's name, case number, admitting/discharge date, doctor's name, referring doctor with electric typewriter.
7. Typing on electric typewriter full lines (6"—6½") including listening to dictation, repeating portion not understood, transcribing, erasing, looking up words in dictionary.
8. Sorting and stapling report if applicable.
9. Arranging reports alphabetically or numerically.
10. Filing reports in patient's chart if applicable.
11. Final review and check of report/chart.

12. Typing and stuffing one envelope per discharge.

The primary workloads needed for the staffing calculations are the number of discharges per day, the number of typed lines per summary, and the number of pages per summary.

The number of lines per summary was found by sampling many discharge summaries. Statistical reliability audits were performed during the data collection to insure a satisfactory confidence level of 90%. The resulting averages were then expressed with a tolerance or accuracy range, e.g., 49 lines plus or minus 10%. To interject a truly conservative nature into the data, the upper end of the tolerance range was used as the workload variable, e.g., 49.0 + 4.9 = 53.9 lines per summary.

The number of pages per summary was found by using the conversion factor of 20 lines for the first page and 40 lines for each succeeding page.

Allowances for personal time (coffee breaks, lavatory, idle), fatigue and unavoidable delays (17%) and productive interaction (8%) were included in the standard times. The formula, presented below, gives the standard time in minutes to perform all the activities mentioned previously for *one* discharge summary:

$$(0.3912)\ (x\ \text{lines/summary}) + (0.425)\ (y\ \text{pages/summary}) + 9.435$$

The next step is to insert the proper values for x and y in the formula and calculate the required man-minutes. The table on the following page presents the results by each nursing unit. The total time required per working day is 1269.3 minutes or 21.2 hours per working day.

25

Required Staffing for Discharge Summary Transcription

Area	Discharges[1] per Day	Lines per Summary	Pages per Summary	Time per[2] Discharge Summary	Total Time per Working Day
B-3	4.4	53.9	2	30.832 min.	135.661 min.
B-4 (I)	1.7	38.6	2	24.999	42.498
B-4 (N)	2.4	48.8	2	28.888	69.331
B-5 (H)	2.3	63.0	3	34.726	79.870
B-5 (R)	1.9	61.4	2	33.691	64.013
B-6	3.3	22.5	2	18.862	62.245
C-2 (G,M,N)	3.4	60.1	2	33.195	112.863
C-2 (D)	8.0	3.0	1	11.004	88.029
C-3 (D)	1.1	59.6	2	33.005	36.306
C-3 (R)	2.0	53.8	2	30.794	61.588
C-3 (FCM)	0.8	50.9	2	29.688	23.750
C-4	1.5	41.2	2	25.990	38.985
C-5	4.8	22.4	2	18.824	90.355
C-6 (G)	3.0	29.1	2	21.378	64.134
C-6 (T)	1.7	28.8	2	21.264	36.149
C-6 (C)	0.2	34.6	2	23.475	4.695
C-6 (P)	0.5	20.6	1	17.713	8.857
A-4 (Gyn)	3.0	27.0	2	20.577	61.731
A-4 (Ot)	4.6	18.6	1	16.950	77.970
A-4 (Op)	1.1	21.6	2	18.519	20.371
F-1	2.3	62.1	2	33.958	78.103
Rehab.	0.2	52.2	2	30.184	6.037
Dental	0.2	49.0	2	28.964	5.793
					1269.334 minutes

[1] From May–June 1973
[2] Hospital Staffing Methodology Manual, MM-7 Medical Records

APPENDIX F

Steno-Pool Discharge Summary

Steno-Pool Discharge Summary Transcription

Hospital Area	Number Done[1] by Steno-Pool	Time/Working Day[2] (in minutes)
B-3	0.05	1.542
B-4 (N)	0.53	15.311
B-5 (H)	0.85	29.517
B-6	0.38	7.168
C-2 (G,M,N)	2.74	90.954
C-3 (D)	0.02	0.660
C-3 (R)	0.66	20.324
C-3 (FCM)	0.02	0.594
C-4	1.12	29.109
C-5	1.04	19.577
C-6 (G)	2.62	56.010
C-6 (T)	1.49	31.683
C-6 (C)	0.17	3.991
C-6 (P)	0.44	7.794
A-4 (Gyn)	1.28	26.339
A-4 (Ot)	1.96	33.222
A-4 (Op)	0.47	8.704
Totals	15.84	382.499

[1] From hospital data for May–June 1973

[2] The time per day was obtained by multiplying the number done by the Steno-Pool (above) by the standard time per summary from Appendix E.

Adjustment:

The 12 month average for discharge summaries typed in the Steno-Pool is 18.604 per day. The above data used only May-June 1973 figures. The average time per summary is 382.499 ÷ 15.84 = 24.148 minutes. The revised staffing to properly reflect the yearly average is, therefore, (24.148) (18.604) = 449.249 minutes per working day.

APPENDIX G

Steno-Pool Operating Room
Note Transcription

Steno-Pool Operating Room Note Transcription

Specialty Area	Notes Per Working Day (May–June 73)	Lines Per Note	Pages Per Note	Lines[1] Per Day	Pages Per Day
General Surgery	4.0	55.4	2	221.6	8.0
ENT	4.7	43.0	2	202.1	9.4
Eye	2.4	36.7	2	88.1	4.8
GU, Urology	2.4	42.9	2	103.0	4.8
Gynaecology	4.7	45.3	2	212.9	9.4
Orthopaedic	4.6	40.0	2	184.0	9.2
Plastic	1.8	30.8	2	55.4	3.6
Neurosurgery	0.4	63.7	3	25.5	1.2
Chest	0.8	55.0	2	44.0	1.6
Cardiology	0.1	56.0	2	5.6	0.2
Dental	1.7	21.2	1	36.0	1.7
Totals	27.4			1178.2	53.9
Average/Note (Total ÷ 27.4)				43.0	2.0

The above averages of 43.0 lines and 2.0 pages per operating note are inserted in the formula shown in Appendix E. The resulting standard is *26.677 minutes per operating note* including all allowances.

[1] From Hospital Staffing Methodology Manual

Index

B

C

D

E

F

G

H

I

K

L

M

N

O